WATER-ROCK INTERACTION

Water Science and Technology Library

VOLUME 40

The titles published in this series are listed at the end of this volume.

WATER-ROCK INTERACTION

edited by

INGRID STOBER

Geological Survey of Baden-Württemberg, Division of Hydrogeology,
Freiburg, Germany

and

KURT BUCHER

Institute of Mineralogy, Petrology and Geochemistry,
University of Freiburg, Germany

SPRINGER SCIENCE+BUSINESS, MEDIA, B.V.

A C.I.P. Catalogue record for this book is available from the Library of Congress.

ISBN 978-94-010-3906-2   ISBN 978-94-010-0438-1 (eBook)
DOI 10.1007/978-94-010-0438-1

Front cover: Title page illustration from the firstprinted publication on "German Spas"
by Hans Foltz, about 1480. It shows the Spa "Wildbad" where people are bathing in pools
that are filled by warm springs flowing from the rocky cliffs to the right. The warm waters
are mineralized by "water-rock-interaction", specifically the interaction of meteoric water
with gneisses and granites of the Black Forest basement. The waters have been utilized
in "Wildbad" for more than 700 years and the bath is still in operation to day.

Printed on acid-free paper

Table of contents

Preface

The chemical interaction of water and rock is one of the most fascinating and multifaceted process in geology. The composition of surface water and groundwater is largely controlled by the reaction of water with rocks and minerals. At elevated temperature, hydrothermal features, hydrothermal ore deposits and geothermal fields are associated with chemical effects of water-rock interaction. Surface outcrops of rocks from deeper levels in the crust, including exposures of lower crustal and mantle rocks, often display structures that formed by interaction of the rocks with a supercritical aqueous fluid at very high pT-conditions. Understanding water-rock interaction is also of great importance to applied geology and geochemistry, particularly in areas such as geothermal energy, nuclear waste repositories and applied hydrogeology. The extremely wide-ranging research efforts on the universal water-rock interaction process is reflected in the wide diversity of themes presented at the regular International Symposia on Water-Rock Interaction (WRI).

Because of the large and widespread interest in water-rock interaction, the European Union of Geosciences organized a special symposium on "water-rock interaction" at EUG10, the biannual meeting in Strasbourg 1999 convened by the editors of this volume. In contrast to the regular WRI symposia addressed to the specialists, the EUG10 "water-rock interaction" symposium brought the subject to a general platform This very successful symposium showed the way to the future of water-rock reaction research.

In this book, the reader will find a selection of papers about groundwater – aquifer rock interaction, origin of solutes in groundwater, reaction of hydrothermal waters in geothermal areas, water – rock interaction in mineral and gas hydrate deposits, groundwater reaction with rocks in active volcanic areas and an experimental study of surface reactions in silica-bearing solutions.

The editors would like to express their thanks to the authors for investing so much time and effort in this venture and for their willingness to share their ideas with the Earth Science community. We also are very grateful to all colleagues who took their time and effort to constructively reviewing the contributions and in doing so helped to significantly improve the quality of the presentations. Without the generous effort of competent reviewers modern science would be impossible.

Reviews were provided by:

M. Azaroual (Orleans), David Banks (Chesterfield), Susanne Barth (Tübingen), Kurt Bucher (Freiburg), Jean Carignan (Nancy), Ian Clark (Ottawa), Joris Gieskes (La Jolla, CA), Ian Hutcheon (Calgary), Margot Isenbeck-Schröter (Heidelberg), Bjørn Jamtveit (Oslo), Gary Klinkhammer (Oregon), Peter Lichtner (Los Alamos), Luigi Marini (Genova), Peter Möller (Potsdam), Christophe Monnin (Toulouse), Eric Oelkers (Toulouse), Ken Osmond (Tallahassee, FL), Joe Pearson (New Bern, NC), Vala Ragnarsdottir (Bristol), R. Shiraki (Davis), Carl Steefel (Livermore, CA), Ingrid Stober (Freiburg), Warren Wood (Reston, VA)

Ingrid Stober and Kurt Bucher

Freiburg, September 27. 2001

Groundwater Evolution in an Arid Coastal Region of the Sultanate of Oman based on Geochemical and Isotopic Tracers

Constanze E. Weyhenmeyer

Institute of Geology, University of Bern,
CH-3012 Bern, Switzerland (weyhen@geo.unibe.ch)

Abstract

In arid and semi-arid regions, conventional hydrological investigations often fail to adequately describe groundwater systems due to a large spatial and temporal variability of hydrological parameters. Alternatively, this study from the Sultanate of Oman utilizes a combination of geochemical and isotopic tracers (O, H, C, Sr) for the identification of aquifer units, recharge areas, groundwater flow paths and residence times in a coastal alluvial aquifer of the Eastern Batinah region in Northern Oman. The hydrochemical investigation of groundwater samples from more than 200 wells and springs clearly indicates that the main recharge areas for the coastal alluvium are the adjacent Oman Mountains, rising up to 3000 m. Groundwater that infiltrates in the high-altitude regions circulates rapidly through the karstified mountains, indicated by close to modern-day tritium activity values measured in springs and wells along the foothills of the mountains. In the piedmont areas, groundwater from the high-altitude regions is diverted by a less permeable ophiolite complex into two geochemically and isotopically distinct corridors (plumes) that stretch through two gaps in the ophiolite across the 50 km wide coastal plain to the Gulf of Oman. Within these plumes the chemical and isotopic signature from the high-altitude regions remains virtually unchanged horizontally as well as vertically to depths exceeding

1

I. Stober and K. Bucher (eds.), Water-Rock Interaction, 1–38.

300 m (deepest wells), suggesting that additional infiltration on the coastal plain is insignificant. Mixing calculations based on strontium isotopes indicate that infiltration on the coastal plain itself accounts for less than 10 % of the total groundwater recharge in the plume areas, which is consistent with the lack of tritium in these groundwater samples. The remaining 90 % of groundwater in the plumes originate in the Jabal Akhdar mountains. In contrast, direct infiltration and recharge on the coastal plain itself is the only source for groundwater in areas adjacent to the two plumes (downstream of ophiolite areas) and in these areas the alluvial aquifer is hydro-chemically layered. Groundwater samples from shallow parts of the aquifer (<50 m) contain tritium indicating modern recharge, whereas groundwater in the deeper aquifer (> 200 m) is of late Pleistocene origin.

1.   Introduction

In arid and semi-arid regions, groundwater is often the only reliable and affordable source of water for rapidly expanding domestic, industrial and agricultural needs. The development of sustainable groundwater management strategies is a prerequisite for a long-term prevention of overexploitation and potential contamination of this precious resource. Effective management, however, requires a detailed assessment of available groundwater resources, which in turn requires a thorough knowledge of aquifer boundaries, recharge areas, groundwater flow paths and residence times. Commonly applied hydrodynamic mass-balance calculations are complicated by an extreme spatial and temporal variability of hydrological input-parameter such as precipitation, evaporation and runoff (Cook *et al.* 1989, Adar and Leibundgut 1995, Froehlich and Yurtsever 1995). Required meteorological and hydrological records are often not available, especially from the more remote areas where the logistical and financial efforts for the necessary long-term monitoring programs often exceed the existing capabilities in these countries.

As an alternative to a purely hydrodynamic approach, geochemical and isotopic investigations of aquifer systems have often proven to be quite successful in determining groundwater recharge, groundwater flow paths and groundwater renewal times in arid regions (Edmunds and Walton 1980, Issar *et al.* 1984, Sharma and Hughes 1985, Chandrasekharan *et al.*

1988, Collerson *et al.* 1988, Simmers 1990, Fontes *et al.* 1991, Edmunds and Gaye 1994, Adar and Leibundgut 1995, Edmunds 1995, Singh *et al.* 1996, Sukhija *et al.* 1996). The advantage of such a hydrogeochemical approach is that the chemical and isotopic groundwater data provide integrated information on all water that enters the reservoir, thereby largely eliminating local differences and short-term variability observed in the hydrological data. Oxygen, hydrogen and carbon isotopes are the most commonly applied isotopes but in recent years a number of studies have demonstrated the great value of strontium isotopes ($^{87}Sr/^{86}Sr$) as a tracer for groundwater flow (Collerson *et al.* 1988, McNutt *et al.* 1990, Johnson and DePaolo 1994, Lyons *et al.* 1995, Neumann and Dreiss 1995, Bullen *et al.* 1996, Katz and Bullen 1996).

This study presents an example from an arid area on the Arabian Peninsula where a combination of various geochemical and isotopic tracers was applied to describe the origin and evolution of groundwater in a coastal aquifer. The investigated alluvial aquifer of the Eastern Batinah coastal plain is one of the most important aquifers in the Sultanate of Oman since it supplies water for the densely populated, cultivated and industrialized capital area. In recent years, overexploitation of this groundwater resource has resulted in a drastic lowering of the groundwater table below sea level, leading to subsequent seawater intrusion into the coastal aquifer, which is a problem commonly encountered in coastal arid regions (Rao *et al.* 1987, Sukhija *et al.* 1989, DeBreuck 1991, Chinn *et al.* 1995, Howard and Mullings 1996). Several different hydrodynamic mass-balance models have been developed and utilized in an attempt to calculate a water balance for the Eastern Batinah region (Tetratech 1980, MacDonald and Partners 1989, Heathcote 1993, Lakey *et al.* 1995). However, as a result of large uncertainties involved in the characterization of model boundary conditions and hydrological model-input parameters, existing models yield different and sometimes contradictory results. In order to improve future model calculations, this study focuses on the characterization of model boundary conditions including the identification of aquifer units, recharge areas and groundwater flow paths. For the characterization of the groundwater flow system in the Eastern Batinah alluvial aquifer, isotope ratios of oxygen ($\delta^{18}O$), hydrogen (δ^2H) and strontium ($^{87}Sr/^{86}Sr$) are used in combination with major ion chemistry. Tritium (^3H) and carbon-14 (^{14}C) data provide additional information on groundwater renewal times.

2.   Study Area

The study area comprises a 10,000 km^2 section of the alluvial aquifer of the Eastern Batinah coastal plain, located northwest of the capital city Muscat, Sultanate of Oman (Figure 1). To the north, the study area is bordered by the Gulf of Oman and the southern limits are defined by the drainage divide of the Jabal Akhdar mountains. The study area includes from west to east the surface wadi catchments of Wadi Bani Ghafir, Wadi Far, Wadi Bani Kharus, Wadi Ma'awil, Wadi Taww and Wadi Samail, including the Al Khwad Fan (Figure 1).

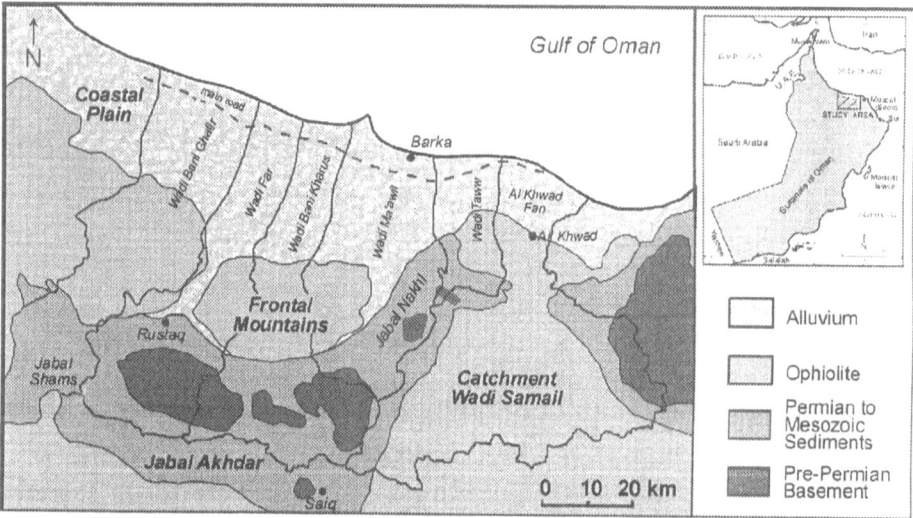

Figure 1. Map of the study area, showing the general location of the Eastern Batinah coastal plain (insert), schematic boundaries of the surface wadi catchment areas as well as four major geological units: 1) coastal alluvial plain, 2) Samail ophiolite, 3) Jabal Akhdar mountains made up of Permian to Mesozoic sediments (mostly carbonates) and 4) pre-Permian basement.

2.1.   GEOLOGY

Within the study area, four geologic (Figure 1) and three major hydrogeologic units can be distinguished: 1) Mesozoic and Permian carbonates of the Jabal Akhdar mountains overlying the pre-Permian basement, 2) coastal alluvium, and 3) Samail ophiolite (Figure 1).

2.1.1. Jabal Akhdar Mountains

The Jabal Akhdar mountains, rising up to 3000 m at the summit of Jabal
Shams (Figure 2), are part of the Oman Mountains that run in a broad arc
parallel to the Gulf of Oman coastline over a distance of nearly 800 km.
The core of the Jabal Akhdar anticline is essentially composed of highly
fractured and faulted pre-Permian siltstone, sandstone and limestone
formations (phyllite, shale, calcite and dolomite), which are exposed in
several topographically low bowls or tectonic windows that have been
created along the crest of the Jabal Akhdar anticline as a result of
extensive erosion. The limbs of the anticline are made up of the Hajar
Supergroup, a sequence of shallow marine carbonates (limestone and
dolomite) and thin sandstone layers, which were deposited between the
Permian and Cretaceous and rest unconformably on the older pre-
Permian basement (Glennie *et al.*, 1974, Robertson *et al.* 1990, Le
Métour *et al.* 1995). The carbonates of the Hajar Supergroup are highly
fractured, and karst features are found throughout most sequences.
Numerous springs within these carbonates provide evidence for
significant, well constrained groundwater circulation. The contribution of
groundwater from these bedrock units to the adjacent alluvial aquifer is,
however, largely unknown and subject to ongoing discussions. Some
investigators consider the contribution of groundwater from the Jabal
Akhdar bedrock units to the alluvial aquifer as 'insignificant'
(MacDonald and Partners 1989, Heathcote 1993, Lakey *et al.* 1995,),
while other studies suggest that most of the groundwater in the alluvium
has its origin in the Jabal Akhdar mountains with only minor recharge on
the coastal alluvial plain itself (Gibb and Partners 1976, Weyhenmeyer
2000, Weyhenmeyer *et al.* in press.).

Several wadis cut deeply into the Jabal Akhdar massif. These wadis are
generally dry with ephemeral streams forming during and after heavy
rainstorm events when wadis are exposed to flash flooding, thereby
transporting eroded rock material from the mountains towards the coast. As
a result, thick alluvial sequences have accumulated in major wadi channels
and on the coastal plain during the late Tertiary and Quaternary.

2.1.2. Coastal Alluvium

The Eastern Batinah coastal plain consists of an alluvial gravel sequence of
late Tertiary to Quaternary origin, underlain by north-dipping Tertiary

carbonates and marls (Young *et al.* 1998). The exact morphology and thickness of this gravel sequence is not well known, but recent transient electromagnetic sounding (TEM) data suggest that the base of the alluvium dips northward from the foothills of the Jabal Akhdar towards the coast (Young *et al.* 1998). At the coast, the thickness of the alluvium exceeds 1,500 m, possibly reaching a depth of 2,000 m. The composition of the alluvium of the Eastern Batinah coastal plain has not been studied in great detail, but on the surface it primarily consists of a mixture of consolidated, partly cemented clastic fragments of Tertiary carbonates together with ophiolitic clasts from the Samail nappe (Glennie *et al.* 1974, Gibb and Partners 1976, Stanger 1986). Although the alluvial sequence is heterogeneous, conventionally it has been divided into two (Lakey *et al.* 1995) or three (Gibb and Partners 1976, Graf 1983, Stanger 1986, Young *et al.* 1998) different hydrostratigraphic units, with decreasing storativity and permeability with increasing depth. The *upper gravel unit* (Gibb and Partners 1976) is made up of boulder beds, uncemented gravels and unconsolidated sands. Its thickness generally varies between 20 and 140 m, but has been found to reach a maximum thickness of 300 m in some coastal sections (Graf 1983, Young *et al.* 1998). The intermediate *clayey gravel unit* consists of bands of clay interlayered with gravel beds and calcrete. The *basal cemented gravels* form the deepest unit of the alluvial sequence and essentially consist of alluvial gravels that are highly cemented by carbonates, mainly high- and low-magnesium calcite (Burns and Matter 1995) and gypsum (Mann *et al.* 1990).

Cemented gravel layers as well as clay layers have been detected throughout the entire gravel sequence (MWR 1993, Young *et al.* 1998). However, seismic and electromagnetic investigations failed to identify any horizontally continuous layer that could potentially form an aquitard (Young *et al.* 1998) and, therefore, the alluvial aquifer of the Eastern Batinah coastal plain is considered an unconfined or semi-confined aquifer system. The groundwater table on the coastal plain lies several tens of meters below the surface, with a maximum depth of 80 m.b.s. in the central part of the plain. In some coastal wells the water table has dropped to several meters below sea level and subsequent saline intrusion effects the aquifer as far inland as 15 km (Lakey *et al.* 1995).

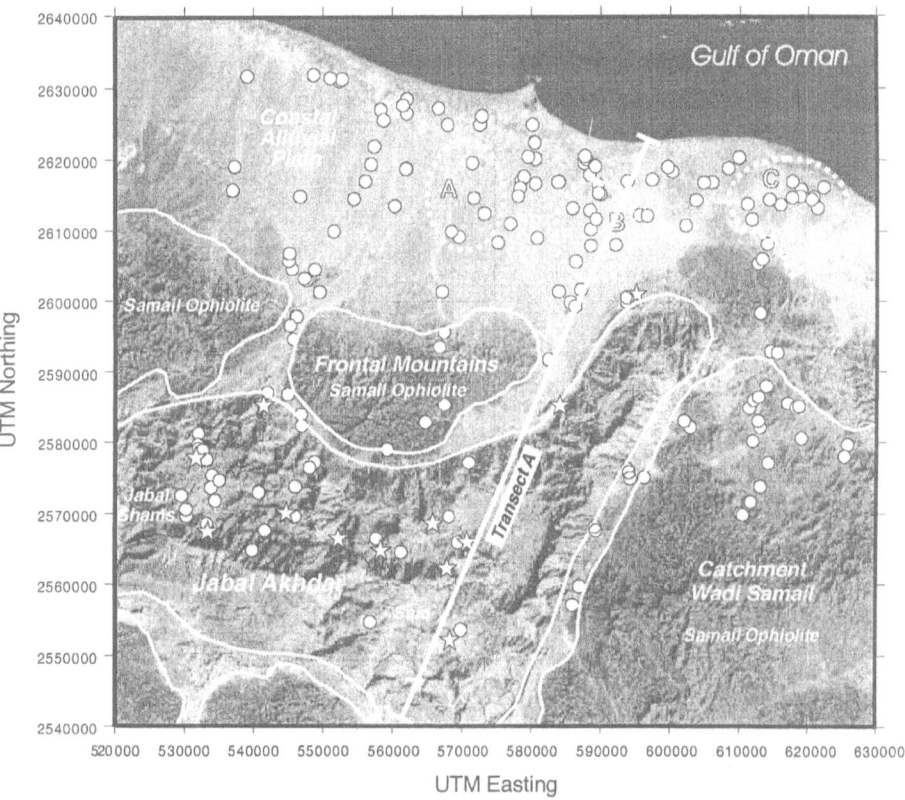

Figure 2: Satellite image of the study area showing the groundwater sampling locations (filled white circles for wells, white stars for springs). Also included is the location of a groundwater transect from the mountains to the coast (Transect A) and three selected areas on the coastal alluvial plain (areas A, B, C). The coordinates are Universal Transverse Mercator (UTM) coordinates where one grid unit equals one meter distance.

2.1.3. Samail Ophiolite

The catchment area of Wadi Samail, situated to the east of the Jabal Akhdar, is geologically different from all other wadi catchments in the study area because it is entirely made up of the Samail ophiolite. The Samail ophiolite, which was obducted during the late Cretaceous, is composed of an oceanic crustal sequence (mainly gabbros, sheeted dykes and pillow lavas) and a mantle sequence composed of a thick, variably serpentinized and tectonized peridotite sequence (mainly foliated harzburgite and dunite) (Lippard *et al.* 1986). As a result of syn- and post-emplacement faulting and folding the ophiolite nappe has been broken into a number of more or less intact structural blocks of which wadi catchment Samail forms the largest single ophiolite block in the area (Figure 1). At the northern foothills of the Jabal Akhdar, another large section of the Samail ophiolite resisted erosion and forms a low-lying mountain range named the Frontal Mountains (Figure 1).

The geology and tectonic setting of the Samail ophiolite has been studied in great detail because it is one of the world's largest and best exposed examples of an oceanic crustal and upper mantle sequence (for reviews see JGR 1981, Lippard *et al.* 1986, Peters *et al.* 1991). However, little attention has focused on the hydrological importance of the ophiolite as a potential aquifer. While primary porosity is probably insignificant, as indicated by several dry boreholes drilled to depths exceeding   150 m., secondary porosity appears to be quite extensive. As a result of tectonic movement the ophiolite became highly fractured and faulted, but it is not known whether existing fractures are effectively hydraulically connected. Earlier studies considered the ophiolite complex an aquitard with groundwater flow being confined to the thin (< 30 m) alluvial deposits overlying the ophiolite. More recent studies, however, suggest significant groundwater circulation within the ophiolite complex (Bhatnagar 1997, Weyhenmeyer *et al.* submitted).

2.2. CLIMATE

The present climate of Northern Oman is classified as 'arid-province' (Schyfsma 1978). Typically the winters are warm with relatively low humidity while the summers are extremely hot and sometimes very humid. The long-term mean annual air temperature is 28.6 °C for the coastal areas (meteorological station at Seeb International Airport; 15 m.a.s.l.) and 17.1 °C in the mountains (meteorological station at Saiq; 1897 m.a.s.l.). As a result of high temperatures and generally low relative humidity values, overall evaporation rates in the area are very high, with long-term mean daily evaporation rates estimated from open pan experiments as 5 to 15 mm d^{-1}, depending on altitude and surface exposure (Gibb and Partners 1976, Stanger 1986).

As in most arid regions, precipitation rates on the Batinah plain are very low and highly erratic in space and time. Most rainfall occurs in the winter months from Mediterranean frontal systems that approach the area from the northwest, mainly between December and March. Additional rainfall occurs during the very hot summers when local convective storm cells form over the mountains, causing short and heavy rainfall mostly in the mountains (orographic rain). A third source for precipitation are tropical cyclones that approach the area occasionally from the southeastern Arabian Sea or the Bay of Bengal, causing heavy precipitation for several days. Because of the strong monsoonal air stream that parallels the Arabian Sea coastline during summer and winter, however, southeastern air masses normally do not reach the northern Arabian Coast. The frequency of southern cyclonic rainfall is estimated as once every 2 to 5 years in southern and central Oman and once every 5 to 10 years in northern Oman (Pedgely 1969, Taylor *et al.* 1990). Mean annual precipitation rates vary between 60 and 100 mm yr^{-1} in coastal regions (80 mm yr^{-1} at Seeb), and exceed 200 mm yr^{-1} in the mountainous regions above 1000 m. Areas located above 2000 m receive more than 300 mm of rain per year (Meteorological Reports; Ministry of Communications, Sultanate of Oman).

3.   Methods

3.1.   SAMPLING SITES, EQUIPMENT AND PROCEDURES

During five field campaigns between 1996 and 1998 more than 200 wells, evenly distributed over the study area, were sampled, including shallow dug wells and boreholes. Furthermore, several springs were sampled in the Jabal Akhdar mountains and along the foothills (Figure 2).

All wells and boreholes were pumped with a submersible pump for a duration representing at least three borehole volumes to retrieve a representative groundwater sample from the aquifer and avoid contaminated or evaporated water from the borehole itself. Three to five groundwater samples were taken during the pumping period and later analyzed for major ion chemistry and stable isotopes, allowing us to monitor changes in groundwater composition during the pumping period. Figure 3 shows two examples of changes in major ion chemistry that occurred during a two hour pumping interval of wells with a discrete, narrow screen interval, illustrating the necessity of sufficient pumping time to obtain a representative aquifer sample. Water level, temperature, electrical conductivity, pH, and dissolved oxygen concentrations were measured continuously during pumping and final groundwater samples were taken only when these parameters were stable for at least 15 minutes. At the end of each sampling period, several samples were taken for chemical and isotopic analyses in glass and polyethylene bottles. For ^{14}C analysis, which requires at least 3 g of carbon, dissolved inorganic carbon (DIC) was precipitated in the field as $BaCO_3$ from 80 to 160 liters of groundwater, depending on the alkalinity. Precipitation was effected by adding about 50 g NaOH dissolved in one liter of groundwater to 80 liters of groundwater collected in a large plastic canister. The addition of NaOH increased the pH of the water to > 11. Furthermore, 100 g $BaCl_2$ dissolved in one liter of groundwater were added to the container which was then quickly sealed from the atmosphere and left in the field for at least eight hours. During this time the dissolved carbon precipitated as $BaCO_3$ into a removable one liter sample bottle attached to the bottom of the large plastic canister.

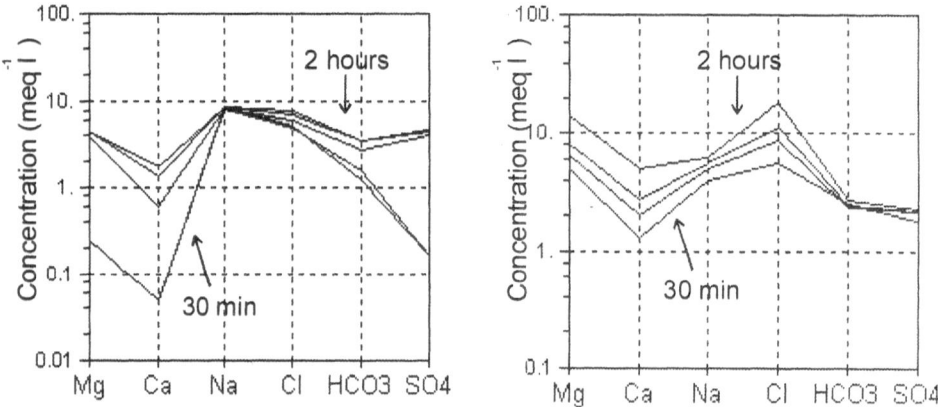

Figure 3. Schoeller plots showing changes in major ion chemistry during a two hour pumping interval of two different wells with discrete screen interval located on the Al Khwad Fan (left: Well 21/6 [screen: 289-297 m]; right: Well RGS-2L [screen: 319-325 m]). In both cases an increase in total mineralization is observed during the pumping period but in well 21/6 chemical changes become smaller, almost reaching a chemical steady state towards the end of the pumping time. However, in both cases pumping time was insufficient to reach a steady state with the aquifer water.

3.2.   LABORATORY ANALYSES

3.2.1.   Ion Chemistry

Major and minor ions were analyzed at the Chemistry Laboratory of the Ministry of Water Resources in Muscat. Anions were analyzed by ion chromatography and cations were measured by standard ICP (inductively coupled plasma spectrometer) methods. Details of the analytical methods are described elsewhere (MWR 1997). Chemical analyses were only accepted when the overall ionic charge balance was better than ± 5%. In addition to ICP measurements of strontium, Sr-concentrations were also determined by isotope dilution (first introduced by Tomlinson and Das Gupta 1953, Moore *et al.* 1973) during the course of strontium isotope analysis by thermal ionization mass spectrometry at the Institute of Mineralogy and Petrology, University of Bern. There was an excellent agreement between the two different methods with an average deviation of < 0.1 ppm.

3.2.2.   Stable Isotope Analyses (O, H, Sr)

Stable isotope ratios of oxygen and hydrogen were analyzed at the Geological Institute of the University of Bern, according to standard gas mass spectrometer techniques (Epstein and Mayeda 1953, Coleman *et al.* 1982). Oxygen isotope analyses on 2-ml water samples were carried out using an Isoprep automated, online-equilibration bench. The samples were equilibrated with a CO_2 standard gas for 5 hours and the equilibrated CO_2 sample was then directly inlet into a VG- Prism II isotope ratio mass spectrometer (IRMS) and analyzed for $\delta^{18}O$ of CO_2.

For hydrogen analyses, 4 µl of water were reduced to hydrogen gas by reaction with a zinc reagent in sealed Pyrex tubes at 500 °C for 30 minutes. At this temperature the zinc reacts quantitatively with H_2O, thus producing H_2 with an isotopic composition equal to that of the water. The samples were then directly inlet into the mass spectrometer. Most samples were analyzed again using a new online-equilibration technique with platinum-coated 'Hokko beads' as a hydrogen reaction catalyst. For the analyses, 20-40 mg of Hokko beads are added to 3 ml of water and then equilibrated with a H_2 standard gas in the Isoprep equilibration bench. The Hokko beads break the surface tension of the water and water-gas equilibration occurs very rapidly, thus requiring an equilibration time of only one hour.

Most water samples were analyzed in duplicate with an overall analytical accuracy of ± 0.1 ‰ for $\delta^{18}O$ and ± 1.0 ‰ for δ^2H with the zinc method and ± 0.6 ‰ using the online-equilibration technique. Isotopic results for oxygen and hydrogen are expressed in ‰ in the δ notation with respect to Vienna Standard Mean Ocean Water (VSMOW) (Gonfiantini 1978).

Measurements of $^{87}Sr/^{86}Sr$ ratios were carried out on a VG-Sector thermal ionization mass spectrometer (TIMS) at the Institute of Mineralogy and Petrology, University of Bern. Standard cation exchange column chemistry was used to separate strontium from 5 to 10 ml of water (Faure and Powell 1972). Extracted samples were loaded onto oxidized tantalum filaments and subsequently measured on the TIMS. Typically, 200 individual isotope ratios were measured per sample with an average standard error (SE) of the mean value of ± 0.000015. For standardization, the NBS 987 standard was run after each measurement cycle, yielding a mean value of 0.710228 ± 0.000012 (n = 7) for the measurement period.

3.2.3.   Radioactive Isotope Analyses (^3H, ^{14}C)

Groundwater samples, stored in glass bottles, were analyzed for tritium (^3H) at the laboratory of Hydroisotop GmbH in Munich, Germany. Tritium activities were measured by β⁻-decay counting in a liquid scintillation counter. To increase the analytical precision, water samples were electrolytically enriched with ^3H before counting. Generally, the precision (2σ) of the tritium analyses is between 0.5 and 1.0 TU, where one tritium unit (TU) corresponds to one ^3H atom per 10^{18} atoms of hydrogen which is equivalent to a radioactivity of 0.12 Bq l⁻¹.

Measurements of radioactive carbon (^{14}C) were carried out at the Physics Institute of the University of Bern by liquid scintillation counting on benzene (C_6H_6) prepared from carbon dioxide liberated from the precipitated $BaCO_3$. As a test for the reproducibility of field and laboratory extraction and measurement techniques, five different wells were sampled twice in two consecutive years. There was an excellent agreement between the different samples with a deviation of less than 0.5 percent modern carbon (pmc) for the individual sample duplicates, where 100 pmc is defined as 95 % of the ^{14}C activity of an oxalic acid NBS standard and equals 13.56 dpm (decays per minute) per 1 g of carbon. The good reproducibility of the ^{14}C measurements provides confidence in our field extraction techniques as well as laboratory methods.

4.   Results

4.1.   GROUNDWATER CHEMISTRY

The chemical composition of groundwater in the Eastern Batinah alluvial aquifer changes from the Jabal Akhdar mountains across the coastal plain towards the Gulf of Oman as illustrated in Figure 4. In the high-altitude regions of the Jabal Akhdar groundwaters classify predominantly as Ca-Mg-HCO_3 type waters (TDS = total dissolved solids: 300-550 mgl^{-1}) while groundwater samples extracted from coastal wells are predominantly Na-Mg-Ca-Cl-SO_4-HCO_3 type groundwaters (TDS: >1000 mgl^{-1}). Associated with this change in chemistry is a general increase in chloride and, to a lesser degree, in sodium and sulfate concentrations from the mountains towards the coast. Concentrations of other ions, including calcium, magnesium and strontium, remain fairly constant across the alluvial plain to within a few kilometers of the coastline where most ion concentrations sharply increase (Table 1). Vertically, the groundwater chemistry varies with depth in some areas of the coastal plain, while in other areas the chemical composition is fairly homogeneous to depths exceeding 300 m (deepest wells) (Figure 5). In area B of the central Batinah plain (Figure 2) groundwater samples from all depths are chemically very similar and classify as Na-Mg-Ca-HCO_3-Cl-SO_4 type waters (TDS: 400-700 mgl^{-1}) (Figure 5, Table 2). In two areas that are located downstream of large ophiolite outcrops (areas A and C in Figure 2), however, there is a distinct difference in the chemical composition of groundwater samples from shallow, intermediate and deep wells (Figure 5). In these areas, groundwater extracted from wells deeper than 200 m classifies as Na-Mg-Ca-Cl-HCO_3-SO_4 type (TDS: 500-1,000 mgl^{-1}), while the shallow groundwater is predominantly Mg-Na-Ca-HCO_3-Cl-SO_4 type groundwater (TDS: 300-500 mgl^{-1}) (Table 2).

Calcium-Magnesium ratios are variable across the Eastern Batinah region, ranging from < 0.01 to 10 (molar). The highest Ca/Mg values of > 1.0 are found in groundwater samples from the Jabal Akhdar mountains and along two corridors that stretch across the coastal plain towards the coast (Figure 6). Lower values of < 0.5 are measured in samples from the two ophiolite areas (Frontal Mountains, catchment Wadi Samail) and from wells located downstream of the ophiolite areas (Figure 6).

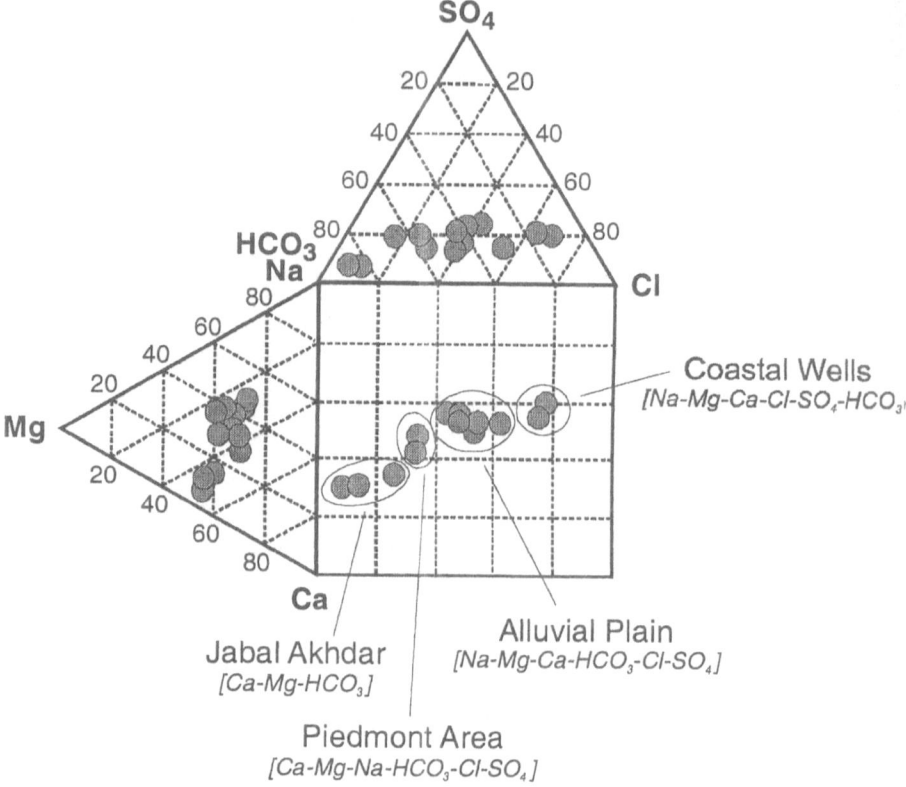

Figure 4. Durov plot showing the chemical composition and general water types of groundwater samples located along transect A from the mountains to the coast (for sampling locations and sample details see Figure 2 and Table 1).

4.2. ISOTOPE RATIOS OF OXYGEN ($\delta^{18}O$) AND HYDROGEN (δ^2H)

The isotopic compositions of groundwater samples range from -4.0 to +1.0 ‰ in $\delta^{18}O$ and from -20 to +5 ‰ in δ^2H. The most isotopically depleted water samples come from wells located in the high-altitude regions of the Jabal Akhdar (Figures 7 and 8). At an average sampling altitude of 1,700 masl the mean oxygen and hydrogen isotope values of 21 groundwater samples are -3.4 ‰ and -16 ‰, respectively. However, isotopically depleted groundwater samples (< -2.5 ‰ $\delta^{18}O$; < -10 ‰ δ^2H) are not only found in the mountainous regions but also in springs and wells located along the foothills of the Jabal Akhdar and in two areas that stretch from the foothills across the plain towards the coast defined as eastern and western plume, respectively. Groundwater samples in the eastern plume, which has been previously described by Macumber *et al.* (1997), are slightly more depleted in $\delta^{18}O$ and δ^2H than groundwater samples in the western plume (Figures 7 and 8). Along the two plumes,

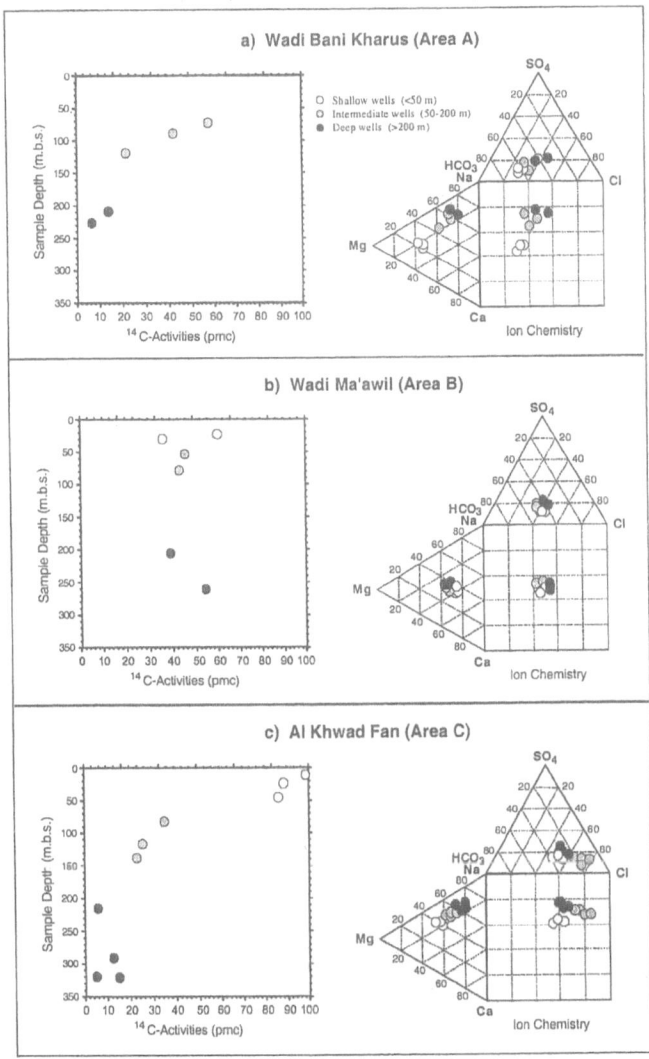

Figure 5: ^{14}C activity values and chemical compositions of groundwater samples from different depths intervals and three different regions (see Figure 2 and Table 2).

isotope ratios do not significantly change from the mountains towards the sea and even in coastal wells, isotope values of less than -2.5 ‰ $\delta^{18}O$ and -10 ‰ δ^2H were measured (Figures 7 and 8, Table 1). Vertically, the isotopic groundwater composition in these plume areas is also very homogeneous to depths exceeding 300 m (Tables 1 and 2).

In contrast to the two plume areas, groundwater samples from the adjacent areas, which are areas within and downstream of ophiolite outcrops (Frontal Mountains and wadi catchment Samail), are significantly enriched in $\delta^{18}O$ and δ^2H, with isotope ratios ranging from -1.5 ‰ to 0 ‰ $\delta^{18}O$ and from -5 to 5 ‰ δ^2H (Figures 7 and 8). As in the plume areas, there are no systematic variations of oxygen and hydrogen isotope ratios with sample depth in these ophiolite areas (Table 2).

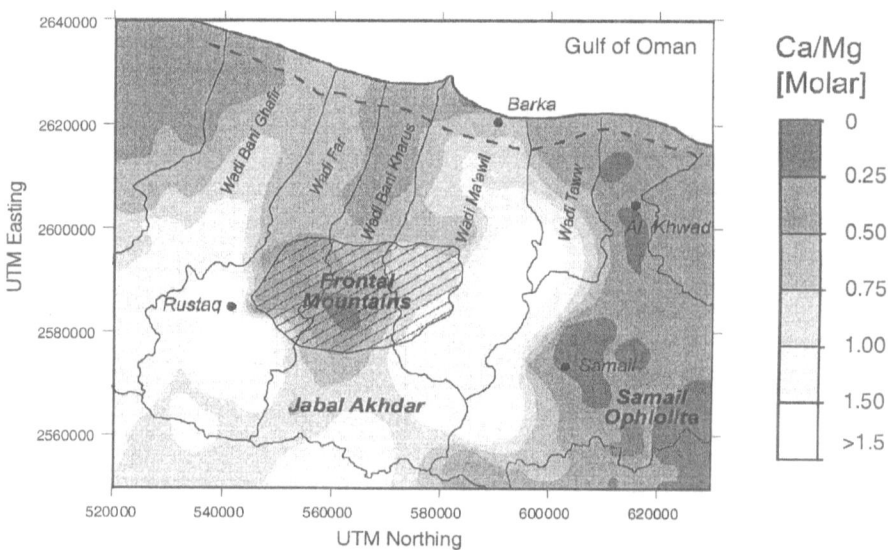

Figure 6: Contour plot of molar Ca/Mg ratios in groundwater samples. The contour lines are based on interpolation of data from all 220 groundwater samples (shown in Figure 2), using the gridding method "Kriging" (Surfer 7.0).

4.3. STRONTIUM ISOTOPE RATIO ($^{87}Sr/^{86}Sr$)

Strontium isotope ratios in groundwater samples range from 0.7067 in the Samail ophiolite catchment to 0.7131 in some areas of the Jabal Akhdar mountains. Sr-isotope ratios in groundwater samples from the Jabal Akhdar fall into two distinct groups: wells and springs located north or upstream of the pre-Permian windows have isotope ratios of around 0.7084 (n = 7), while those samples collected downstream of the pre-Permian windows have an average isotope ratio of 0.7118 (n = 9). Groundwater samples from the foothills of the Jabal Akhdar mountains have Sr-isotope values between 0.7100 and 0.7120, virtually unchanged from groundwater samples from the high altitude regions downstream of the pre-Permian windows (Figure 9, Table 1). On the coastal plain, Sr-isotope ratios of > 0.7095 are measured in groundwater samples from two distinct corridors or plumes that stretch from the mountains towards the coast. Isotopically, groundwater in these two areas can be distinguished from the adjacent regions (i.e., downstream of the ophiolite areas Wadi Samail and Frontal Mountains) where groundwater samples are characterized by lower Sr-isotope ratios of less than 0.7090 (Figure 9). Vertically, strontium isotope ratios within the plume areas are fairly uniform at least to a depth of 300 m (Table 2).

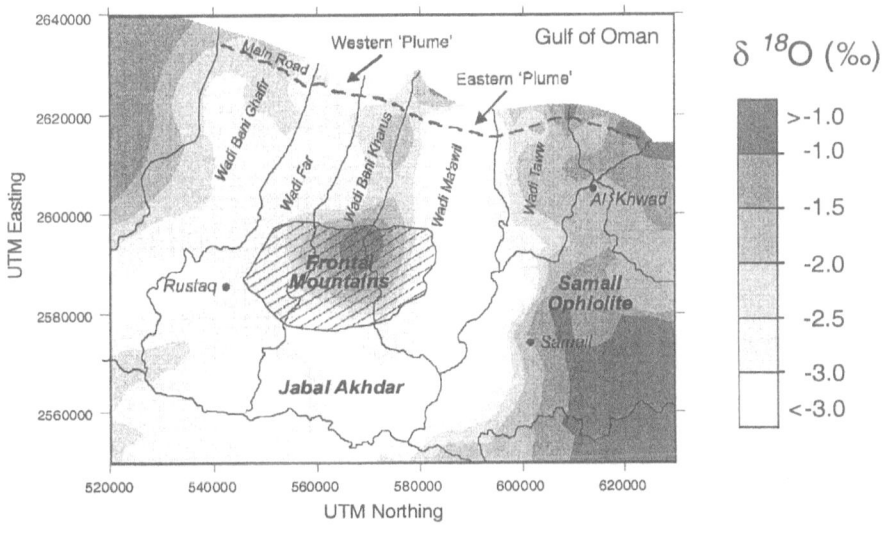

Figure 7: Contour plot of oxygen isotope ratios ($\delta^{18}O$) in groundwater samples based on interpolation of all samples shown in Fig. 2, using "Kriging" (Surfer 7.0).

Sr-isotope ratios of groundwater samples collected within the Samail ophiolite and downstream of the two main ophiolite areas (Frontal Mountains, catchment Wadi Samail) range from 0.7080 to 0.7088 and are, therefore, significantly lower than isotope ratios of groundwater samples from the Jabal Akhdar and the two plume areas (Figure 9). In the ophiolite dominated areas, strontium isotope ratios are fairly homogeneous with depth with the exception of the Al Khwad Fan, located downstream of wadi catchment Samail, where two groundwater samples from the deepest parts of the aquifer have distinctively lower isotope ratios of around 0.7075 (Table 2).

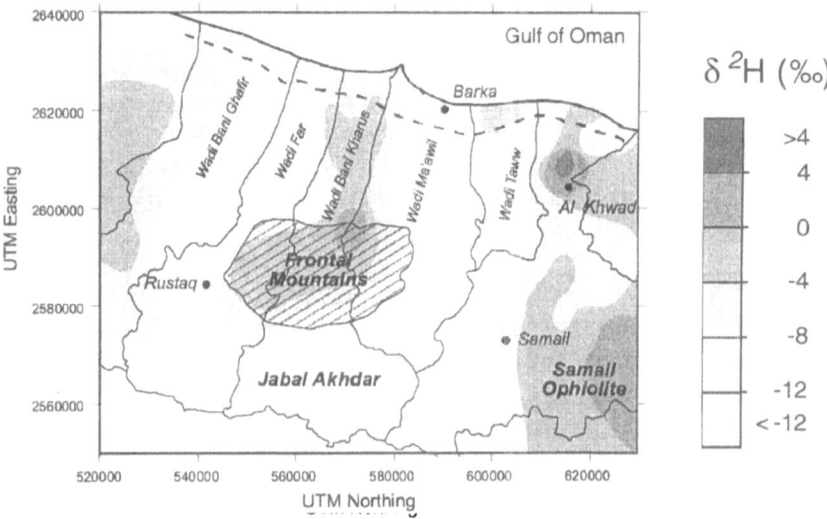

Figure 8: Contour plot of hydrogen isotope ratios (δ^2H) in groundwater samples based on interpolation of all samples shown in Figure 2, using "Kriging" (Surfer 7.0).

4.4. TRITIUM (^{3}H)

Tritium activities in groundwater samples from the high-altitude regions of the Jabal Akhdar mountains vary between 2.1 and 5.6 TU (Figure 10), which is only slightly lower than tritium activity values in modern rainfall which range from 5 to 10 TU (Weyhenmeyer 2000). Springs and wells sampled along the foothills of the Jabal Akhdar also have elevated tritium levels of > 3.0 TU. A few kilometers onto the coastal plain, however, tritium concentrations fall, with few exceptions, below the detection limit of 0.5 to 1.0 TU. The exceptions are shallow groundwater samples from areas downstream of ophiolite outcrops (areas A and C in Figure 2). In contrast, deeper wells in these downstream-ophiolite areas do not contain measurable amounts of tritium (Table 2).

4.5.   RADIOACTIVE CARBON (^{14}C)

^{14}C activity values are only presented for selected wells from three different areas on the central coastal plain to assess the vertical structure of the alluvial aquifer. ^{14}C activities in these wells range from nearly 100 percent modern carbon (pmc) in the upper layers of the aquifer in Areas A and C to less than 5 pmc in the deepest wells (> 300 m) (Figure 5, Table 2). ^{14}C values systematically decrease with depth in areas downstream of the two ophiolite blocks (areas A and C in Figure 2), while in the eastern plume area (Area B) ^{14}C activities vary between 36 and 60 without showing any systematic trends of ^{14}C activity values with sample depth (Figure 5, Table 2). A more detailed presentation and discussion of results from ^{14}C analyses and estimated groundwater residence times is presented elsewhere (Weyhenmeyer 2000).

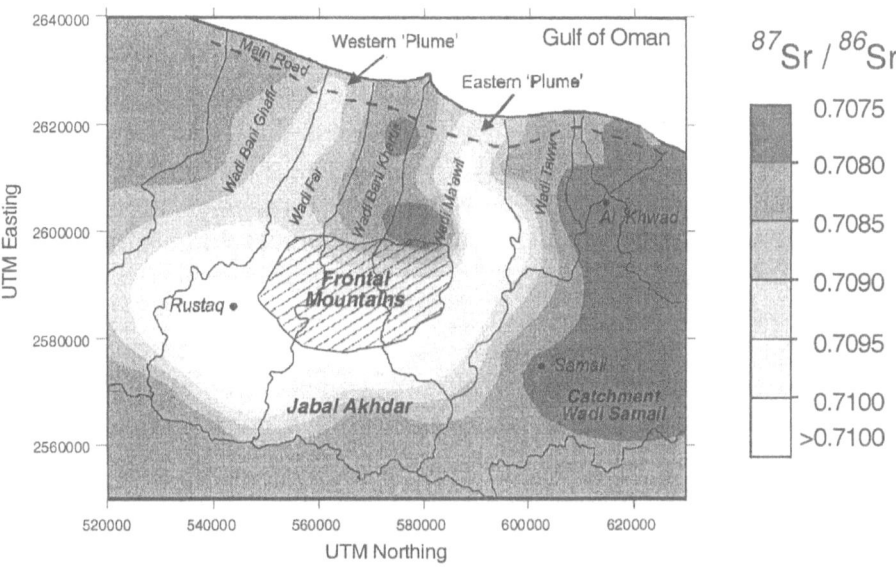

Figure 9: Contour plot of strontium isotope ratios (^{87}Sr/^{86}Sr) in groundwater samples. The contour lines are based on an interpolation of 120 groundwater samples, evenly distributed over the study area, using the gridding method "Kriging" (Surfer 7.0).

5. Discussion

5.1. *RECHARGE AREAS AND GROUNDWATER FLOW PATHS*

Systematic changes in the isotopic composition of precipitation with changes in elevation (i.e., altitude effect) allow an identification of the altitude at which groundwater recharge took place. In the Eastern Batinah region where the potential recharge areas span 3000 m in altitude from the mountains to the coast, oxygen and hydrogen isotope ratios are therefore extremely valuable tracers for groundwater recharge. The altitude dependent isotopic fractionation in the Batinah region results in a -0.15 to -0.3 ‰ decrease in $\delta^{18}O$ of rainfall per 100 m increase in altitude, starting at oxygen isotope ratios of approximately 0 to -1 ‰ $\delta^{18}O$ in the coastal regions (Stanger 1986, Weyhenmeyer *et al.* in press.). As expected from the rainfall data, groundwater samples with the most isotopically depleted isotope values of < -2.5 ‰ $\delta^{18}O$ and   < -10 ‰ δ^2H are found in the high-altitude regions of the Jabal Akhdar. However, groundwater samples from springs and wells located along the foothills of the mountains also have depleted isotope values and close to modern-day tritium activity values (Figures 7, 8 and 10, Table 1), indicating that these springs and wells are fed by water that originated at higher altitudes and circulated rapidly through the karstified Jabal Akhdar mountains. Surprisingly, depleted isotope values that are virtually unchanged from the mountains are also found along two distinct corridors (eastern and western plumes) that stretch across the coastal plain towards the coast. These two isotopically distinct plumes suggest that the high-altitude groundwater component is diverted by an ophiolite area (Frontal Mountains) at the foothills of the mountains into two groundwater streams that stretch through gaps in the ophiolite block across the 50 km wide Batinah plain towards the coast. The two plumes, therefore, can be considered as preferential pathways for groundwater that originated at higher altitudes (Figures 7 and 8). Along the eastern plume of isotopically depleted groundwater from the Jabal Akhdar mountains, the high-altitude isotope signal is maintained horizontally as well as vertically, down to a depth of at least 300 m (Figure 5, Table 2). Even in wells located near the coast, oxygen and hydrogen isotope values vary from -3 to -3.5 ‰ and from -10 to -20 ‰, respectively, suggesting that there is neither significant infiltration of the isotopically more enriched low-altitude rainfall nor another influence (e.g., seawater intrusion, brine mixing) along this groundwater flow path. If there was significant recharge on the coastal plain itself, a progressive enrichment in oxygen and hydrogen isotope ratios from the mountains towards the coast should be observed. As an example, the addition of 20 % low-altitude rainwater with an average isotopic signal of -0.5 ‰ $\delta^{18}O$ to a groundwater component with a high-altitude signal of -3.0 ‰ $\delta^{18}O$ would result in an isotopic shift of 0.5 ‰ towards more enriched values; a shift that is not seen in the groundwater data (Figure 7, Table 1). An upper limit of 20 % low-altitude recharge to the high-altitude groundwater component in the two plume areas can also be verified by the interpretation of tritium activity values. Assuming tritium activity values of 5 to 10 TU in present-day precipitation (Weyhenmeyer 2000), the addition of 20 % rainwater to a tritium-free groundwater component would result in an overall groundwater tritium concentration of 1 to 2 TU which is above the general detection limit of 0.5 to 1.0 TU. However, except for a few wells located in the downstream ophiolite areas (Areas A and C in Figure 2)

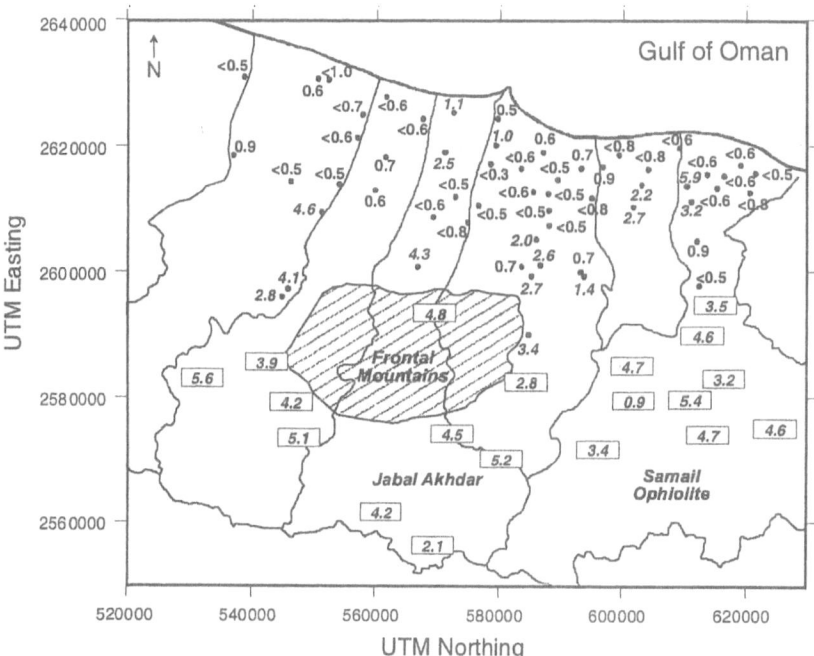

Figure 10: Tritium activities (in TU = tritium units, where one TU corresponds to one ^3H atom per 10 [8] atoms of hydrogen) for selected wells and springs of the Eastern Batinah coastal plain. White boxes around data are only drawn for clearer presentation of the tritium data in the mountain areas.

virtually none of the groundwater samples from the Eastern Batinah alluvial aquifer contain tritium (Figure 10). Even groundwater samples from wells located along the active wadi channels are tritium-free, demonstrating that within the plume areas infiltration of modern precipitation on the coastal plain does either not occur at all or occurs only very slowly (> 50 years). In either case, the low-altitude recharge component must be small compared to the component that has its origin in the Jabal Akhdar mountains.

In contrast to the two plume areas, local infiltration on the coastal plain itself appears to be the only source for groundwater recharge in adjacent areas, which are ophiolite areas (Frontal Mountains and wadi catchment Samail) or areas downstream of the ophiolites. In these areas, groundwater samples are isotopically enriched by about 2 to 3 ‰ in $\delta^{18}O$ and by 10 to 15 ‰ in δ^2H compared to samples from the Jabal Akhdar mountains and the two plume areas (Figures 7 and 8). These higher isotope ratios of -1.5 to 0 ‰ $\delta^{18}O$ and -5 to 5 ‰ δ^2H strongly suggest that low-altitude rainfall is the only source for groundwater in the ophiolite areas. Local recharge is also indicated by elevated tritium activity values measured in some of the shallow wells within and downstream of the ophiolites (Table 2, Figure 10). The large isotopic difference of groundwater samples collected upstream and downstream of the Frontal Mountains confirms the hypothesis that this ophiolite block forms a barrier to groundwater that infiltrates in the high-altitude regions of the Jabal Akhdar mountains. As a result, the high-altitude groundwater is more or less confined to the two plume areas.

The two plumes of groundwater that originates in the Jabal Akhdar mountains can also be distinguished from locally infiltrated groundwater in the adjacent areas by Ca/Mg ratios of groundwater samples. Molar Ca/Mg ratios within the eastern plume range from 1 to 1.5 (Figure 6), reflecting water-rock interaction with the limestone and dolomite formations of the Jabal Akhdar Mountains. In contrast, in areas adjacent to the plumes, which are areas within or downstream of ophiolite complexes, Ca/Mg ratios are significantly lower with values well below 0.5, indicating water-rock interaction of this groundwater component with magnesium-rich minerals from ophiolitic rocks (e.g., olivine, pyroxene, serpentine). Along the eastern and particularly the western plume, Ca/Mg ratios slightly decrease from the mountains towards the coast (Figure 6), either as a result of an addition of a second groundwater component that has interacted with the magnesium-rich ophiolite minerals or as the result of actual water-rock interaction of the high-altitude groundwater with ophiolitic gravel clasts from the alluvium. Since neither oxygen or hydrogen isotopes nor tritium activity values show any progressive change across the coastal plain, water-rock interaction appears to be the more likely explanation for the observed changes in groundwater chemistry. Likewise, the progressive increase in chloride, sodium and sulfate concentrations from the mountains to the coast is more likely the result of groundwater mixing with saline pore fluids rather than the addition of an evaporated rainwater component that would also have to affect the stable isotope ratios of the resulting groundwater mixture.

Strontium isotope ratios provide additional strong support for the lack of significant groundwater recharge on the coastal plain itself as well as for the presence of two preferential flow paths of groundwater that originated in the Jabal Akhdar mountains. Groundwater samples from the Jabal Akhdar mountains that are located upstream of the pre-Permian windows have strontium isotope ratios of around 0.7084, which reflects water-rock interaction with the carbonate rocks of the Hajar Super Group (Weyhenmeyer *et al.* submitted). Groundwater samples from areas located downstream of the pre-Permian windows, on the other hand, have higher Sr-isotope ratios between 0.7110 and 0.7131, which reflects water-rock interaction with the pre-Permian clastic sediments, because these are the only rocks in the region with Sr-isotope ratios > 0.7095 (Weyhenmeyer *et al.* submitted). The distinctive Sr-isotope signature from the pre-Permian rocks of the Jabal Akhdar is vertically and horizontally maintained along the two groundwater plumes that stretch across the coastal plain to the Gulf of Oman (Figure 9, Tables 1 and 2). A detailed transect through the eastern plume reveals a slight decrease in strontium isotope ratios from ~0.7110 to ~0.7100 (Table 1) near the foothills of the mountains, suggesting the addition of a second groundwater component or continued water-rock interaction.

In contrast to the high Sr-isotope ratios in the Jabal Akhdar mountains and two plume areas, groundwater samples from the Samail ophiolite catchment and the Frontal Mountains as well as from areas downstream of these ophiolite complexes, have significantly lower strontium isotope ratio between 0.7080 and 0.7088. These lower isotope ratios are an indication of water-rock interaction with the Tertiary gravel and calcite and magnesite lined fractures within the ophiolite (Weyhenmeyer *et al.* submitted) and, therefore, provide further evidence that the alluvial aquifer units downstream of the ophiolitic Frontal Mountains receive no recharge from the mountainous regions of the Jabal Akhdar but have only locally infiltrated rainfall sources.

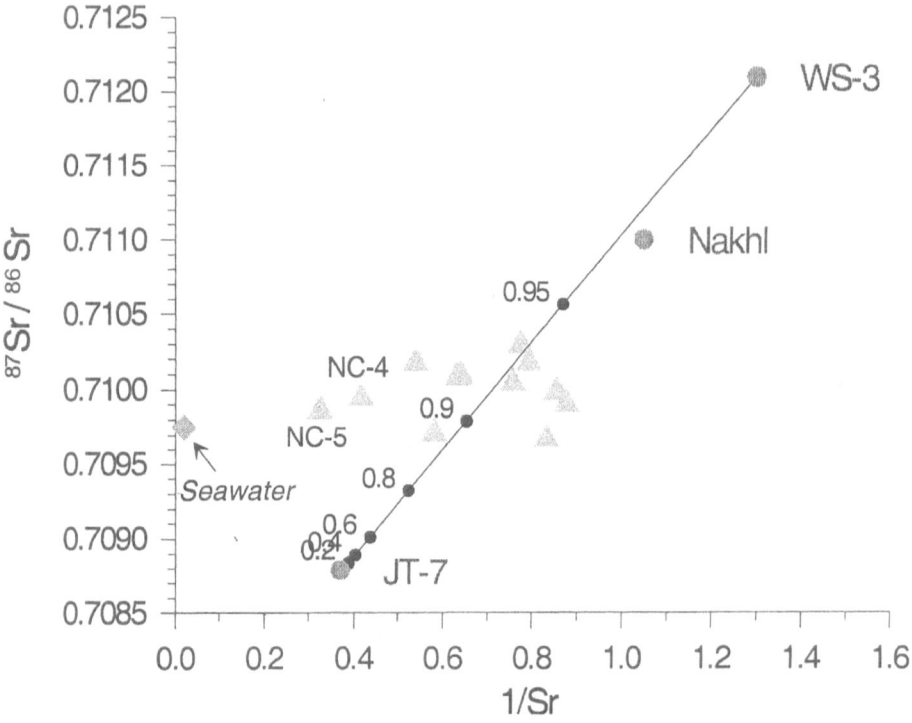

Figure 11: Theoretical mixing line between a groundwater component from the Jabal Akhdar (WS-3) and the coastal alluvial plain (JT-7) based on strontium concentrations and isotope ratios. The black dots represent mixing ratios between the two components (i.e., 0.9 means that 90 % of the groundwater component comes from WS-3 and 10 % from JT-7). The triangles mark groundwater samples from the 'eastern plume' that are listed in Table 1. The strontium value for seawater is represented by the rhomb.

In addition to providing qualitative information on groundwater flow, strontium isotope ratios can also be used to quantitatively estimate groundwater mixing between different groundwater components because strontium isotope ratios are considered a conservative tracer. This means that there is no isotopic fractionation during chemical and physical processes such as the dissolution or precipitation of minerals, evaporation or degassing and therefore general mass-balance mixing calculations can be performed based on strontium concentrations and isotope ratios (Faure 1986). In this study, such mixing calculations can, for example, be used to verify the upper limit of 20 % local recharge to the two high-altitude plumes that was established based on oxygen isotope values and tritium activities. As a representative mixing end-member for the high-altitude groundwater component from the Jabal Akhdar mountains, groundwater from well WS-3 was selected because it best represents the average chemical and isotopic composition of all groundwater samples collected in the mountains downstream of the pre-Permian windows. As a mixing end-member for groundwater that recharged locally on the coastal alluvial plain, well JT-7 was chosen. This well is located in the piedmont areas of the ophiolitic Frontal Mountains just a few kilometers to the west of the eastern groundwater plume. According to the mixing calculations, which are graphically illustrated in Figure 11, approximately 90 % of the groundwater within the eastern plume originates in the Jabal Akhdar mountains, and only the remaining 10 % can be attributed to recharge on the coastal plain itself. The calculations also show that the spring at Nakhl, located at the foothills of the Jabal Akhdar, is almost entirely fed by groundwater that has interacted with the pre-Permian sediments which form the core of the mountains. The calculations also reveal that additional groundwater infiltration only occurs in the piedmont areas and not across the entire coastal plain. If indeed significant amounts of rainwater would infiltrate over the entire coastal plain, this should result in a progressive decrease in strontium isotope ratios and associated changes in mixing ratios along the eastern and western plumes, which is not observed in the groundwater data (Figures 9 and 11, Table 1). Minor deviations of groundwater samples from the theoretical mixing line in Figure 11 can be explained by small amounts of seawater intrusion (< 5 %), particularly in two wells located near the coast (NC-4 and NC-5; Table 1).

Results from strontium mixing calculations are consistent with results from groundwater chemistry, stable isotope ratios and tritium activities and strongly support the hypothesis that the Jabal Akhdar is the main recharge area for large parts of the coastal alluvial aquifer. The combined chemical and isotopic evidence suggests that the groundwater from the high-altitude regions circulates rapidly through the karstified Jabal Akhdar to the foothills of the mountains where the water is diverted by the ophiolitic Frontal Mountains into two isotopically distinct plumes that stretch across the coastal plain towards the coast. Within these plumes, additional infiltration on the coastal plain only accounts for approximately 10 % of the total groundwater recharge. Assuming homogeneous rainfall rates across the coastal plain, the results suggest that recharge rates to the two plume areas are at least five to ten times higher than recharge rates in the adjacent areas. Interestingly, the two plumes of groundwater from the high altitude regions of the Jabal Akhdar do not follow the present-day active wadi channels (Figures 6-9). Instead, the western plume stretches across two surface wadi catchments (Wadi Bani Ghafir and Wadi Al Far), suggesting that in some areas there is significant groundwater flow across surface wadi catchment boundaries, which needs to be considered when assigning boundary conditions to hydrodynamic groundwater models.

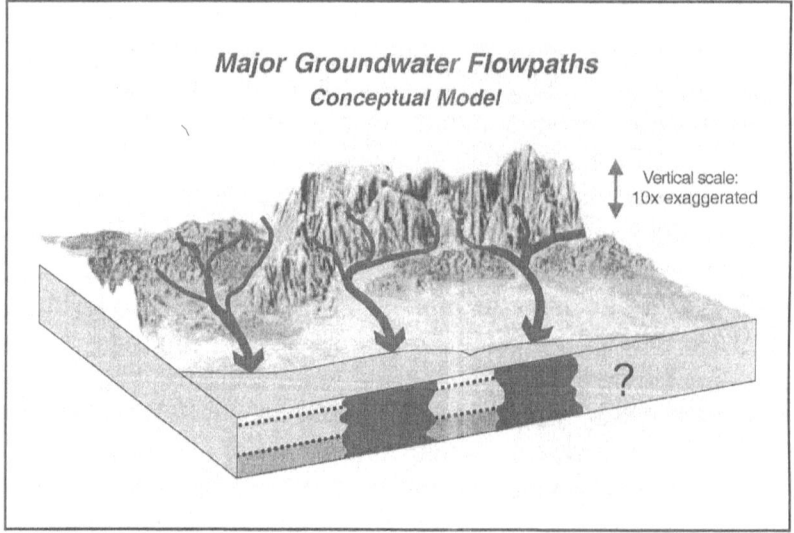

Figure 12: Simplified conceptual model showing the major groundwater flowpaths and aquifer units across the Eastern Batinah coastal plain as identified by geochemical and isotopic investigation. The black arrows mark the major groundwater flowpaths identified in this study. The vertical geochemical homogeneity of the two plumes of groundwater that were recharged mainly in the high altitude regions of the Jabal Akhdar mountains is indicated by the black blocks. In the adjacent areas the alluvial aquifer is geochemically stratified as indicated by the three different gray layers. Despite the high density of sampling, vertical and horizontal aquifer boundaries are not yet well established, mainly due to the lack of sufficiently detailed geological information. Aquifer boundaries are therefore only schematic. The vertical scale is approximately 10 x exaggerated.

5.2. *VERTICAL STRUCTURE OF THE ALLUVIAL AQUIFER*

Information on the vertical structure of the alluvial aquifer is very limited in the Eastern Batinah region due to a lack of wells with deep screen intervals. There are, however, three areas on the central Batinah plain where observation wells were drilled to depths of approximately 300 m (Areas A, B and C in Figure 2). A comparison of groundwater samples from shallow (< 50 m), intermediate (50 to 200 m) and deep (> 200 m) wells in these areas reveals a large variability in the chemical and isotopic composition of groundwater samples between the three different areas. In areas located downstream of ophiolite areas (Frontal Mountains and Samail Ophiolite; see Figure 2) there is chemical and isotopic evidence for a hydrochemical stratification of the alluvial aquifer. Groundwater samples from the different depth intervals are chemically distinct (Figure 5, Table 2), suggesting that the groundwater in these parts of the alluvial aquifer is not vertically well mixed. In addition, groundwater samples from shallow wells contain tritium, indicating modern recharge, while intermediate and deep groundwater samples are tritium-free (Table 2). ^{14}C activity values systematically decrease with depth to values of < 5 pmc (Figure 5), which corresponds to groundwater residence times of greater than 15,000 years (Weyhenmeyer *et al.* 2000), suggesting that this groundwater must have infiltrated sometime during the Late Pleistocene (15,000 to 24,000 yr BP). From the combined chemical and isotopic evidence we conclude that there is little effective vertical groundwater mixing in these downstream-ophiolite areas. The geochemical layering of the aquifer could either be the result of different groundwater flow dynamics (local versus regional groundwater flow), or could be caused by the existence of confining layers such as clay zones or cemented gravel units. Alternatively, the vertical layering could be the result of discrete major recharge events occurring at widely separated time intervals as it has been described in other arid regions (Calf 1978, Pearson and Swarzenski 1974). With the existing data it is impossible to completely resolve this issue. However, because data from seismic, transient electromagnetic sounding (TEM) and from borehole logs do not provide evidence for continuous layers that could function as aquitards and because the decrease in ^{14}C activities appears to be fairly continuous (Figure 5), different groundwater flowdynamics or discrete recharge events appear to be the more likely explanation for the observed geochemical layering of the aquifer.

In contrast to the areas downstream of the ophiolites, groundwater samples from the eastern plume of high-altitude water (Area B in Figure 2), are isotopically and chemically rather homogeneous throughout the entire depth range, suggesting that groundwater within the plume is well mixed to a depth of at least 300 m. In this area, there is no geochemical evidence for a layered aquifer system as has been previously proposed for the entire Batinah coastal plain (Gibb and Partners 1976, Graf 1983, Stanger 1986, Lakey *et al.* 1995, MacDonald and Partners 1989). Radiocarbon (^{14}C) data also support the lack of vertically distinct aquifer units in this area since there is no systematic change in ^{14}C activities with depth (Figure 5, Table 2). The observed vertical and horizontal differences in ^{14}C activity values within the eastern plume rather reflect the overall heterogeneity of the alluvium, particularly with respect to differences in the degree of cementation (Young *et al.* 1998).

To date, it is not clear why in certain areas of the coastal alluvial aquifer the groundwater appears to be well mixed to depths exceeding 300 m while in other areas the aquifer is geochemically stratified with depths. Deep borehole logs, seismic data and transient electromagnetic sounding indicate that the degree of cementation of the alluvial gravels is highly variable and that indeed the eastern and western plume areas may be associated with less cemented gravel units (MWR 1993, Young *et al.* 1998). Whether cementation in these areas did not take place, or whether previously cemented zones may have experienced extensive dissolution over time, thereby creating these preferential groundwater channels with higher hydraulic conductivities is as yet unknown. For the development of regional groundwater models, however, it is crucial to consider the described vertical and horizontal heterogeneity of the coastal alluvial aquifer, particularly for the definition of model boundary conditions.

6.   Summary and Conclusions

The combination of a variety of different geochemical and isotopic tracers has provided a more detailed three-dimensional picture of groundwater flow in the alluvial aquifer of the Eastern Batinah coastal plain than previously possible based on hydrological investigations alone (Figure 12). Chemical groundwater data as well as isotopic ratios of oxygen, hydrogen and strontium ($\delta^{18}O$, $\delta^{2}H$ and $^{87}Sr/^{86}Sr$) clearly indicate that the high altitude regions of the Jabal Akhdar mountains are the most significant recharge areas for large sections of the adjacent coastal alluvial aquifer. Groundwater that infiltrates in the Jabal Akhdar mountains circulates rapidly through the karstified Permian and Mesozoic limestone and dolomite formations to the base of the Oman Mountains as indicated by high tritium activity values in groundwater samples from wells and springs located along the foothills. The contribution of this high-altitude water to the coastal alluvial aquifer is, however, not homogeneous across the entire study area. A less permeable ophiolite complex at the foothills of the mountains diverts this high-altitude groundwater into two isotopically and chemically distinct groundwater streams (eastern and western plume) that stretch through two gaps in the ophiolite across the coastal plain to the Gulf of Oman. Interestingly, these two groundwater streams do not follow the present-day active wadi surface channels, suggesting significant cross-surface catchment flow in the subsurface. This was not generally considered in previous hydrodynamic models of individual catchment areas.

The chemical and isotopic composition of groundwater within the two plumes of groundwater with high-altitude origin is vertically and horizontally rather homogeneous. It is characterized by relatively high Ca-Mg ratios (>1) and by strontium isotope ratios above 0.7095, which is a clear indication of water-rock interaction with the pre-Permian basement rocks that make up the core of the Jabal Akhdar mountains. Furthermore, depleted oxygen and hydrogen isotope ratios that are characteristic for high-altitude rainfall confirm that groundwater within these plume areas originates in the Jabal Akhdar mountains. Mixing calculations based on strontium concentrations and isotope ratios suggest that within the eastern plume 90% of the groundwater has its origin in the high-altitude regions of the mountains while only approximately 10% of the total recharge occurs on the coastal plain itself. Low infiltration rates on the coastal plain are consistent with the lack of tritium in these groundwater samples.

In contrast to the plume areas, direct infiltration and recharge on the coastal plain itself is the only source for groundwater recharge in areas adjacent to the two plumes, downstream of ophiolite regions, as suggested by chemical and isotopic data. In these regions, groundwater is characterized by lower Ca/Mg ratios (< 0.5) and lower strontium isotope ratios (0.7080-0.7088), reflecting rock water-interaction with the limestone and ophiolite clasts that make up the alluvium. The groundwater is further characterized by enriched oxygen and hydrogen isotope ratios that are characteristic for low altitude rainfall.

The vertical structure of the alluvial aquifer is not as homogeneous across the Batinah coastal plain as had been previously suggested. While groundwater within the two plume areas has a fairly homogeneous chemical and isotopic signature down to a depth of at least 300 m (deepest wells), suggesting that groundwater is well mixed, there is evidence for geochemical layering in those aquifer units that are located downstream of the two main ophiolite areas (i.e., Frontal Mountains, Wadi Samail Catchment). This layering is evident from systematic changes in groundwater chemistry and ^{14}C activity values with depth. Groundwater in the upper parts of the aquifer (< 50 m) contains tritium indicating modern recharge while water in the deeper aquifer (> 200) is of Late Pleistocene origin.

Overall, the results from this geochemical study indicate considerable vertical and horizontal heterogeneity of the alluvial aquifer that needs to be taken into account when establishing regional water budget models for the area. This study, therefore, provides important new boundary conditions, which will help to better constrain future hydrodynamic models, thus allowing a better quantitative assessment of the available groundwater resources of the Eastern Batinah region.

Acknowledgements

The author wants to thank the Ministry of Regional Municipalities, Environment and Water Resources in the Sultanate of Oman for their permission to conduct this fieldwork, for their logistic and scientific support and for carrying out the chemical analyses of all water samples. In addition, I want to thank the Ministry for their authorization to publish this manuscript. Special thanks go to Zaher Al Suleimani, former Director of Research, to Ahmed Al Malki, Acting Director of Research, and to Dr. Phil Macumber, former Manager of the Groundwater Section for their highly appreciated assistance during various stages of this project. Thank you also to several other people from the Ministry for their invaluable help in the field, namely Mohammed Niwas, Khalifa Al Hinai, Salim Al Khanbashi, Salim Al Ma'shari and Hashim Al Balushi. Special gratitude is expressed towards my Ph.D. supervisors Professor Albert Matter, Dr. Niklaus Waber and Dr. Stephen Burns who initiated this research and provided essential scientific support during various stages of this project. The valuable manuscript reviews by F.J. Pearson and W.W. Wood were highly appreciated.

References

Adar, E. M., and C. Leibundgut, Application of tracers in arid zone hydrology, *IAHS Series of Proceedings and Reports, 232*, 452 pp., IAHS Press, Wallingford, 1995.

Bhatnagar, G. C., The hydrogeology of Wadi Samail upper catchment, *Report*, Ministry of Water Resources, Muscat, 1997.

Bullen, T. D., Krabbenhoft, D. P. and C. Kendall, Kinetic and mineralogic controls on the evolution of groundwater chemistry and ^{87}Sr / ^{86}Sr in a sandy silicate aquifer, northern Wisconsin, USA, *Geochim. Cosmochim. Acta*, *60*(10), 1807-1821, 1996.

Burns, S. J., and A. Matter, Geochemistry of carbonate cements in surficial alluvial conglomerates and their paleoclimatic implications, Sultanate of Oman, *J. Sed. Res.*, *A65*(1), 170-177, 1995.

Calf, G. E., The isotope hydrology of the Mereenie Sandstone Aquifer, Alice Springs, Northern Territory, Australia, *J. Hydrol.*, *38*, 343-355, 1978.

Cansult, Origin and age of groundwater in Oman - A study of environmental isotopes, *Report PAWR 86-7*, Public Authority for Water Resources, Muscat, 1986.

Chandrasekharan, H., S. V. Navada, S. K. Jain, S. M. Rao, and Y. P. Singh, Studies on natural recharge to the groundwater by isotope techniques in arid western Rajasthan, India, in *Estimation of Natural Groundwater Recharge, Series C(222)*, Mathematical and Physical Sciences, edited by I. Simmer, pp. 205-221, Reidel Publishing Company, Dordrecht, 1988.

Chinn, B. D., Davis, T. E., Hurst, R. W. and K. C. Leslie, Evaluation of seawater intrusion and mixing in the Dominguez Gap area, Los Angeles County, California, in *Diversity in Engineering Geology and Groundwater Resources*, AEG-GRA 1995 Annual Meeting, Association of Engineering Geologists, United States, 41 pp., 1995

Clark, I. D., and P. E. Fritz, *Environmental Isotopes in Hydrogeology*, CRC press LLC, New York, 311 pp., 1997.

Coleman, M. L., T. J. Shepherd, J. J. Durham, J. E. Rouse, and G. R. Moore, Reduction of water with zinc for hydrogen isotope analysis, *Anal. Chem.*, *54*, 993-995, 1982.

Collerson, K. D., Ullman, W. J. and T. Torgersen, $^{87}Sr/^{86}Sr$ ratios in the Great Artesian Basin, Australia, *Geology*, *16*, 59-6, 1988.

Cook, P. G., G. R. Walker, and I. D. Jolly, Spatial variability of groundwater recharge in a semiarid region, *J. Hydrol.*, *111*(1-4), 195-212, 1989.

Dansgaard, W., Stable isotopes in precipitation, *Tellus*, *16*(4), 436-468, 1964.

DeBreuck, W., Hydrogeology of Salt Water Intrusion, International Contributions to Hydrogeology, *A Selection of SWIM Papers, vol. 11*, IAHS press, Wallingford, 1991.

Edmunds, W. M., Characterization of groundwaters in semi-arid and arid zones using minor elements, in *Groundwater Quality*, edited by H. Nash, and G. J. McCall, pp. 19-30, Chapman & Hall, London, United Kingdom, 1995.

Edmunds, W. M., and C. B. Gaye, Estimating the spatial variability of groundwater recharge in the Sahel using chloride, *J. Hydrol.*, *156*, 47-59, 1994.

Edmunds, W. M. and N. R. G. Walton, A geochemical and isotopic approach to recharge evaluation in semi-arid zones; past and present, in *Arid-Zone Hydrology, Investigations with Isotope Techniques*, edited by J. C. Fontes, pp. 47-68, Int. At. Energy Agency, Vienna, 1980.

Epstein, S., and T. K. Mayeda, Variations of 18O/16O ratio in natural waters, *Geochim. Cosmochim. Acta*, *4*, 213, 1953.

Faure, G. E. and J. L. E. Powell, Strontium Isotope Geology, *Series: Minerals, Rocks and Inorganic Materials, Monograph Series of Theoretical and Experimental Studies 5*, 188 pp., Springer Verlag, Berlin, 1972.

Faure, G. E., *Principles of Isotope Geology*, 2nd edition, 589 pp., John Wiley and Sons, New York, 1986.

Fontes, J. C., Andrews, J. N., Edmunds, W. M., Guerre, A. and Y. Travi, Paleorecharge by the Niger River (Mali) deduced from groundwater geochemistry, *Water Resour. Res.*, *27*(2), 199-214, 1991.

Froehlich, K., and Y. Yurtsever, Isotope techniques for water resources in arid and semiarid regions, in *Application of Tracers in Arid Zone Hydrology*, edited by E. M. Adar, and C. Leibundgut, pp. 3-12, IAHS press, Wallingford, 1995.

Gibb, Sir A. and Partners, Water resources survey of northern Oman, *Report*, Ministry of Water Resources, Muscat, 1976.

Glennie, K. W., M. G. A. Boeuf, M. W. Hughes Clarke, M. Moody-Stuart, W. F. H. Pilaar, and B. M. E. Reinhardt, *Geology of the Oman Mountains*. Verh. Kon. Ned. Geol. Minjnb. Gen., 1974.

Gonfiantini, R., Standards for stable isotope measurements in natural compounds, *Nature*, *271*, 534-536, 1978.

Gonfiantini, R., Environmental isotopes in lake studies, in *Handbook of Environmental Isotope Geochemistry*, vol. 2, edited by P. Fritz, and J. C. Fontes, pp. 113-168, Elsevier, New York, 1986.

Graf, C. G., The Hydrology of the Sultanate of Oman, *Report*, Public Authority for Water Resources, Muscat, 1983.

Heathcote, J. A., Conceptual models for Eastern Batinah catchments, *Report by Hydrotechnica Ltd.*, Ministry of Water Resources, Muscat, 1993.

Howard, K. W. F., and E. Mullings, Hydrochemical Analysis of groundwater flow and saline intrusion in the Clarendon Basin, Jamaica, *Groundwater*, *34*(5), 801-810, 1996.

Issar, A., Nativ, R., Karnieli, A. and J. R. Gat, Isotopic evidence of the origin of groundwater in arid zones, in *Isotope Hydrology*, pp. 85-104, International Atomic Energy Agency, Vienna, 1984.

JGR, Special Issue on the Samail Ophiolite, *J. Geophys. Res.*, 86(4), 2495-2782, 1981.

Johnson, T. M. and D. J. DePaolo, Interpretation of isotopic data in groundwater-rock systems; model development and application to Sr isotope data from Yucca Mountain, *Water Resour. Res.*, *30*(5), 1571-1587, 1994.

Katz, B. G. and T. D. Bullen, The combined use of $^{87}Sr/^{86}Sr$ and carbon and water isotopes to study the hydrochemical interaction between groundwater and lakewater in mantled karst, *Geochim. Cosmochim. Acta*, *60*(24), 5075-5087, 1996.

Lakey, R., P. Easton, and H. Al Hinai, Eastern Batinah Resource Assessment - Numerical Modeling, *Report*, Ministry of Water Resources, Muscat, 1995.

Le Métour, J., J. C. Michel, F. Béchennec, J. P. Platel, J. E. and J. Roger, *Geology and Mineral Wealth of the Sultanate of Oman*, Directorate General of Minerals, Ministry of Petroleum and Minerals, Muscat, 1995.

Lippard, S. J., A. W. Shelton, and I. E. Gass, *The Ophiolite of Northern Oman, Memoir, 11*, Geological Society London, 178 pp., 1986.

Lyons, W. B., Tyler, S. W., Gaudette, H. E. and D. T. Long, The use of strontium isotopes in determining groundwater mixing and brine fingering in a playa spring zone, Lake Tyrrell, Australia, *J. Hydrol.*, *167*(1-4), 225-239, 1995.

MacDonald, M. and Partners, Groundwater recharge schemes for the Barka - Rumais area, *Report*, Ministry of Water Resources, Muscat, 1989.

Macumber, P. G., J. M. Niwas, A. Al Abadi, and R. Seneviratne, A new Isotopic Water Line for Northern Oman, *Proceedings of 'The Third Gulf Water Conference'*, Muscat, 1997.

McNutt, R. H., Frape, S. K., Fritz, P., Jones, M. G. and I. M. MacDonald, The $^{87}Sr/^{86}Sr$ values of Canadian Shield brines and fracture minerals with applications to groundwater mixing, fracture history, and geochronology, *Geochim. Cosmochim. Acta*, *54*(1), 205-215, 1990.

Mann, A., S. S. Hanna, and S. C. Nolan, The post-Campanian tectonic evolution of the central Oman Mountains: Tertiary extension of the Eastern Arabian Region. *Special Publication, 49*, Geological Society, London, 1990.

Moore, L. J., Moody, J. R., Barnes, I. L., Gramich, T. J., Murphy, T. J., Paulsen, P. J. and W. R. Shields, Trace determination of rubidium and strontium in silicate glass standard reference materials, *Analytical Chem.*, *45*, 2384-2387, 1973.

MWR, Drilling completion report contract 92-21 Eastern Batinah, *Report MWR-93-49*, Ministry of Water Resources, Muscat, 1993.

MWR, Water Quality Laboratory: Methods of Analyses, *Report*, Ministry of Water Resources, Muscat, 1997.

Neumann, K. and S. Dreiss, Strontium87/strontium86 ratios as tracers in groundwater and surface waters in Mono Basin, California, *Water Resour. Res.*, *31*(12), 3183-3193, 1995.

Pearson, F. J. Jr., and W. V. Swarzenski, ^{14}C Evidence for the Origin of Arid Region Groundwater, Northeastern Province, Kenya, in *Isotope Techniques in Groundwater Hydrology Vol. II: Symposium Proceedings*, International Atomic Energy Agency, Vienna, 95-109, 1974.

Pedgely, D. E., Cyclones along the Arabian Coast. *Weather*, *24*, 456-486, 1969.

Peters, Tj., A. Nicolas, and R. G. E. Coleman, Ophiolite Genesis and Evolution of the Oceanic Lithosphere, *Series: Petrology and Structural Geology, vol. 5*, 903 pp., Kuwer Academic Publishers, Dordrecht, 1991.

Rao, S. M., S. K. Jain, A. R. Navada, and K. Shivana, Isotopic studies on sea water intrusion and interrelationships between water bodies: some field examples, in *Isotope Techniques in Water Resources Development, Proceedings of a Symposium*, pp. 403-425, Int. At. Energy Agency, Vienna, 1987.

Robertson, A. H. F., M. P. Searle, and A. C. E. Ries, The Geology and Tectonics of the Oman Region, *Special Publication, 49*, 845 pp., Geological Society London, 1990.

Schyfsma, E., Climate, in *Quaternary Period in Saudi Arabia 1: Sedimentological, Hydrogeological, Hydrogeochemical, Geomorphological and Climatological*

Investigations in Central and Eastern Saudi Arabia, edited by S. S. Al-Sayari and J. G. Zotl, pp. 31-44, Springer Verlag, 1978.

Sharma, M. L., and M. W. Hughes, Groundwater recharge estimation using chloride, deuterium and oxygen-18 profiles in the deep coastal sands of western Australia, *J. Hydrol.*, 81, 93-109, 1985.

Simmers, I., Natural groundwater recharge estimation in (semi-)arid zones; some state-of-the-art observations, in *Proceedings of the Sahel forum on the state-of-the-art of hydrology and hydrogeology in the arid and semi-arid areas of Africa, edited by G. E. Stout, and M. Demissie, pp. 373-386, 1990.

Singh, D., Y. P. Singh, S. P. Bairwa, C. P. Porwal, and K. M. Mathur, Isotopic and hydrochemical study of groundwater in Shahgarh Bulge area of Jaisalmer. *J. Appl. Hydrol.*, 9, 84-87, 1996.

Stanger, G., The Hydrogeology of the Oman Mountains, Ph.D. Thesis, Open University, U.K, 1986.

Sukhija, B., D. Reddy, and I. Vasanthakumar-Reddy, Droughts as means of delineation of areas prone to seawater intrusion, in *International Workshop on Appropriate Methodologies for Development and Management of Groundwater Resources in Developing Countries, New Delhi*, pp. 733-741, Oxford and IBH Publishing Company, 1989.

Sukhija, B. S., P. Nagabhushanam, and D. V. Reddy, Groundwater recharge in semi-arid regions of India; an overview of results obtained using tracers. *Hydrogeol. J.*, 4(3), 50-71, 1996.

Surfer 7.0, Software for *Contouring and 3D Surface Mapping for Scientists and Engineers,* Golden Software Inc., Colorado, 1991.

Taylor, D., P. D. Jones, and T. M. L. Wigley, Rainfall in Oman: Data Acquisition, Statistics and Climatology, *Report*, Ministry of Water Resources, Muscat, 1990.

Tetratech, International Inc., Evaluation of alternative groundwater development schemes for the Wadi Samail Aquifer, *Report,* Ministry of Water Resources, Muscat, 1980.

Tomlinson, R.H. and A. K. Das Gupta, The use of isotope dilution in determination of geologic age of minerals, *Can. J. Chem.*, 31, 909-914, 1953.

Weyhenmeyer, C. E., S. J. Burns, H. N. Waber, W. Aeschbach-Hertig, R. Kipfer, H. H. Loosli, and A. Matter, Cool glacial temperatures and changes in moisture source recorded in Oman groundwaters, *Science*, 287, 842-845, 2000.

Weyhenmeyer, C. E., Burns, S. J., Waber, H. N., and A. Matter, Isotope study of moisture sources, recharge areas and groundwater flow paths within the Eastern Batinah Coastal Plain, Sultanate of Oman, *Water Resour. Res., in press.*

Weyhenmeyer, C. E., Waber, H. N., Burns, S. J., Kramers, J. and A. Matter, Strontium isotopes (^{87}Sr/^{86}Sr) as a tracer for groundwater movement and mixing in a coastal alluvial aquifer of Northern Oman, *submitted*

Weyhenmeyer, C. E., Origin and evolution of groundwater in the alluvial aquifer of the Eastern Batinah coastal plain, Sultanate of Oman – A hydrogeochemical approach, Ph.D. thesis, Geological Inst., Univ. of Bern, Switzerland, 2000.

Young, M. E., R. G. M. De Bruijn, and A. S. Al-Ismaily, Exploration of an alluvial aquifer in Oman by time-domain electromagnetic sounding. *Hydrogeol. J.*, 6, 383-393, 1998.

Well No.	Location	Coast dist. [km]	Screen [m.b.s.]	Temp. [deg C]	pH	Cond. [µS/cm]	Cl [mg/l]	Na [mg/l]	Ca [mg/l]	Mg [mg/l]	SO4 [mg/l]	Sr [mg/l]	HCO3 [mg/l]	TDS [mg/l]	δ18O [‰]	δ2H [‰]	87Sr/86Sr	Tritium [TU]
Spring (S-1)	Mountains	71	—	24.0	7.7	626	14	10	66	35	24	0.4	375	524	-3.58	-15.0	0.70854	—
JA-5	Mountains	69	open	24.5	7.8	636	26	13	69	38	22	0.3	326	494	-3.88	-16.2	0.70823	2.1 ± 1.0
Spring (S-146)	Mountains	50	—	24.6	7.8	514	40	21	47	31	53	0.7	175	368	-3.35	-13.2	0.71232	5.2 ± 0.8
Spring (Nakhl)	Piedmont	38	—	38.6	7.4	633	62	44	59	24	48	0.8	238	476	-3.08	-12.0	0.71100	2.8 ± 0.8
428PW	Piedmont	31	20-60	34.6	7.4	940	88	69	69	47	85	1.1	322	681	-2.99	-13.2	0.70990	3.4 ± 1.0
422PW	Alluvial Plain	21	open	33.2	7.5	1276	167	121	71	58	119	1.8	316	854	-3.06	-12.2	0.70170	2.6 ± 0.5
421PW	Alluvial Plain	17	open	35.0	7.3	1122	150	102	72	46	103	1.2	262	736	-3.08	-13.0	0.71009	<0.5
21/1	Alluvial Plain	15	280-288	36.5	7.5	1118	136	103	62	43	124	1.6	274	796	-3.33	-14.9	0.71014	<0.5
JT-5	Alluvial Plain	11	23-34	35.2	7.5	949	124	85	65	39	104	1.3	212	636	-3.45	-14.8	0.71004	<0.5
419PW	Alluvial Plain	10	open	34.4	7.4	1000	144	77	70	42	116	1.7	209	660	-3.24	-13.1	—	<0.8
ADW-7	Alluvial Plain	8	70-87	36.1	7.6	1096	223	80	69	48	72	1.1	200	726	-3.36	-16.2	0.71017	<0.5
NC-4	Coast	6	5-103	33.8	7.4	2230	534	242	123	72	209	2.4	190	1372	-3.29	-15.4	0.70994	0.7 ± 0.5
NC-5	Coast	5	53-103	34.2	7.4	2042	449	171	125	74	182	2.7	198	1202	-3.51	-14.7	0.70984	0.6 ± 0.6
WS-3	Mountains	46	38-350	33.6	7.4	721	59	61	55	31	71	0.8	246	524	-2.82	-12.0	0.71210	4.5 ± 1.0
JT-7	Alluvial Plain	23	117-140	34.6	8.1	1237	168	128	66	63	89	2.7	372	889	-2.85	-10.3	0.70879	2.7 ± 1.0

Table 1. Geochemical and isotopic properties of selected groundwater samples from wells and springs located along Transect A (see Figure 2). Wells WS-3, located along the transect and JT-7, located to the west of the transect are used for mixing calculations based on strontium concentrations and isotopes.

Well No.	Screen [m.b.s.]	Temp. [°C]	pH	Cond. [µS/cm]	Cl [mg/l]	Na [mg/l]	Ca [mg/l]	Mg [mg/l]	SO_4 [mg/l]	Sr [mg/l]	HCO_3 [mg/l]	TDS [mg/l]	Water Type	$\delta^{18}O$ [‰ VSMOW]	δ^2H [‰ VSMOW]	$\delta^{13}C_{DIC}$ [‰ VPDB]	$^{87}Sr/^{86}Sr$	^3H [TU]	^{14}C [pmc]
Area A - Wadi Bani Kharus																			
Shallow Wells (<50m)																			
UBKD-3	11-28	36.3	7.5	437	47	22	21	29	27	0.3	148	294	$Mg\text{-}Na\text{-}Ca\text{-}HCO_3\text{-}Cl\text{-}SO_4$	-1.72	-6.9	-12.4	0.70867	4.3 ± 1.0	—
JT-57	24-35	35.0	7.9	416	44	26	16	26	27	0.5	143	282	$Mg\text{-}Na\text{-}Ca\text{-}HCO_3\text{-}Cl\text{-}SO_4$	-1.74	-7.2	-11.8	0.70856	2.5 ± 0.9	—
Intermediate Wells (50-200m)																			
JT-11	71-82	37.5	7.8	860	117	118	30	27	86	0.7	188	566	$Na\text{-}Mg\text{-}Ca\text{-}HCO_3\text{-}Cl\text{-}SO_4$	-1.97	-5.6	-10.8	—	<0.8	58.2 ± 0.3
JT-24	86-97	37.9	7.8	451	56	42	19	23	26	0.3	147	313	$Na\text{-}Mg\text{-}Ca\text{-}HCO_3\text{-}Cl\text{-}SO_4$	-1.75	-7.4	-9.6	—	<0.5	42.0 ± 0.1
JT-68	116-140	38.0	8.1	500	44	70	11	15	41	0.2	156	337	$Na\text{-}Mg\text{-}Ca\text{-}HCO_3\text{-}Cl\text{-}SO_4$	-2.37	-9.1	-10.6	0.70855	<0.6	21.4 ± 0.2
Deep Wells (>200m)																			
21/3	210-218	38.4	7.6	974	155	142	27	27	97	0.6	165	613	$Na\text{-}Mg\text{-}Ca\text{-}HCO_3\text{-}Cl\text{-}SO_4$	-1.93	-14.2	-9.7	—	<0.5	13.1 ± 0.2
21/9B	218-226	38.5	7.6	708	90	97	10	29	65	0.4	188	479	$Na\text{-}Mg\text{-}Ca\text{-}HCO_3\text{-}Cl\text{-}SO_4$	-2.00	-14.8	-11.6	0.70843	<0.5	7.0 ± 0.1
Area B - Wadi Ma'awil (Eastern Plume)																			
Shallow Wells (<50m)																			
JT-5	23-34	35.2	7.5	949	124	85	65	39	104	1.3	212	636	$Na\text{-}Mg\text{-}Ca\text{-}HCO_3\text{-}Cl\text{-}SO_4$	-3.45	-14.8	-10.4	0.71004	<0.5	60.2 ± 0.3
MD-1	25-50	35.0	7.5	729	116	64	46	29	46	0.9	176	477	$Na\text{-}Mg\text{-}Ca\text{-}Cl\text{-}HCO_3\text{-}SO_4$	-3.52	-17.6	-9.4	0.70948	<0.6	36.1 ± 0.2
Intermediate Wells (50-200m)																			
RE-4P	51-101	36.2	7.7	811	109	72	48	32	66	1.1	211	538	$Na\text{-}Mg\text{-}Ca\text{-}HCO_3\text{-}Cl\text{-}SO_4$	-3.39	-14.5	-9.7	0.70998	<0.6	45.5 ± 0.2
ADW-7	70-87	36.1	7.6	1096	223	80	69	48	72	1.1	200	726	$Na\text{-}Mg\text{-}Ca\text{-}HCO_3\text{-}Cl\text{-}SO_4$	-3.36	-16.2	-10.7	0.71017	<0.5	43.2 ± 0.2
Deep Wells (>200m)																			
21/1	280-288	36.5	7.5	1118	136	103	62	43	124	1.6	274	796	$Na\text{-}Mg\text{-}Ca\text{-}HCO_3\text{-}Cl\text{-}SO_4$	-3.33	-14.9	-10.5	0.71014	<0.5	55.8 ± 0.2
21/2	202-210	37.4	7.3	548	71	44	32	20	38	0.6	142	347	$Na\text{-}Mg\text{-}Ca\text{-}HCO_3\text{-}Cl\text{-}SO_4$	-3.96	-19.0	-10.3	0.70945	<0.6	39.8 ± 0.5
Area C - Al Khwad Fan																			
Shallow Wells (<50m)																			
SLU-2B	15-23	35.2	7.6	610	67	90	24	14	61	0.3	167	423	$Na\text{-}Mg\text{-}Ca\text{-}HCO_3\text{-}Cl\text{-}SO_4$	-2.05	-7.2	-13.3	0.70807	3.5 ± 0.9	98.8 ± 0.3
RGS-2U	22-201	31.8	7.9	744	89	73	19	43	82	1.3	194	500	$Mg\text{-}Na\text{-}Ca\text{-}HCO_3\text{-}Cl\text{-}SO_4$	-1.79	-6.9	-9.6	0.70867	6.7 ± 1.1	89.3 ± 0.5
RGS-5F	< 50	33.3	7.7	1093	183	94	28	65	97	0.6	189	656	$Mg\text{-}Na\text{-}Ca\text{-}HCO_3\text{-}Cl\text{-}SO_4$	-1.13	-3.3	-9.0	0.70871	1.1 ± 1.0	87.1 ± 0.3
Intermediate Wells (50-200m)																			
BZ-4	70-78	35.3	7.5	1446	305	194	31	47	112	0.5	217	906	$Na\text{-}Mg\text{-}Ca\text{-}Cl\text{-}HCO_3\text{-}SO_4$	-0.74	-6.3	-8.2	0.70859	<0.6	35.1 ± 0.3
21/7	120-135	36.0	7.7	1397	215	211	30	36	182	0.7	224	898	$Na\text{-}Mg\text{-}Ca\text{-}Cl\text{-}HCO_3\text{-}SO_4$	-0.77	-7.1	-9.0	0.70844	<0.6	24.6 ± 0.2
JT-31	140-148	35.2	7.6	2240	509	265	79	74	173	0.8	192	1292	$Na\text{-}Mg\text{-}Ca\text{-}Cl\text{-}HCO_3\text{-}SO_4$	-0.81	-5.7	-8.4	—	<0.5	23.8 ± 0.2
Deep Wells (>200m)																			
KWD-1	321-366	33.6	9.4	1005	196	207	1	5	15	0.1	117	541	$Na\text{-}Mg\text{-}Ca\text{-}HCO_3\text{-}SO_4$	-1.53	-13.8	-9.8	0.70818	<0.5	17.3 ± 0.2
21/6	289-297	33.4	7.9	1136	180	192	15	19	87	0.3	247	740	$Na\text{-}Mg\text{-}Ca\text{-}HCO_3\text{-}SO_4$	-0.37	-6.3	-8.0	0.70860	<0.6	13.2 ± 0.1
RGS-2L	319-325	32.6	7.8	1218	187	200	61	65	137	0.4	135	785	$Na\text{-}Mg\text{-}Ca\text{-}HCO_3\text{-}SO_4$	+0.01	-2.4	-11.1	0.70769	<0.8	4.6 ± 0.2
KWD-3L	216-366	34.5	8.1	1506	255	197	35	54	220	0.3	218	979	$Na\text{-}Mg\text{-}Ca\text{-}HCO_3\text{-}SO_4$	-1.18	-12.9	-9.8	0.70748	<0.6	3.8 ± 0.1

Table 2. Selected chemical and isotopic properties of groundwater samples from different depths intervals and three different areas corresponding to areas A, B and C in Figure 2.

A combined isotopic tool box for the investigation of water-rock interaction: An overview of Sr, B, O, H isotopes and U-series in deep groundwaters from the Vienne granitoid (France)

Ph. Négrel [1], J. Casanova [1], W. Kloppmann [1], J. F. Aranyossy [2]

[1]BRGM, Water Department, BP 6009, F 45060 Orléans Cedex 2, France
[2]ANDRA, 1 rue J. Monnet, F 92298, Châtenay-Malabry, France

Abstract

An "isotopic toolbox" comprising strontium, stable boron, oxygen and hydrogen isotopes as well as the uranium series has been used to place constraints on the origin deep groundwaters from the crystalline basement of the Vienne region in France.

For deep groundwaters from the Vienne granitoid, the conclusions drawn from Sr and B isotope compositions converge towards a marine origin modified by water rock interaction (WRI). The B isotope ($\delta^{11}B$) compositions of the most saline waters lie close to those of present-day seawater while the $^{87}Sr/^{86}Sr$ ratios are slightly higher than those of the Jurassic ocean (i.e. the last transgressive episode) as a result of subsequent WRI. This is in agreement with a model developed to determine the $^{87}Sr/^{86}Sr$ ratio of water after interaction with granitoid. The stable O-H isotope data suggest that, as is the case for waters from many crystalline shields, WRI under low temperature-low porosity conditions has modified the ^{18}O and/or ^{2}H compositions of the liquid phase. Example where ^{234}U has been preferentially mobilised relative to ^{238}U is presented and discussed in relation to two main processes of U mobility: preferential solution related to oxidative paleo-events in the upper zone (0 - 400m depth) of the site and alpha-recoil induced solution in the lower zone (400 - 900m).

I. Stober and K. Bucher (eds.), Water-Rock Interaction, 39–59.

1. Introduction

As summarised by Fritz and Frape (1987), a large variety of geochemical tools have been applied over the last few decades in order to investigate the origin of highly saline waters in deep geological formations. Highly saline waters from crystalline basement rocks have a controversial origin. The salinity of groundwater may be autochthonous, originating from water-rock interaction processes, or may allochthonous, with the salinity derived from seawater intrusion, residual evaporite brines, or evaporite dissolution and migration of secondary fluids. The origin of salinity in deep waters has long been a matter of debate (Fritz and Frape, 1987, Bottomley et al., 1994, 1999, Beaucaire et al., 1999, and references therein).

The combination of individual isotopic tools in groundwater studies helps to constrain hypotheses regarding the nature of water-rock interactions and possible end-members involved in mixing processes. A tool-box containing strontium, boron, oxygen and hydrogen isotopes has been tested in several geological environments in order to elucidate the origin of salinity (Casanova et al., 1998, 1999a, b, Kloppmann et al., 1999a, b).

Due to the large relative mass difference between ^{10}B and ^{11}B and the high chemical reactivity of boron, isotope fractionation accompanying natural chemical reactions produces large variations in the $^{11}B/^{10}B$ ratios of natural samples from different geological settings(Barth, 1993). This results in a large isotopic contrast between potential mixing sources, but also in process-specific changes of the isotopic signature. Boron is omnipresent in groundwaters and may behave as a conservative tracer (e.g. Aggarwal et al., 2000, Barth, 2000). Boron isotopes are, for instance, insensitive to redox-processes unlike sulphur and oxygen isotopes. In other cases (e.g. high clay contents in the solid matrix or coprecipitation with solid phases such carbonates and evaporites) a partial phase change can lead to significant isotopic fractionation (Vengosh et al., 1991a, b, c, Palmer et al., 1987).

Strontium isotopes exhibit no detectable fractionation by any natural process. Given the relatively short timescale of the processes studied, the observed differences in $^{87}Sr/^{86}Sr$ ratios can be interpreted as due to the addition of Sr derived from various sources with different isotopic compositions (Faure, 1986, Négrel et al., 1993, Gaillardet et al., 1997). Thus, $^{87}Sr/^{86}Sr$ ratio variations within a hydrosystem can provide information about those sources, the mixing proportions of groundwater components, or the degree of water-rock interaction (Albarède and Michard, 1987, Andersson et al., 1994, Négrel et al., 1999a).

If no evaporation or exchange with dissolved gases occurs in the groundwater body, the stable isotopes of oxygen and deuterium in the water molecule behave as conservative tracers which reflect mixing of different recharge components in the groundwater. Under particular conditions (low water-rock ratios, long residence times or high reservoir temperatures) water-rock interactions may modify the stable isotope composition (Frape and Fritz, 1982) of groundwaters.

Uranium-series disequilibrium studies may provide further information because radioactive disequilibria observed in crystalline basement rock always reflect low-temperature water-rock interaction (Suksi et al., 2001). Finally, uranium concentration and activity ratios ($^{234}U/^{238}U$) in groundwaters can be used as potential signals indicating changes in redox conditions, in order to document recent fluid circulation in the crystalline basement (Osmond and Ivanovich, 1992).

In this paper, we present a synthesis of results obtained using the "isotopic tool-box" of strontium (Négrel et al., 1997, Négrel et al., in press), boron (Négrel et al., 1999b, Casanova et al., in press), stable oxygen and hydrogen (Kloppmann et al., in prep), and U-series (Casanova and Aranyossy, 1998) in deep groundwaters collected in the Vienne granitoids (West France) during a geological characterisation program. The research was undertaken to assess the feasibility of an underground laboratory in granitoid basement. As part of the Nuclear Waste Management Program managed by ANDRA, 17 boreholes were drilled in the Vienne granitoids, and the deep groundwaters from these holes were sampled for further geochemical investigation.

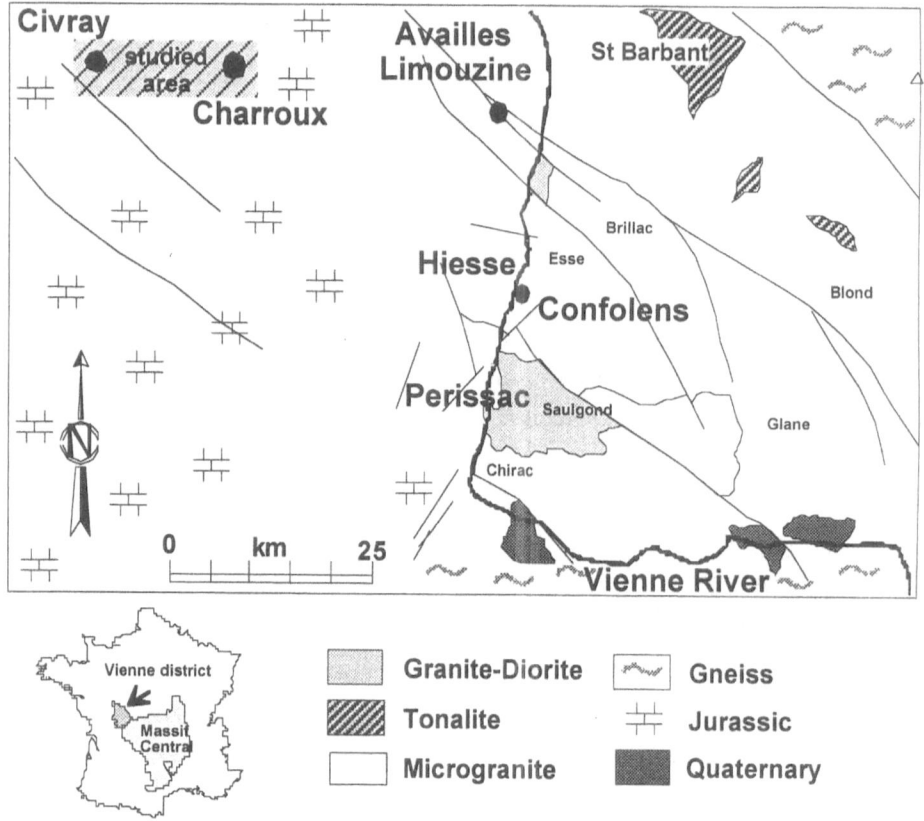

Figure 1a: General location of the Vienne granitoids

2. Site Location, Materials and Methods

The Vienne granitoids are located west of the French Massif Central (figure 1a). On both sides of the Vienne river, Devonian metamorphic units (gneisses, migmatites, amphibolites) and Carboniferous granitic terranes form an old peneplained and weathered bedrock (Capdeville et al., 1983, Mourier et al., 1989). Further west in the Vienne district, the site studied by ANDRA is composed of several magmatic intrusions comprising several mineralogical and chemical facies, partially covered by sedimentary deposits. The site framework, illustrated in figure 1b, shows the main bedrock facies (calc-alkaline, peraluminous, sub-alkaline) and the associated boreholes. The Jurassic sedimentary cover (Hantzpergue et al., 1997) consists of 150 m of bioclastic limestones (Dogger) containing an open aquifer, argillaceous limestones and marls, dolomite, sands and sandstones (Lower Liassic: Infra-Toarcian) containing a confined aquifer.

Deep groundwater samples from the granitoid basement, shallow from the two overlying sedimentary aquifers (Dogger/Infra-Toarcian) were collected from the different boreholes (CHA106, 107, 113, 115, 117, 212 and 312, figure 1b), during a hydrogeological survey to a maximum depth of 650 m (Matray et al., 1998). Mineral spring waters emerging from the granitoid outcrops in the Limousin region (namely Availles springs) and mineral spring waters from the crystalline basement of the Massif Central and Pyrenees represent various degrees of water-rock interaction (Krimissa, 1995, Mossadik, 1997, Négrel et al., 1997b, c) and were used for comparative purposes.

Several features have been identified by previous geochemical investigations of these various waters (Matray et al., 1998):

- waters from sedimentary aquifers are dilute, with a total dissolved solid contents (TDS) generally lower than 0.5 mg/l,

- $Ca-HCO_3$ and $Ca-Na-HCO_3$ groundwater types are observed in the Dogger and Infra-Toarcien aquifers, respectively.

- deep groundwaters are more saline, with TDS contents ranging from 0.5 to 10 g/l and having a $Na-Cl-SO_4$ composition.

Chemical analyses of the water samples were performed by ICP-MS for Sr, B, U and Th. Standard mass spectrometric techniques (Kloppmann et al., in prep) were applied for the stable isotope determinations. The $^{87}Sr/^{86}Sr$ and $\delta^{11}B$ measurements were performed by solid-source mass spectrometry (Casanova et al., in press, Négrel et al., in press), as were $^{234}U/^{238}U$ isotope ratio measurements (Casanova and Aranyossy, 1998).

3. Results and Discussion

3.1. Salinity origin in deep groundwaters constrained by B isotopes

High boron contents are observed in the deep groundwaters (range of 2-6 mg/l) and the highest boron contents surpass the seawater

44 PHILIPPE NÉGREL ET AL.

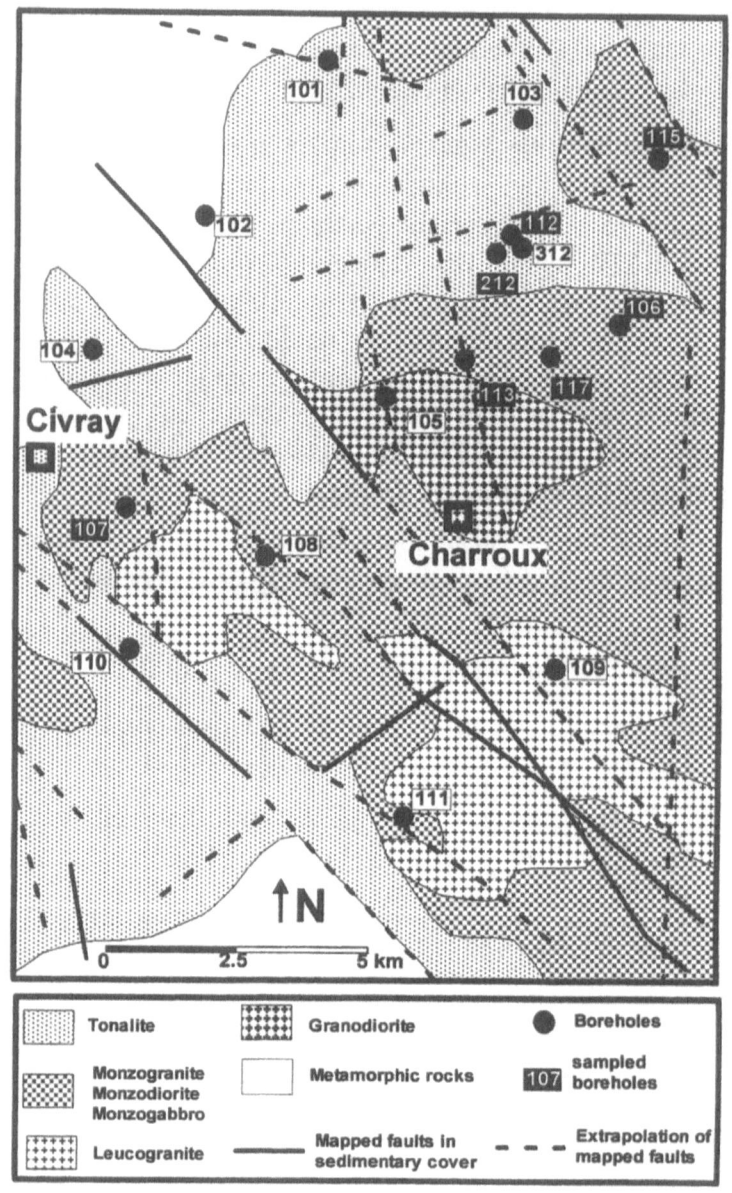

Figure 1b: Sample locations (overview Fig. 1a); Calc-alkaline series (i) tonalites - leucogranites CHA101-103, CHA112, CHA212 and (ii) diorites CIV104 and CHA110. Leucogranite CHA109. Sub-alkaline facies; (i) monzodiorites and monzogranites CHA106, CHA108, CHA111, CHA113, CHA117 and CIV107 and (ii) monzogabbrodiorites (CHA115).

concentration (4.6 mg/l, Bruland, 1983). The deep groundwaters display high $\delta^{11}B$ values $(+24.9 < \delta^{11}B < +36.1‰)$, slightly ^{11}B depleted with respect to present-day seawater ($\delta^{11}B \approx +40‰$, Vengosh, 1992, Barth, 1993). Shallow groundwaters from the Infra-Toarcian aquifer have intermediate $\delta^{11}B$ values $(+13.0/+13.2‰)$ whereas shallow groundwaters from the Dogger aquifer show the lowest $\delta^{11}B$ value (-2.4‰), together with low B concentrations (0.006 mg/l).

A plot of $\delta^{11}B$ vs. B content shows two distinct fields (figure 2) without any relationship. The first field is defined by the mineral spring waters from the Massif Central (Mossadik, 1997, Négrel et al., 1997) and the Pyrenees (Krimissa, 1995). These waters, with variable boron contents (1->8 ppm), are characterised by ^{11}B depletion (range of -10 to 0‰) in agreement with the $\delta^{11}B$ signature of the continental crust (Barth, 1993, 2000). The scattering of data in this field must be related mainly to B concentration variations, which may reflect lithological variability of the B-bearing phases interacting with the waters.

The deep groundwaters from the Vienne granitoids, which have similar boron contents to mineral springs from the Massif Central, exhibit more positive $\delta^{11}B$ values (> +25‰) and therefore plot much more closely to the present-day seawater value. The Availles spring waters plot in an intermediate position.

The shallow groundwaters from the Infra-Toarcien and Dogger aquifers plot within a large range of $\delta^{11}B$ values (-2.4 to +13.2‰) associated with very low boron contents (<0.2 mg/l). These characteristics might result from unmodified rainwater input and/or interaction of rainwater with rocks. In the NO_3-rich waters from the Dogger aquifer, the boron isotopic ratio may be related to farming activities in the area, because fertilisers may add boron with negative $\delta^{11}B$ values to local groundwaters (Komor, 1997, Vengosh, 1998).

The distinctive isotopic composition of boron from different water types (i.e. marine or terrestrial, Barth, 1993, Aggarwal et al., 2000) makes it an attractive geochemical tracer for discriminating between a marine and a non-marine origins for the salinity.

The deep groundwaters from the Vienne granitoids display positive $\delta^{11}B$ values that never reach the seawater value. Our data are clearly higher than values proper to water-rock interaction with granites

(Barth, 1993, 2000, Mossadik, 1997, Krimissa, 1995) and suggest the influence of a marine water end-member with a boron isotopic signature close to seawater (Casanova et al., in press).

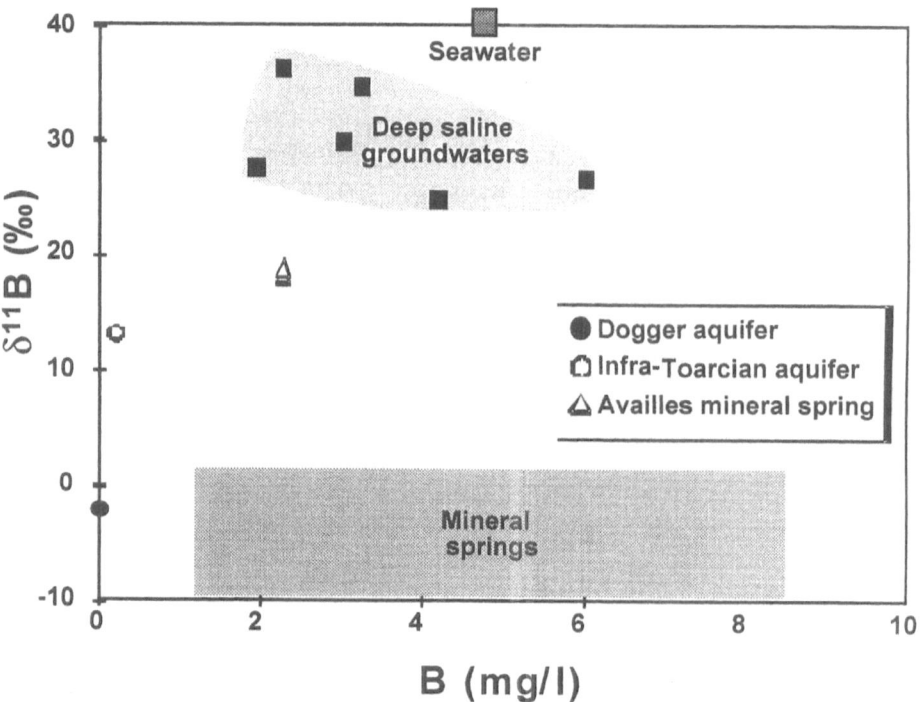

Figure 2. Boron concentrations plotted versus the $\delta^{11}B$ values of the Vienne groundwaters, mineral springs from granitic basement in the Massif Central and in the Pyrenees. Data are presented in Casanova et al., in press.

3.2. Sr isotopes as a WRI tracer in deep groundwaters from the Vienne granitoids

The Sr contents and $^{87}Sr/^{86}Sr$ ratios of deep groundwaters are corrected for contamination by cleaning fluids (Négrel et al., in press). The corrected Sr contents for deep groundwaters in the granitoids vary considerably (range of 1 to 11 mg/l) with a gradual increase from the shallow groundwaters in the sedimentary aquifers towards the deeper groundwaters (figure 3a). The deep groundwaters in the Vienne granitoids rapidly approach the Sr content observed in the Availles mineral springs (range of 4.3 to 13.5 mg/l).

The $^{87}Sr/^{86}Sr$ ratios fluctuate between 0.70781 ± 0.00012 (borehole 115) and 0.70909 ± 0.00019 (borehole 117) and fall in the range of 0.70807-0.70827 (n = 19) in boreholes 106, 107, 113 and 212. There is no direct link with the geological character of the bedrock, as a range of $^{87}Sr/^{86}Sr$ ratios are observed within the same geological lithology. The plot of $^{87}Sr/^{86}Sr$ ratios vs depth (figure 3b) suggests more radiogenic values in the Infra-Toarcian aquifer than in the Dogger aquifer but deep groundwaters in the granitoids have lower $^{87}Sr/^{86}Sr$ ratios than the sedimentary aquifers. The $^{87}Sr/^{86}Sr$ ratios of the deep groundwaters in the granitoids are slightly variable with depth, except for the samples from borehole 117.

One major feature of the $^{87}Sr/^{86}Sr$ variations with depth is the lack of intermediate values between the three zones, suggesting a lack of communication between the different formations. In addition, samples of the deep groundwaters are clustered in five separate groups in the mixing relationship defined by $^{87}Sr/^{86}Sr$ vs. 1/Sr ratios. The lack of direct linear relationships among the samples implies the existence of more than two end-members and a site-specific $^{87}Sr/^{86}Sr$ signature for each bedrock type.

It is generally agreed that the Sr isotopic composition of groundwater is constrained by weathering of potential Sr-bearing phases from the host rocks (Franklyn et al., 1991, Zuddas et al., 1995, Brantley et al., 1998). The Sr content in the water depends on the dissolution of Sr-bearing phases and control by the formation of neogenic phases (Seimbille et al., 1998). Among the minerals typically found in granitic rocks, apatites, feldspars (plagioclase and potassic feldspar) and micas (biotite and muscovite) are the most common minerals that contain a significant amounts of Rb and/or Sr. We have developed a model based on that developed by Zuddas et al., (1995) and Bullen et al. (1997) in order to determine the $^{87}Sr/^{86}Sr$ ratio of water after interaction with granitoid (Négrel et al., in press). This model assumes that Sr is derived from the three minerals: plagioclase, K-feldspar and biotite and that the neoformed phases will contain Sr in isotopic equilibrium with their parent solutions. If dissolution is stopped after the formation of neogenic phases, the initial Sr content of the fluid is modified by the formation of these phases but the $^{87}Sr/^{86}Sr$ will be unchanged.

The calculated $^{87}Sr/^{86}Sr$ ratio of the water after equilibration with the minerals (Négrel et al., in press) yields low values for the tonalite (0.70463) and for the monzogranite (0.70704). Higher $^{87}Sr/^{86}Sr$ ratios are observed in the deep groundwaters than are implied by the calculated $^{87}Sr/^{86}Sr$ ratios. This divergence indicates that the $^{87}Sr/^{86}Sr$ ratios of the deep groundwaters analysed within the Vienne hydrosystem cannot be directly linked to the weathering of tonalite or monzogranite and another source of Sr, with higher $^{87}Sr/^{86}Sr$, must be invoked. We therefore assume that the deep groundwaters from the Vienne district originated from marine intrusions during the Jurassic and have subsequently been diluted by mixing with earlier fluids produced by WRI with the granitoids (cf. § 3.2).

3.3. Stable O-H isotopes as a possible WRI tracer

The stable isotopic data, corrected for the contamination by cleaning fluids (Kloppmann et al., in prep), are represented in figure 4 in a δ^2H vs. $\delta^{18}O$ diagram. The deep groundwaters from the Vienne granitoids are isotopically shifted to the left of the global meteoric water line (GMWL). The maximum shift to the left of the GMWL is to values of relative deuterium excess RDE (e.g. RDE = δ^2H - a $\delta^{18}O$ - b) close to 10.3 ‰ and a relative ^{18}O depletion ROD [e.g. ROD = $(\delta^2H$ -b)/a - $\delta^{18}O$] close to 1.3 ‰, with respect to the local meteoric waterline LMWL (δ^2H = a·$\delta^{18}O$ + b), is observed for the waters from the borehole106. The other fluids define a linear relationship with a slope >8. The most depleted values plot close to shallow groundwaters from the Infra-Toarcian aquifer, which are themselves depleted with respect to shallow groundwaters from the Dogger aquifer.

The stable O-H isotope data for deep groundwaters in the Vienne granitoid suggest that, similarly to other crystalline shields, water-rock interaction under low-temperature low-porosity conditions has modified the ^{18}O and/or ^2H contents in the liquid phase (Kloppmann et al., in prep). If water rock interaction induced the isotopic shift, either the mineral or a group of minerals contained in the bedrock, or matrix or fracture minerals may have been involved. Either type of mineral reactant will act on the isotopic composition of the water in much the same way provided that some general conditions (low WR ratio, low temperature) are fulfilled.

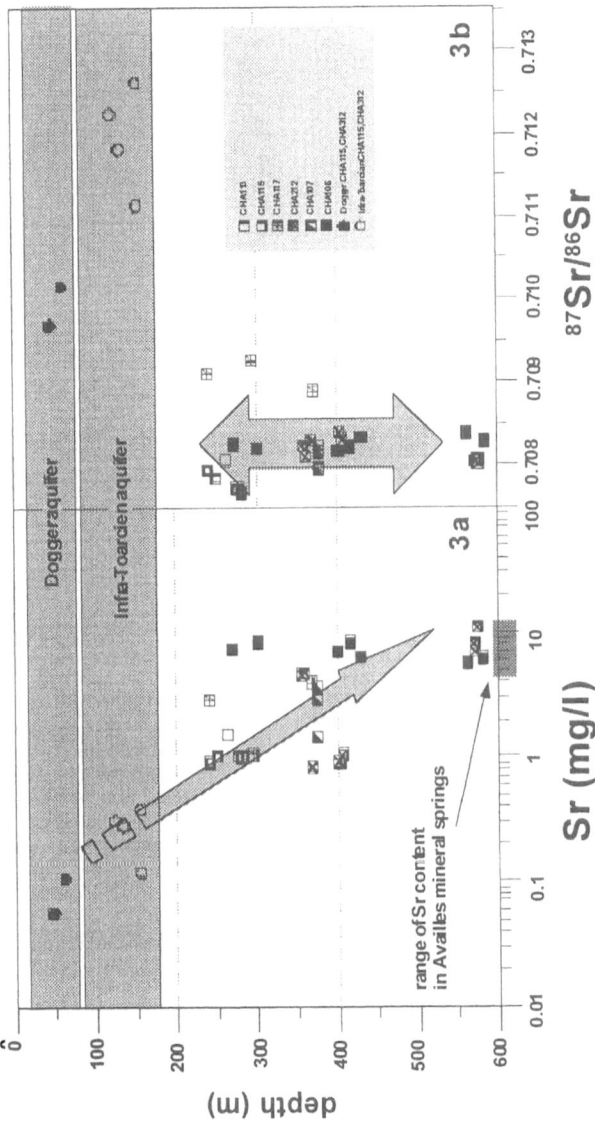

Figure 3. Strontium contents (3a) and $^{87}Sr/^{86}Sr$ ratios (3b) of deep groundwaters in granite and the overlying sedimentary aquifers plotted as function of depth. Data are presented in Négrel et al., in press.

In deep groundwaters from the Vienne granitoid, the isotopic shift occurs at salinities < 10g/l and the RDE vs. TDS relationship (not shown) defines a steeper mixing line than most of the Canadian (Frape and Fritz, 1982) and Scandinavian (Blomqvist et al., 1989) shield brines.

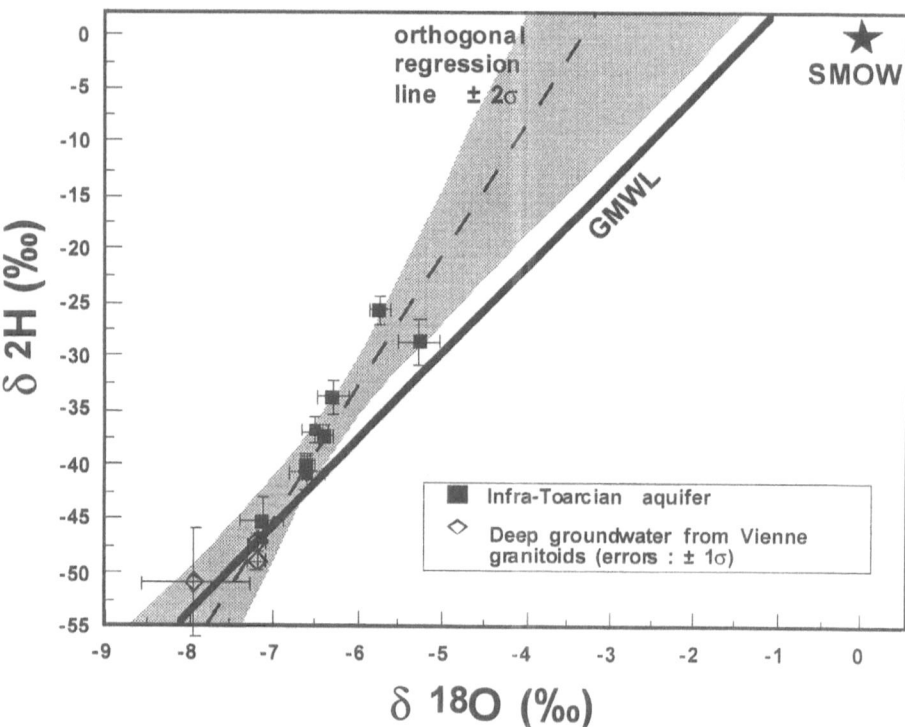

Figure 4. Simulations of the isotopic effect of sea water interaction with fracture minerals of the Vienne granites. The measured isotopic compositions of deep saline groundwaters are represented together with the cumulative errors related to analytical precision and correction for contamination by drilling fluids, the orthogonal regression line and the 95% confidence interval of the regression. The hypothetical mixing endmember has been calculated by extrapolation to seawater salinity along the linear relationship of measured isotope contents and TDS. Data are presented in Kloppmann et al., in prep.

Modelling of these processes (Kloppmann et al., in prep) takes into account the dissolution of calcite with precipitation of dolomite, the dissolution of smectite with precipitation of kaolinite, and the dissolution of feldspars with precipitation of kaolinite-goethite. These processes produce, within certain limits of temperature and water-rock ratio, modelled water isotopic compositions corresponding to the observed δ^2H vs. $\delta^{18}O$ correlation. The measured isotopic composition in the deep groundwaters of the Vienne granitoids are thus compatible with a model whereby the original fluid was seawater, modified by the dissolution/precipitation of mineral phases and subsequently diluted by meteoric waters.

3.4 U-series as a redox indicator

The uranium contents of the Vienne groundwaters range between 23.7 ± 0.001 ppb and 0.022 ± 0.001 ppb. Uranium has two relatively stable oxidation states, immobile +4 and mobile +6, which means that its geochemical behaviour is sensitive to redox conditions (Langmuir, 1978). Although it is evident that U has been mobile in the system (Casanova and Aranyossy, 1998), the lack of clear depth-dependant trend (figure 5) reflects primarily the uranium content of the rocks (0.1<[U $_{MATRIX}$]<10 ppm). The thorium contents of the groundwaters range between 0.008 ± 0.001 ppb and 0.001 ± 0.001 ppb. Thorium is extremely insoluble in a wide range of natural conditions (Langmuir and Herman, 1980) and the tendency of increasing Th contents with depth (figure 5) might be related to preferential Th adsorption on clay minerals and hydroxides in shallow oxidising waters.

The U activity ratio ($^{234}U/^{238}U$) for dissolved uranium in the groundwaters is greater than unity (figure 5). The highest disequilibria (3.87 < $^{234}U/^{238}U$ < 4.55) are observed in shallow groundwaters together with uranium contents up to 24 ppb. The isotopic fractionation takes place as selective chemical release which dominates over direct physical α recoil. This preferential ^{234}U release depends on the valence contrast between the U isotopes, ^{238}U occurring in +4 form and ingrown ^{234}U, due to oxidising microenvironment, in +6 form (Suksi et al., 2001). The results of a uranium series isotopic study of fracture infill materials from the Vienne granitoids are presently interpreted in terms of two different zones (Casanova and Aranyossy, 1998). In the upper zone (0 - 400 m), the excess of ^{234}U relative to ^{238}U that occurs in fine-grained minerals (especially Fe-oxyhydroxides) suggests that preferential ^{234}U-solution processes occurred some 102 to 181 ka ago, assuming a closed-system behaviour for these minerals. Activity ratios as a function of distance from fracture surfaces (sample at 342m) show a clear, absolute uranium release up to 5-6 cm from the fracture surface. Both the $^{230}Th/^{234}U$ and $^{234}U/^{238}U$ profiles indicate episodic ^{234}U mass flow events within the last 350 000 years (Casanova and Aranyossy, 1998). The hypothesis relating the observed absolute uranium release and isotopic fractionation between ^{234}U and ^{238}U is based on the radioactive decay-induced oxidation of ^{234}U to U(VI), the more soluble oxidation state of uranium. In such a situation the oxygen concentration in groundwater plays an important role. If there is much oxygen, it can actually oxidise and release the bulk of the uranium without inducing marked isotopic fractionation. If there is only a small amount of oxygen, the already oxidised ^{234}U is easier to remove than ^{238}U, leading to clear isotopic fractionation. Finally, if there is very little or no oxygen, there would be no effect on ^{234}U.

The moderate disequilibrium ($1.87 < {}^{234}U/^{238}U < 3.36$) observed in the deep groundwaters, together with very low uranium concentration, is characteristic for reducing conditions (Andrews et al., 1989, Osmond and Ivanovich, 1992, Osmond and Cowart, 1992). Fractionation between U isotopes which is seen as ^{234}U enrichment in waters and ^{234}U depletion in rocks provides further information of the disequilibration processes. This behaviour has been generally ascribed to α recoil in reducing aquifers (Kronfeld, 1974) and is supported here by disequilibria measured in minerals filling water-conducting fractures and the rock matrix next to the fracture face (Casanova and Aranyossy, 1998). In the lower zone (400 - 900 m), a deficiency of ^{234}U relative to ^{238}U, suggests alpha-recoil induced solution of ^{234}U from the mineral surfaces to the fracture fluids. It is therefore very likely that deep groundwaters with uranium contents as low as 0.022 ppb underwent an activity ratio increase due to the alpha-recoil process. Observed U isotopic fractionation in groundwaters combined with other uranium series disequilibrium measurements provides a tool for locating redox fronts formed as a result of low temperature rock-groundwater interaction.

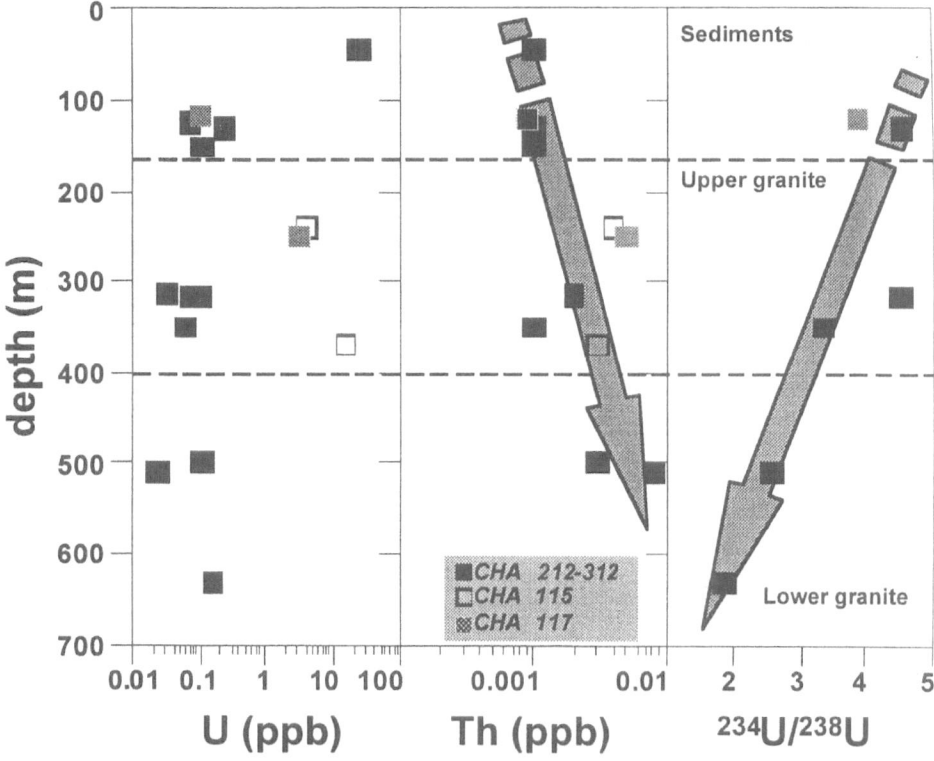

Figure 5. Depth versus dissolved U and Th, and the $^{234}U/^{238}U$ isotopic ratio for the Vienne groundwaters. Data are presented in Casanova and Aranyossy (1998).

4. Conclusions: combined isotopes investigations

In the deep groundwaters from the Vienne granitoids, the conclusions drawn from Sr and B isotopes converge towards a marine origin of the fluids prior to further WRI influence. The B isotopic compositions of the most saline waters lie close to those of present-day seawater and the $^{87}Sr/^{86}Sr$ ratios are slightly higher than those of the Jurassic ocean because of WRI. The stable O-H isotope data suggest that, as is the case for many crystalline shields, water rock interaction under low-temperature low-porosity conditions has modified the ^{18}O and/or ^2H contents of the liquid phase.

A plot of $\delta^{11}B$ vs. $^{87}Sr/^{86}Sr$, avoiding biases linked to relative concentrations, reveals two distinct groups of waters (figure 6): (a) the deep groundwaters from the Vienne granitoids and (b) the mineral spring waters from the Massif Central and Pyrenees. The deep groundwaters display uniform low Sr isotope ratios and pronounced ^{11}B-enrichment. The mineral spring waters show a crustal $\delta^{11}B$ signature and a variable (WRI-induced) $^{87}Sr/^{86}Sr$ signature. Therefore, we conclude that the deep groundwaters from the Vienne district originated from marine intrusions during the Jurassic and have been diluted by mixing with former fluids produced by WRI with the granitoids.

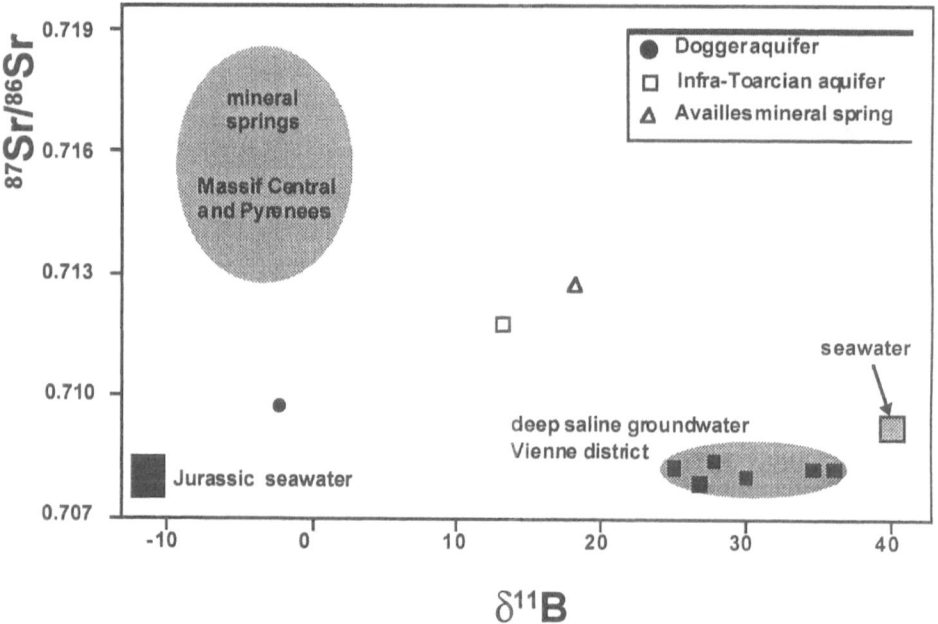

Figure 6. Plot of $^{87}Sr/^{86}Sr$ ratios vs. the $\delta^{11}B$ values of the Vienne groundwaters as well as mineral springs from granitic basement in the Massif Central and the Pyrenees. Data are presented in Casanova et al., in press.

The proportions of such mixing can be estimated using the relationship between the $\delta^{11}B$ and Cl content (figure 7). A representative mixing line lies between the less Cl-rich mineral water sample ([Cl] = 120 µg/l, $\delta^{11}B$ = -8‰) and modern seawater ([Cl] = 20000 mg/l, $\delta^{11}B$ = 40‰). The observed relationship is likely to reflect the simple evolution of a marine intrusion in the granitoids through water-rock interaction specific to the Vienne district. The deep groundwaters lie on this trend and represent 15 to 20% of the seawater end-member. The measured isotopic composition of the fluids of the Vienne granitoids are thus compatible with a model of seawater having dissolved/precipitated fracture minerals and subsequently been diluted by meteoric (paleo-) waters.

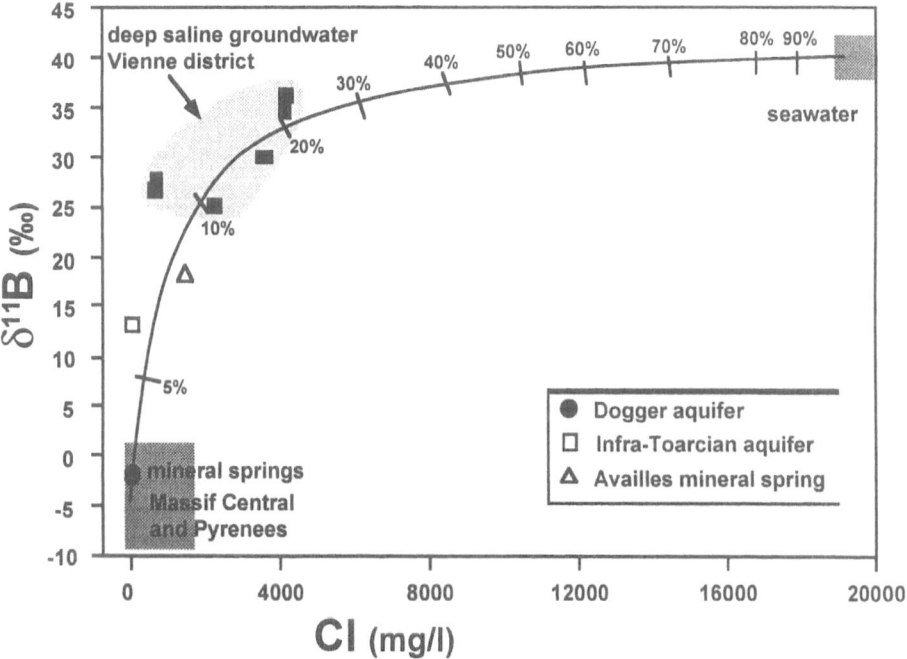

Figure 7. Plot of $\delta^{11}B$ values vs. the chlorine content in the Vienne groundwaters as well as mineral springs from granitic basement in the Massif Central and the Pyrenees. Data are presented in Casanova et al., in press.

Acknowledgements

This study was supported by ANDRA, France. Chemical and isotopic analyses were performed in the geochemistry Laboratory of BRGM, France.

References

Aggarwal, J.K., Palmer, M.R., Bullen, T.D., Arnórsson, S., Ragnarsdóttir, K.V. 2000. The boron isotope systematics of Icelandic geothermal waters : 1. Meteoric water charged system. Geochim. Cosmochim. Acta 64, 579-585.

Albarède, F, Michard, A. 1987. Evidence for slowly changing ^{87}Sr/^{86}Sr in runoff from freshwater limestones of southern France. Chem. Geol., 64, 55-65.

Andersson, P.S., Wasserburg, G.J., Ingri, J. 1994. The sources and transport of Sr and Nd isotopes in the Baltic Sea. Earth and Planet. Sci. Lett., 113, 459-472.

Andrews, J.N., Ford, D.J, Hussain, N., Trivedi D.,.Youngman, M.J. 1989. Natural radioelement solution by circulating groundwaters in the Stripa granite. Geochimica et Cosmochimica Acta. 53, 1791-1802.

Barth, S.R. 2000. Geochemical and boron, oxygen and hydrogen isotopic constraints on the origin of salinity in groundwaters from the crystalline basement of the Alpine foreland. App. Geochem. 15, 937-952.

Barth, S.R. 1993. Boron isotope variations in nature: a synthesis, Geol. Rundsch, 82, 640-651.

Beaucaire, C., Gassama, N., Tresonne, N., Louvat, D. 1999. Saline groundwaters in the hercynian granites (Chardon Mine, France): geochemical evidence for the salinity origin. App. Geochem. 14, 67-84.

Blomqvist, R., Lahermo, P.W., Lahtinen, R., Halonen, S. 1989. Geochemical profiles of deep groundwater in Precambrian Bedrock in Finland. In G.D. Garland (ed.): Proceedings of Exploration '87. Third Decennial International Conference on Geophysical and Geochemical Exploration for Minerals and Groundwater, Ontario Geological Survey, Special Volume 3, 746-757.

Bottomley, D.J., Gregoire, D., Raven, K.G. 1994. Saline groundwaters and brines in the Canadian shield: Geochemical and isotopic evidence for a residual evaporite brine component. Geochim Cosmochim. Acta, 58, 1483-1498.

Bottomley, D.J., Katz, A., Chan, L.H., Starinsky, A., Douglas, M., Clark, I.D., Raven, K.G. (1999) The origin and evolution of Canadian Shield brines: evaporation or freezing of seawater? New lithium isotope and geochemical evidence from the Slave craton. Chem. Geol., 155, 295-320.

Brantley, S.L., Chesley, J.T., Stillings, L.L. 1998. Isotopic ratios and release rates of strontium measured from weathering feldspars. Geochim. Cosmochim. Acta, 62, 1493-1500.

Bullen, T., White, A., Blum, A., Harden, J., Schulz, M. 1997. Chemical weathering of a soil chronosequence on granitoid alluvium: II. Mineralogic and isotopic constraints on the behaviour of strontium. Geochim. Cosmochim. Acta, 61, 291-306.

Bruland, K.W. 1983. Trace elements in seawater. In Riley, J.P. and Chester, R (Eds.). Chemical Oceanography, vol8. Academic Press (London), 157-220.

Capdeville, J.P., Floc'h, J.P., Lougnon, J., Recoing, M. 1983. Carte géologique de la France a 1/50000, feuille Confolens. Notice de la carte géologique, 32p. BRGM Ed., Orléans.

Casanova, J. and Aranyossy, J.F. (1998). Uranium series isotopic data of fracture infill materials from the potential underground laboratory site in the Vienne granitoids, France. In Water-rock Interactions, Arehart & Hulston (eds.), Balkena, Rotterdam. WRI9, 965-967.

Casanova, J., Négrel, Ph., Kloppmann, W., Aranyossy, J.F. In press. Origin of deep saline groundwaters in the Vienne Granitoids (France): Constrains inferred from Boron and Strontium Isotopes. Geofluids.

Casanova, J., Négrel, Ph., Kaija, J., Blomqvist, R. 1998. Constraints added by the strontium and boron isotopes on the geochemical characterization of the Palmottu hydrosystem. Goldschimdt Conference, Toulouse, 1998. Mineralogical Magazine, volume 62A, 278-279.

Casanova, J., Négrel, Ph., Frape, S., Kaija, J., Blomqvist, R. 1999a. Multi isotopes geochemistry of the Palmottu hydrosystem (Finland). In Geochemistry of the Earth's Surface (Armannsson Ed.), Balkema, Rotterdam. 483-486.

Casanova, J., Machard de Gramont, H., Kloppmann, W., Négrel, Ph. 1999b. Boron and strontium isotopic geochemistry of the Wadi Ahin catchment (Sultanate of Oman). European Union of Geosciences Strasbourg, 1999, J. of Conference Abstracts, vol 4, n°1, p 557.

Faure, G. 1986. Principles of Isotope Geology. John Wiley & Sons. 589 p.

Franklyn, M.T, McNutt, R.H, Kamineni, D.C, Gascoyne, M, Frape, S.K. 1991. Groundwater $^{87}Sr/^{86}Sr$ values in the Eye-Dashwa Lakes pluton, Canada : Evidence for plagioclase-water reaction. Chem. Geol. (Isotope Geoscience Section), 86, 111-122.

Frape, S.K., Fritz, P. 1982 The Chemistry and Isotopic Composition of Saline Groundwaters from the Sudbury Basin, Ontario. Canadian Journal of Earth Sciences 19, 4, 645-661.

Fritz, P. and Frape, S.K., 1987. Saline waters and gases in crystalline rocks. Geological Association of Canada, Special Paper 33, 245p.

Gaillardet, J., Dupré, B., Allegre, C.J., Négrel, Ph. 1997. Chemical and physical denudation in the Amazon River Basin. Chem. Geol., 142, 141-173.

Hantzpergue, P., Branger, P., Ducloux, J., Lemordant, Y., Joubert, J.M., Tournepiche, J.F. 1997. Carte géologique de la France a 1/50000, feuille Civray. Notice de la carte géologique, 41p. BRGM Ed., Orléans.

Kloppmann, W., Négrel, Ph., Casanova, J. 1999a. A combined isotopic tool for water-rock interaction studies : B, Sr, O, H isotopes in groundwater. In International Symposium on Isotope Techniques in Water Resources Development and Management, Vienne, Autriche, 10-14 mai 1999, IAEA-SM-361/38, 84-86.

Kloppmann, W., Négrel, Ph., Casanova, J., Guerrot, C. 1999b. Boron and Strontium isotopes in saline groundwaters in the North German Basin (Gorleben diapir). European Union of Geosciences Strasbourg, 1999, J. of Conference Abstracts, vol 4, n°1, p 584.

Kloppmann, W., Girard J.P., Aranyossy J.F. Exotic stable isotope compositions in saline waters and brines from the crystalline basement: Literature review and a new example from French granites. Submitted to Geofluids.

Komor, S.C. 1997. Boron contents and isotopic compositions of hog manure, selected fertilizers and water in Minnesota. J. Environ. Qual., 26, 1212-1222.

Krimissa, M. 1995. Application des méthodes isotopiques à l'étude des eaux thermales en milieu granitique (Pyrénées, France). Ph.D. Thesis, Paris XI, 248p.

Kronfeld, J. 1974. Uranium deposition and Th-234 alpha-recoil : an explanation for extreme U-234 fractionation within the Trinity aquifer. Earth Planet. Sci. Let., 21, 327-330.

Langmuir, D., 1978. Uranium solution-mineral equilibria at low temperatures with applications to sedimentary ore deposit. Geochim. Cosmochim. Acta, 42, 547-569.

Langmuir, D. and Herman, J.S., 1980. The mobility of thorium in natural waters at low temperatures. Geochim. Cosmochim. Acta, 44, 1753-1766.

Matray, J.M., Gadalia, A., Aquilina, L., Kloppmann, W., Lemière, B., Négrel, Ph. 1998. Site de la Vienne, synthèse des reconnaissances hydrogéochimiques. DRP0ANT 97-067A, 117p.

Mossadik, H. 1997. Les isotopes du bore, traceurs naturels dans les eaux : Mise au point de l'analyse en spectrométrie de masse à source solide et applications à différents environnements. PhD Thesis, Univ. Orleans, 224p.

Mourier, J.P., Floc'h, J.P., Coubes, L. 1989. Carte géologique de la France a 1/50000, feuille L'Isle-Jourdain. Notice de la carte géologique, 73p. BRGM Ed., Orléans.

Négrel, Ph., Allegre, C.J., Dupré, B., Lewin, E.. 1993. Erosion sources determined from inversion of major, trace element ratios and strontium isotopic ratio in river water : the Congo Basin case. Earth and Planet. Sci. Lett., 120, 59-76.

Négrel, Ph., Fouillac, C., Brach, M. 1997a. A strontium isotopic study of mineral and surface waters from the Cézallier (Massif Central, France): implications for the mixing processes in areas of disseminate emergences of mineral waters. Chem. Geol., 135, 89-101.

Négrel, Ph, Casanova, J., Aranyossy, J.F., 1997b. Strontium Isotopic Characterization of Groundwaters and Calcites from the Potential Underground Laboratory Site in the Vienne Granitoids (France). Goldschimdt Conference, Tucson, 1997. J. of Conference Abstracts, 1, p 149.

Négrel, Ph., Casanova, J., Guerrot, C., Cocherie, A., Azaroual, M., Fouillac, Ch. 1999a. Isotopes geochemistry in thermo-mineral waters in the Massif Central

(France). In Geochemistry of the Earth's Surface (Armannsson Ed.), Balkema, Rotterdam. 531-534.

Négrel, Ph., Casanova, J., Kloppmann, W., Aranyossy, J.F. 1999b. Origin of deep saline groundwaters in the Vienne granitoids (France); constraints inferred from strontium and boron isotopes. European Union Strasbourg, 1999, J. of Conference Abstracts, vol 4, n°1, p 520.

Négrel, Ph., Casanova, J., Aranyossy, J.F. In press. Strontium isotope studying the origin of fluids in the Vienne Granitoids, France. Chem. Geol.

Osmond, J.K., Ivanovich, M.1992. Uranium-series mobilization and surface hydrology. In M. Ivanovich & R.S. Harmon (eds), Uranium series disequilibrium: application to environmental problems, 259-289, Clarendon Press, Oxford.

Osmond, J.K., Cowart, B. 1992. Ground water. In M. Ivanovich & R.S. Harmon (eds), Uranium series disequilibrium: application to environmental problems, 290-333, Clarendon Press, Oxford.

Palmer, M.R., Spivack, A.J., Edmond, J.M. 1987. Temperature and pH controls over isotopic fractionation during adsorption of boron on marine clay, Geochim. Cosmochim. Acta 51, 2319-2323.

Seimbille, F., Zuddas, P., Michard, G. 1998. Granite-hydrothermal interaction : a simultaneous estimation of the mineral dissolution rate based on the isotopic doping technique. Earth and Planet. Sci. Lett., 157, 183-191.

Suksi, U.J., Pitkänen, P., Ruskeeniemi, T., Rasilainen, K., Casanova, J. 2001. Uranium-series disequilibrium indicting oxygen intrusion in rocks. Tenth International Symposium on Water Rock Interaction, Villasimius, Italy, June 10-15, 2001, submitted.

Vengosh, A., Chivas, A.R, McCulloch, M.T., Starisnky, A., Kolodny Y. 1991a. Boron isotope geochemistry of Australian salt lakes, Geochim. Cosmochim. Acta 55, 2591-2606

Vengosh, A., Kolodny, Y., Starisnky, A., Chivas ,A.R., McCulloch, M.T. 1991b. Coprecipitation and isotopic fractionation of boron in modern biogenic carbonates, Geochim. Cosmochim. Acta, 55, 2901-2910.

Vengosh A., Starisnky A., Kolodny Y., Chivas A.R. 1991c. Boron isotope geochemistry as a tracer for the evolution of brines and associated hot springs from the Dead Sea, Geochim. Cosmochim. Acta 55, 1689-1695

Vengosh, A., 1992. Boron isotope variations during brine evolution and water-rock interactions. In Water-rock Interactions, Kharaka & Maest (eds.), Balkena, Rotterdam. WRI6, 965-967.

Vengosh, A., 1998. Boron isotopes and groundwater pollution. Water & Environ. News, 3, 15-16.

Zuddas, P., Seimbille, F., Michard, G. 1995. Granite-fluid interaction at near equilibrium conditions : experimental and theoretical constraints from Sr contents and isotopic ratios. Chem. Geol., 121, 145-154.

Water-rock reaction experiments with Black Forest gneiss and granite

Kurt Bucher
Institute of Mineralogy, Petrology and Geochemistry,
University of Freiburg,
Albertstr. 23b, 79104 Freiburg, Germany
bucher@ruf.uni-freiburg.de

Ingrid Stober
Geological Survey of Baden-Württemberg,
Albertstr. 5, 79104 Freiburg, Germany
stober@lgrb.uni-freiburg.de

Abstract

Five characteristic samples of crystalline rocks of the Black Forest basement, three gneisses from the Kinzig Valley and two granites (Triberg & Bärhalde) from the Variscan basement of the Central Black Forest have been experimentally reacted with water in a batch reactor under a series of different experimental conditions in order to better understand the composition and evolution of groundwater in the crystalline basement.

Experiments with fine-grained powders ($< 10\mu m$) showed a high mobility of chloride: gneiss contains up to 48 mg dissolvable chloride per kg rock, granite up to 300 mg Cl / kg rock. The salinity of leachates is proportional to the amount of preserved fluid inclusions which is higher in Qtz-rich granites than in feldspathic gneisses with little Qtz. Experiments with CO_2-saturated water did not distinctively increase Cl, although TDS increased by a factor of 6. In coarsely crushed gneiss

I. Stober and K. Bucher (eds.), Water-Rock Interaction, 61–95.

and granite (grain size > 1 mm) 5 and 8 mg Cl was extractable per kg rock, respectively. Average measured Cl/Br ratio was about 100 (on a ppm basis), a value typical of primary high-T fluids trapped in fluid inclusions. The Cl/Br-ratios of leachates from granite and gneiss were very similar and independent of temperature.

In both, granite and gneiss, more Mg than Ca is extractable. However, Ca + Mg is higher by an order of magnitude in gneiss leachates. Leachates from both rocks are strongly alkali dominated and K > Na under most experimental conditions. Sulfate and nitrate are low (5 mg/l) compared with Cl which is typically higher by a factor of 30 in granite. Experiments at 50°C resulted in higher TDS and bulk reaction rate. Na/Ca ratio increased from 4 at 25°C to 13 at 50°C. Ion exchange on the fine rock powder and reaction products (clays) is important and increases K+(Na) relative to Ca+(Mg) in the water.

The origin of the solutes is related to alteration of biotite and secondary chlorite (K, Mg), alteration of plagioclase (Ca, Na) and opened fluid inclusions (Na, Cl). The dominate primary anion is chloride which resides in fluid inclusions and in halite on grain boundaries (in contrast to HCO_3 which is atmospheric). The composition of the leachates is clearly related to the composition and nature of the minerals present in the rocks. The amount of water extractable solutes is very different for granites and gneisses, which reflects the petrographic and mineral compositional differences between the two rock types. These results correspond well with the observation of two different types of deep groundwater in the Black Forest basement.

1. Introduction

Groundwater in crystalline rocks of the continental basement has some characteristic compositional features that are related to the reaction of water with the rock matrix exposed along the fracture walls. Although other sources of solutes are known, such as fossil seawater and formation water from the sedimentary cover, typical crystalline waters draw their solutes from water-rock reactions.

Composition data of water in basement rocks have been reported from commercial and scientific wells, from deep underground mines and other sources (e.g. Frape and Fritz, 1987; Gascoyne et al., 1987; Lahermo and Lampén, 1987; Banks et al., 1995; Möller et al., 1997a; Möller et al., 1997b; Pauwels, 1997; Lodemann et al., 1998; Stober and Bucher, 1999a, Bucher and Stober, 2000). On a world wide basis there are perhaps three different waters types that can be found in the fracture porosity of the basement (Bucher and Stober, 2000).

In the Variscan crystalline basement of the Black Forest, water is found in the fracture-porosity throughout. Also deep drillholes in the basement (4450 m in Urach, 2000 m in N-Switzerland) found water-filled interconnected fracture porosity (Stober and Bucher, 2000). The chemical composition and characteristics of Black Forest basement waters has been described and discussed in several papers (Bucher and Stober, 2000 and references therein).

Deep Black Forest groundwater (drill holes to 2000m) occurs in two distinct populations; one is of the Na-Cl type and the other of the Ca-HCO$_3$ type (Fig. 1). The typical Na-Cl water contains about 10g/l NaCl and shows a pH of around neutral. This water type commonly has an elevated temperature suggesting a deep origin. The maximum well head temperature is 70°C and the water is drawn from 4000 m deep reservoirs (e.g. Stober et al., 1999). The Ca-HCO$_3$ type water has a TDS of up to 2-3 g/l and a pH around 6 due to high concentrations of dissolved CO_2 gas (2-3 g/l). Also, sulfate is often high in this type of water.

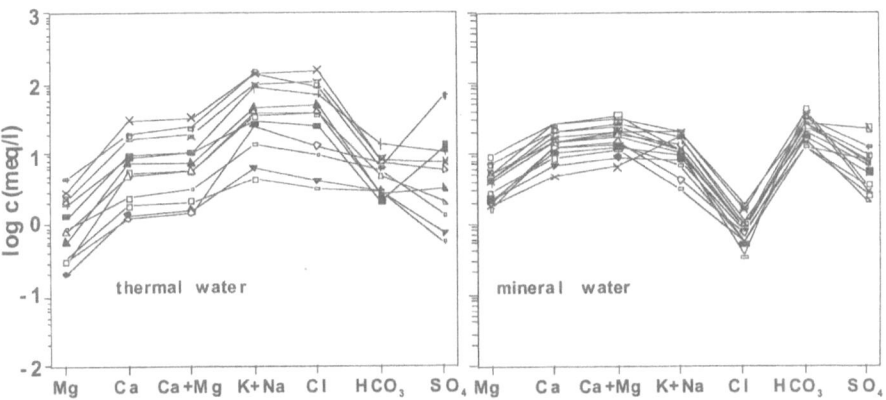

Figure 1: Composition patterns of Black Forest water (Bucher and Stober, 2000).

Whereas the high total of dissolved solids (TDS) can be attributed to passive desiccation (Bucher & Stober, 2000), the source of the solids and the composition pattern is linked to the reaction of meteoric recharge water with the minerals of the basement rock exposed along the fracture walls. Also, the water takes up salty fluids from broken up fluid inclusions in minerals of the basement rocks.

The water composition of granitic and gneissic fracture aquifers has attracted much research effort (Garrels, 1967; Garrels and MacKenzie, 1975; Frape and Fritz, 1987; Grimaud et al., 1990; May et al., 1996; Bottomley et al., 1999) and numerous experimental studies led to a better understanding of the genesis of deep groundwater in the crust (e.g. Garrels and Howard, 1959; Nesbitt et al., 1991; Azaroual and Fouillac, 1997; White et al., 1999). Still, much is unknown, particularly, halogen data from granites and gneisses are needed to explain the origin of high salinity in crustal fluids. Chloride – bromide ratios of the waters are widely used as a diagnostic parameter to differentiate the origin of anions in the waters. Cl-Br data from rocks and minerals are very scarce, however (Kamineni, 1987). Halogen data from fluid inclusions in granite minerals are available

(Böhlke and Irwin, 1992; Yardley et al., 1992; Savoye et al., 1998), but the data base is thin.

In this paper we report experimental data of leaching experiments carried out in batch reactors with powders of typical Black Forest basement rocks (Liegl, 1998; Liegl et al., 1999; Treß, 1999). The experimental conditions varied widely and a total of five different gneisses and granites were used as experimental staring materials. Peters (1986) has carried out similar experiments and determined the amount of extractable chloride from a granite sample but did not analyze the leachates for other halogens. The experimental data presented here give new insights into the genesis of deep groundwater in the crystalline basement.

2. Regional geological setting

In the Black Forest area the crystalline basement of the European continent is exposed in an about 130 km long and 40 km wide N-S trending outcrop representing an erosional window through the cover sediments. The outcrop has been generated by the rifting in the Rhine river valley and the subsequent uplift of the Graben shoulders mainly in the Oligocene (Thury et al., 1994).

The basement consists predominantly of a number of carboniferous granitic plutons (Finger et al., 1997; Emmermann, 1977; Schleicher and Fritsche, 1978) intruding an older basement composed of high-grade metasediments and migmatitic orthogneisses (Waldeck, 1970; Stenger, 1982; Hofmann, 1989; Mazurek and Peters, 1992). Other rock types, such as amphibolites and ultramafic rocks are rare (Kalt et al., 1995). Carbonate rocks (marbles) are virtually absent.

The basement rocks experienced the last major tectono-thermal reworking during the Variscan orogeny. During the Mesozoic a major hydrothermal phase of unknown origin pervasively altered much of the rocks (Lippolt and Kirsch, 1994). Barite-fluorite mineralization and Ag-Cu-Zn vein deposits are related to this phase (Hofmann, 1989). Tertiary rifting, Graben formation and uplift caused a relatively high fracture density in the basement (Illies and Greiner, 1978). The fracture-related permeability is relatively high (Stober, 1996).

3. Sample characterization

The rock samples for the leaching experiments were collected in quarries and in the Clara mine, a deep (700m) underground barite and fluorite mine (Table 1; Figure 2). The samples are believed to be typical and representative of Black Forest basement. This means that water in the basement normally resides in and interacts with rocks of the type used in the reported leaching experiments. There are three sources of solutes that can be taken up by water reacting with the rock matrix: a) the minerals constituting the rock, b) fluid inclusions in the minerals and c) thin films of solids coating the grain boundaries of the matrix minerals. The group c) phases are distinguished from the matrix minerals by the circumstance that the group c) phases cannot be studied by ordinary techniques commonly used in mineralogy such as petrographic microscope and electron microprobe. The contribution of the grain boundary films and fluid inclusions to the bulk rock composition is small and below the sensitivity of conventional XRF analyses. The bulk rock composition reflects the mode and composition of the matrix minerals.

Table 1: Samples of Black Forest rocks

sample	num.	locality	rock type	leaching data from:	probe data
CM	kb810a	Clara Mine (14 level)	banded gneiss	Bucher	*
CM	kb810b	Clara Mine (14 level)	banded gneiss	Bucher	*
PS	kb832	Pauli-Schänzle	gneiss	Liegl	*
AB	kb833	Artenberg	gneiss	Treß	*
TB	kb837	Hornberg	Triberg granite	Liegl	*
BH	kb839	Caritash.-Farnwitte	Bärhalde granite	Treß	*

microprobe data of all samples produced by: Bucher

4. Petrography of analyzed samples

CMa: Banded gneiss from the Clara Mine (14th level)
The sample with a greenish color and a weak chlorite foliation contains the minerals: quartz, plagioclase, K-feldspar, chlorite, relic biotite, secondary muscovite, ilmenite, rutile, apatite, pyrite and magnetite.
Biotite in the sample is strongly altered and mostly replaced by chlorite and ilmenite, plagioclase is altered to albite and mica. Muscovite is an early replacement product of biotite, pseudomorphs and intergrowth of chlorite + K-feldspar + rutile indicate the second-stage reaction Bt + Ms ⇒ Chl + Kfs. Abbreviation of mineral names after Kretz (1983).

CMb: Banded gneiss from the Clara Mine (14th level)
Similar sample as above with a weak banding and foliation. The sample contains abundant fresh primary biotite. Minerals identified include: quartz, plagioclase, K-feldspar, biotite, some chlorite, white mica, much apatite, zircon, rutile, pyrite and dolomite. The sample displays some interesting feldspar and biotite alteration textures. The biotite alteration to chlorite + dolomite is particularly remarkable. The overall alteration reaction can be written as plagioclase + biotite ⇒ albite + chlorite + K-feldspar + dolomite ± quartz.

PS: Garnet-biotite-gneiss, Pauli-Schänzle quarry
Grey-brown gneiss with irregular banding and gneissic structure. All grains are strongly deformed and stressed. The mineral content of the gneiss includes: quartz, plagioclase, K-feldspar, biotite, garnet, chlorite, apatite and calcite. Quartz contains abundant fluid inclusions. Plagioclase shows various degrees of alteration and some calcite veining. Biotite and garnet is locally somewhat altered to chlorite and magnetite. A quite a large amount of secondary calcite is present in parallel lenses in biotite.

AB: Biotite-gneiss, Artenberg quarry
Biotite-rich gneiss with thin mica bands and lenses of deformed K-feldspar. Identified minerals include: K-feldspar (40%), quartz (30%), plagioclase, biotite (20%), chlorite, ilmenite, magnetite, apatite, pyrite and titanite. Plagioclase shows complex alteration textures. Main product is albite and a porous residual feldspar body. Biotite is fairly fresh and only locally altered to chlorite and opaque minerals.

TB: Triberg-granite, Hornberg quarry
Coarse granite with a clear brownish alteration stain. Minerals identified: K-feldspar, plagioclase, quartz, biotite, muscovite, chlorite, ilmenite, magnetite, apatite, Ce-monazite and zircon. Both feldspars are strongly altered and much secondary albite is present. Locally primary plagioclase is preserved in patches and polysynthetc twinning is visible. Subordinate muscovite appears to be secondary. Plagioclase is present as spongy aggregates with large porosity and mica in the pores.

BH: Bärhalde-granite, Caritashaus-Farnwitte
Coarse-grained, gray, fresh granite with large Carlsbad-twinned K-feldspars. Minerals: K-feldspar, plagioclase, quartz, muscovite, biotite, apatite, topaz, magnetite and zirconolite. Both feldspars are strongly altered to albite and mica. Muscovite in the rock is predominantly primary, however. K-feldspar is much coarser and more abundant than plagioclase. Biotite is always partly altered to chlorite and opaques (magnetite mostly).

4.1. Composition of Black Forest gneiss and granite

The bulk composition of two typical gneisses and granites of the Black Forest basement were analyzed by XRF (Philips 2404 instrument). The data shown in Table 2 are representative for typical granitoid continental basement. The bulk compositional features reflect the mineralogical composition of the two types of rocks. K-feldspar dominated granite is rich in potassium, biotite-rich gneiss contains more Fe, Mn, Ti and Mg than granite. The most important difference between gneiss and granite is the calcium content which is very low in the later. Ca in gneiss is stored in plagioclase which contains some anorthite-component. Ca in the Bärhalde granite is mostly tied to apatite. Phosphate is present in similar amounts to Ca in the granite and it is much more abundant than Mg in the Bärhalde granite. The magnesium content of both granites is remarkably low. Selected trace element concentrations are given in Table 2. Barium and vanadium, trace elements stored in biotite, are much higher in gneiss than in granite.

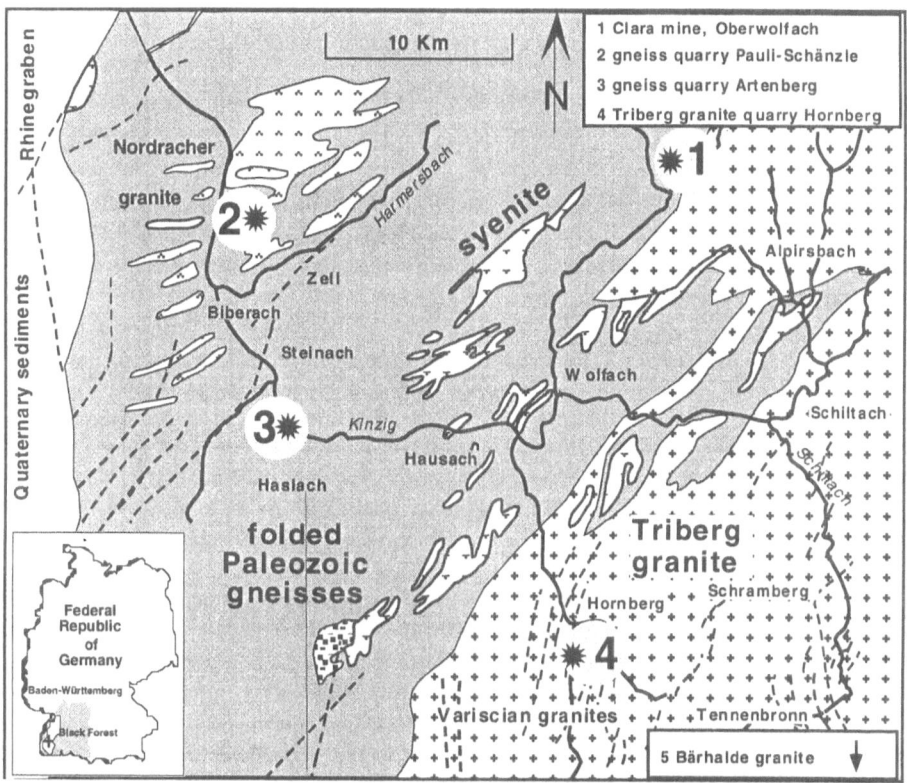

Figure 2: Geologic map of the Kinzig valley area, Central Black forest (Schleicher and Fritsche, 1978). Sample localities of four analyzed samples are shown, Bärhalde granite sample (BH) is in the South outside the map area.

4.2. Mineral chemistry

The composition of minerals from Black Forest basement samples were measured on a Cameca SX100 electron microprobe using natural and synthetic standards. The microprobe was also used to identify very fine-grained accessory minerals and to investigate the alteration products and the retrograde reactions of feldspar and biotite modification. All samples show feldspar and biotite alteration to variable degree. Plagioclase typically is replaced by secondary albite and a fine grained mass of "clay", presumably illite. Secondary Ca-minerals have not been found, except for rare calcite and dolomite. The altered feldspar is very porous (spongy). The formation of this porosity is secondary and related to the alteration process. Biotite alteration takes place by a number of different processes, the most prominent is chlorite + magnetite/ilmenite replacement. However, also muscovite + chlorite forms from biotite as well as chlorite + K-feldspar from biotite and the early alteration product muscovite. The composition of the analyzed minerals is briefly described below:

Plagioclase: All samples contain plagioclase that is altered to various extent. In the gneiss samples the anorthite content varies between An_{19} and An_{32}. Three gneisses contain An_{max} of 27-32 one An_{21}. The Artenberg gneiss contains three populations of plagioclase characterized by An_{19}, An_{27}, and An_{32}. Hence, the gneisses typically contain andesin and may be designated andesin-gneisses. Typically, in gneisses, primary oligoclase or andesine is replaced by secondary albite which can be found in all samples. Granite contains albite and the anorthite component of Bärhalde granite plagioclase is very low (An_{03}).

K-feldspar: In the samples 810A and 810B from the Clara mine K-feldspar is intimately intergrown with albite and all transitions between Kfs and albite can be measured. The totals are low due to the porous structure of the feldspar. K-feldspar of gneiss has two different compositions: an Ab-rich alkali feldspar representing a primary phase and small grains of pure K-feldspar that coexists and forms together with late chlorite. Kfs in granite contains Ab_{10} and Ab_{14}. Kfs in the Hechtsberg pegmatite is zoned with Ab-rich cores and Ab-poor rims.

Biotite: All samples contain biotite. The composition of which varies widely and differs dramatically between gneiss and granite. Biotite in gneiss has an X_{Fe} between 0.48 and 0.68, Ti is high and Mn is present in distinct amounts. Fluorine is present in two of the gneisses and absent in two other samples. In F-poor biotite chlorine is present in small amounts. Biotite in gneiss is not oxidized and all iron is present as Fe^{2+}. Biotite in both granites contains much fluorine and small amounts of Cl (~0.2 wt.%). In the Bärhalde granite, three types of biotite with different composition are present. The highest F-content measured is about 3 wt.% corresponding to 1/3 replacement of the OH-sites by F. Ti and F appear to be inversely correlated. Particularly remarkable is the extremely high iron content of the granite biotite. Bärhalde granite contains biotite with $X_{Fe} > 0.9$. Manganese in biotite is nearly as high as magnesium. The biotite of the granite, particularly of the Bärhalde sample, is remarkably oxidized. Based on stoichiometry, more than 50% of the total iron could be in the trivalent state.

Muscovite: Primary muscovite is present in granite and pegmatite. Muscovite in all samples shows some tschermak variation, X_{Fe} is always lower than in the coexisting biotite. Like biotite, the muscovite of the Bärhalde granite contains much fluorine (2.4 wt.%).

Illite/smectite or other clay minerals are present in all samples as a result of feldspar alteration. No reliable composition data could be obtained by microprobe analysis, except for the Pauli-Schänzle gneiss. The clay contains between 3 and 4 wt.% K and between 0.2 and 0.3 wt.% F.

Granet: Three samples contained garnet, although in very different textures. Slightly chloritized primary garnet of the Pauli-Schänzle gneiss is very rich in almandine and contains 8 mole% grossular component. Both samples from Hechtsberg, gneiss and pegmatite, contain very small (~40 μm) grains of young garnet growing from the alteration products of biotite and feldspar. This garnet is rich in manganese.

Chlorite has an X_{Fe} similar to the biotite from which it grew. Chlorite, however, does not incorporate the Ti of the biotite, which is then found in rutile (or ilmenite), nor does chlorite contain fluorine or chlorine. In most samples different types of chlorite are present. As an example, chlorite in the Artenberg gneiss contains variable Fe/Mg and one chlorite population is highly oxidized with > 60% of total Fe present as Fe^{3+}.

Apatite is present in all samples, in Bärhalde granite it contains most of the Ca present in the rock. All analyzed apatite is very rich in fluorine and chlorine is present in small amounts. Some of the specimens are pure fluorapatite. In the Bärhalde granite fluorapatite occurs together with abundant topaz.

Carbonate: In the Artenberg gneiss, magnesian calcite low in Fe and Mn has been identified. The Clara gneiss contains secondary dolomite with equal amounts of siderite and rhodochrosite (3-4 mole% each).

Table 2a: composition of gneiss and granite of Black Forest basement (wt%, XRF).

locality rock	Clara mine gneiss	Pauli-S gneiss	Artenberg gneiss	Triberg granite	Bärhalde granite
SiO_2	64.37	70.87	65.77	73.89	74.48
TiO_2	n.d.	0.30	0.68	0.09	0.06
Al_2O_3	14.43	14.31	16.39	13.57	14.25
Fe_2O_3 (tot)	5.55	3.25	5.03	1.79	1.30
MnO	0.08	0.06	0.1	0.01	0.02
MgO	3.33	0.89	2.62	0.36	0.09
CaO	2.89	1.44	2.29	0.15	0.34
Na_2O	3.29	3.23	4.43	2.57	3.96
K_2O	2.77	4.11	2.45	5.58	4.74
P_2O_5	n.d.	0.12	0.16	0.04	0.28
H_2O	2.19	1.08		1.14	
Ba (ppm)	642		565		16
Zr (ppm)			55		527
Cr (ppm)			77		97
V (ppm)			107		5
Σ	98.90	99.66	100.05	99.19	99.59

Additional data: CM: Zn 55; Rb 128; Sr 188; CO_2: 8905; S 1050, FeO 2.89 %

Table 2b: modal compositions of Black Forest basement rocks (vol%)

locality rock	Clara mine gneiss	Pauli-S gneiss	Artenberg gneiss	Triberg granite	Bärhalde granite
quartz	17	20	18	50	30
plagioclase	36	35	31	9	20
K-feldspar	6	8	5	27	40
biotite	35	33	37	3	3
muscovite				11	5
chlorite	5	1.5	4		0.5
garnet		1.5			
apatite	1	0.8	1	<1	1.5

Summary of composition features of rocks and minerals

The major elements of the rocks that were used in the leaching experiments are mainly present in aluminous silicates and quartz. Alkalis, K and Na, dominate over Ca and Mg. Ca is mainly stored in plagioclase, apatite and garnet. In granite, biotite and chlorite is extremely rich in iron and fluorine is very abundant in sheet silicates (3 wt.%). Chlorine is present in biotite and muscovite in small amounts.

5. Experimental setup

The rock samples were washed, weathering crusts removed and then crushed. Two powder grain sizes were prepared for the experiments: a) a powder with the characteristic grain size (CG) of the sample with the intention to minimize opening fluid inclusions in quartz. The CG parameter was obtained from petrography and is 0.5 - 1.0 mm for the gneisses and 2 - 4 mm for the granites. The crushed rock was sieved to these grain size fraction and adhesive dust removed with pressurized air. b) powder with the grain size < 5 μm. After breaking the rock to the CG, the material was ground for 4 min in a laboratory ball mill. After sieving into three fractions, the fraction < 0.5 mm was then further ground to < 5 μm in a ZrO_2-ball mill. In this powder most of the fluid inclusions were opened and became available to the leachate. A quantitative grain size distribution analysis of the powder showed a size range from 0.1 to 50 μm and a bi-modal size distribution. The larger peak at 2 μm represents the brittle phases, mostly feldspar and quartz, and the smaller peak at 15 - 20 μm represent the ductile phases, mainly mica and chlorite. The powder with two grain size fractions from the samples were then used for the leaching experiments. During preparation, all samples were kept dry at all times.

All experiments were carried out in a glass batch reactor. 100 g of powder was funneled into the reactor, the cleaned and dry electrodes for measuring electrical conductivity, pH and Redox potential were attached to the reactor. Finally the stirrer was mounted and 800 ml bi-distilled water was added to the reactor and the clock measuring the reaction time started. The stirrer speed was kept constant in all experiments (856 rpm). The natural CO_2-content of the clean water used has not been extracted. Some experiments were carried out at CO_2-saturation by streaming the charge with pure CO_2 gas. The pH and Eh electrodes were inserted into the reactor for the actual measurement only during the long run experiments to avoid contamination of the leachate with electrolyte from the electrode (see also Azaroual and Fouillac, 1997). After fixed reaction time of 2, 5, 10, 50, (60), 100 min. the leachate was removed from the reactor, filtered and centrifuged.

The composition of the leachate was analyzed using a Dionex DX-120 ion-chromatograph for the solutes: Li, Na, K, Ca, Mg, F, Cl, Br, SO_4, NO_3. Iodine was measured by a Metrohm IC. The elements Al, Ba, Sr, and Mn were measured by Perkin-Elmer 4110 ZL graphite furnace atomic absorbtion, Fe and Rb by Perkin-Elmer 3030 flame atomic absorbtion. Si was determined on a Merck SQ2000 photometer.

6. Reaction progress and kinetics

The chief interest of the present study was to determine the nature and amount of easily water extractable solids from granite and gneiss powders. However, the first data and observations acquired in each experiment were related to the dissolution process itself. These data are now the first to be presented and discussed below. Although the kinetic data obtained from these experiments cannot be directly used to model groundwater evolution in natural environments, they are important for understanding the time dependence of the composition of the waters and for deciding the run duration of the batch experiments.

6.1. Electrical conductivity patterns

The progress of the dissolution process can be monitored by measuring the electrical conductivity (ec) as a function of reaction time. In principle, at the beginning of the reaction the ec of the solution is that of pure water or of CO_2-saturated water in the case of the CO_2-saturated experiments. However, because the various actions that need to be performed to initiate the experimental run, the first reading of ec indicated that very soluble material dissolved very quickly resulting in a rapid and irregular increase of ec during the initial wetting period of pure water with the powder charge. Consequently, $ec° \neq 0$ at $t = 0$ in the rate equations derived below. After about 30 - 60 seconds the powder is suspended in the reaction fluid and kept in suspension by the stirrer during the experiment.

In dilute multi-solute solutions involved in the experiments the total of dissolved solids (TDS) can be approximated by: 20 ec (in $\mu S\ cm^{-1}$) = TDS (in μeq

kg^{-1}). This ignores small amounts of uncharged species present in the solution, notably SiO_{2aq}.

The electrical conductivity regularly and systematically increases during all experimental series. None of the experiments reached an equilibrium or a steady state. This is not surprising, of course, as all solutions remained out of equilibrium with at least some of the primary minerals of granite and gneiss, notably biotite and plagioclase. The detailed behavior of the dissolution process with time depends on the material leached, on the temperature, grain size and CO_2-pressure. Some example reaction series will be presented and discussed in detail below.

The electrical conductivity evolution of the fine powder of the gneiss from the Clara mine (CM) can best be described by a power function of the form $ec = ec° \, t^n$ (Fig. 3). After an initial very rapid increase in ec, the release of solutes to the fluid slows down during further reaction. The reaction rate $(d \, ec \, / \, d \, t) = ec° \, n \, t^{(n-1)}$ or more conveniently expressed as $(d \, TDS \, / \, d \, t) = TDS° \, n \, t^{(n-1)}$ decreases from about 200 µeq kg^{-1} min^{-1} at 1 min reaction time to less than 2.4 µeq kg^{-1} min^{-1} at 100 min reaction time. After about 30 min reaction time, the electrical conductivity evolution is better described by a linear function $ec = k \, t + ec°$ or $TDS = k't + TDS°$ (Fig. 3). The reaction rate of the bulk dissolution reaction is constant. This rate law of the bulk reaction corresponds to a zero'th order reaction kinetics (Lasaga, 1981). The rate constant of $k = 0.265$ corresponds to an ec increase of 0.265 µS cm^{-1} min^{-1} or, which is equivalent, an increase of TDS of 5.2 µeq kg^{-1} min^{-1} or 86.6 neq kg^{-1} s^{-1}.

The initial power law describes the data considerably better than the $ec = k \, t^{0.5} + ec°$ square root function that has been often observed in other experiments (Wollast, 1967; Busenberg and Clemency, 1976; Luce et al., 1972). Yet, the functual dependency suggests that highly soluble material such as salt from fluid inclusions, secondary calcite and ultra-fine grained silicate dust, dissolves very quickly and then the slow process of silicate dissolution proceeds at a constant, slow rate.

An experiment using the fine powder of the Clara gneiss (CM) as above but keeping $P_{CO2} = 1$ bar constant (CO_2-saturation, Fig. 4), resulted in a TDS that was 6 times higher after 100 min reaction time compared with the experiment without CO_2. The TDS evolution of this experiment is best described by a power function with a much higher exponent (~ 0.1) and the reaction rate decreases with the inverse of time. However, also here after about 30 minutes reaction time, the TDS increase follows a linear relationship typical of zero'th order reaction kinetics. The slope of the straight line on Fig. 4 corresponds to a bulk reaction rate of 40 µeq kg^{-1} min^{-1} or 660 neq kg^{-1} s^{-1}. This rate is higher than the rate in the CO_2-free experiment at the same temperature and grain size by a factor of 7.6. The dramatic difference in the TDS evolution is mainly the effect of different pH. The CO_2-free experiment evolved to a pH of 9.58 after 100 min reaction time while the pH in the CO_2-saturated experiment was 6.12 after the same time.

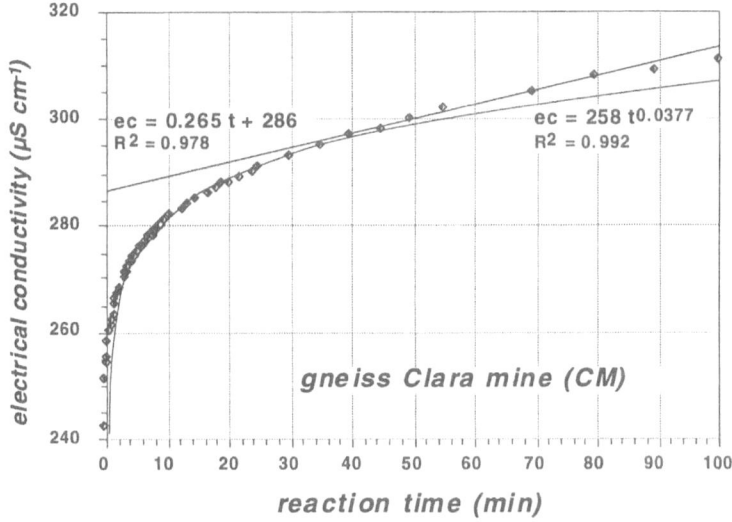

Figure 3: leaching fine powder of gneiss from Clara mine in pure water, ec evolution during a 100 min experiment at 25°C. Linear trend established after ca. 30 min reaction time.

Leaching experiments with coarse grain size charges at 25°C (Fig. 5) showed a sqrt t relationship of the TDS (ec) evolution. After about 30 min reaction time, also in these experiments the square root in time relation grades into a linear ec=f(t) function. The TDS of the leachate of the 1-2 mm grain size charge (characteristic grain size of the sample) reached 660 µeq kg^{-1} after 30 min reaction time and the 0.5 - 1 mm size sample had a TDS of 700 µeq kg^{-1} after the same time. The < 10 µm powder (Fig. 3) reached a TDS of 6000 µeq kg^{-1} at 30 min.

Experiments at elevated temperature (50°C) were carried out with some of the samples. A clear and consistent temperature dependence of the bulk water-rock reaction was found in all experiments. (see also White et al., 1999). Using the Pauli-Schänzle gneiss (PS, Fig. 6), for example, the linear ec versus t portion of the ec evolution shows a bulk reaction rate that is nearly three times higher at 50°C compared to the rate at 25°C (6.94 µeq kg^{-1} min^{-1} at 50°C; 2.52 µeq kg^{-1} min^{-1} at 25°C). Similar consistent data follow from the experiments with the Artenberg gneiss (AB, Fig. 7). The corresponding rates were: 11.8 µeq kg^{-1} min^{-1} at 25°C, 18.0 µeq kg^{-1} min^{-1} at 50°C and 43.8 µeq kg^{-1} min^{-1} at 25°C and CO_2 saturation. The different rate of the two gneisses is most likely related to differences in mineral

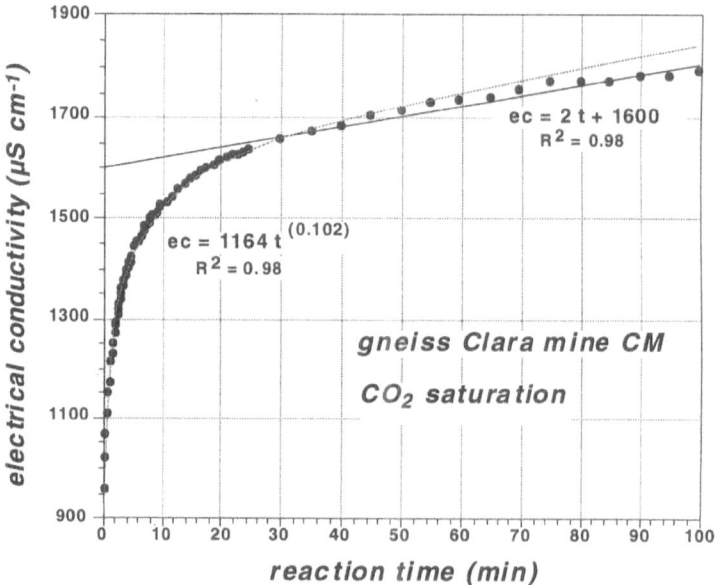

electrical conductivity ($\mu S\ cm^{-1}$)

$ec = 1164\ t^{(0.102)}$
$R^2 = 0.98$

$ec = 2\ t + 1600$
$R^2 = 0.98$

gneiss Clara mine CM

CO_2 saturation

reaction time (min)

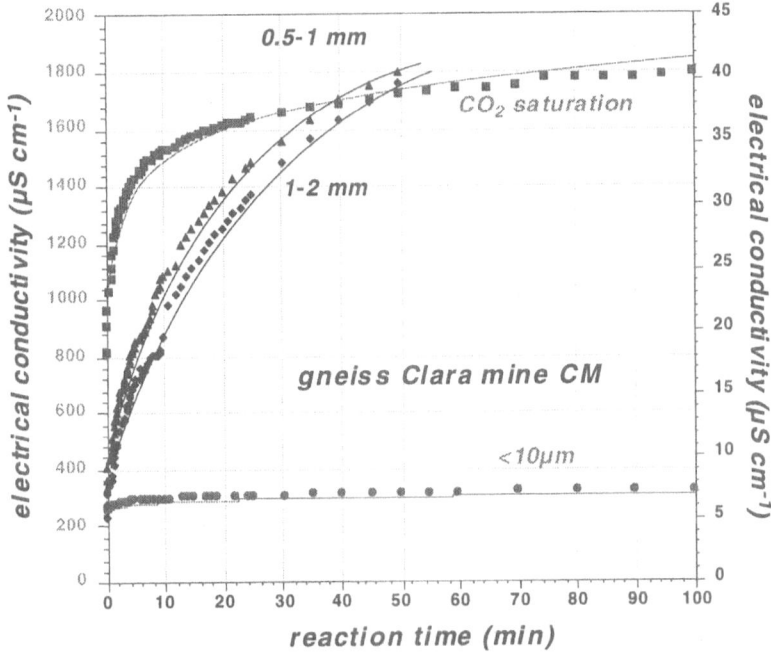

Figure 5: All leaching data of CM gneiss. Scale of fine powder experiments on the l.h.s, for coarse grained experiments scale on the r.h.s.

patterns (Fig. 9). Again, the main reason is probably more calcic plagioclase and more magnesian biotite in gneiss. The grain size distribution pattern shows a slight shift towards finer grains in brittle feldspar-rich granite compared with gneiss rich in ductile biotite. This favors high reaction rates in granite powders. The observation of the contrary suggests that the possible minimal smaller average grain size in granite powders is outweighed by the effects of mineral compositions. This effect is very clearly seen by comparing all CO_2-saturated experiments of two gneisses and two granites. The electrical conductivity curves of Triberg and Bärhalde granite are nearly indistinguishable (Fig. 9). This also shows that the experimental data are significant and consistent. The same grain size spectrum, the same experimental procedure and two different rocks with similar composition and mineralogy produce very similar leachates. The gneisses are more variable with respect to chemical and modal composition and consequently show more variation in leachate composition as well, but gneisses are consistently much more soluble than granites (Fig. 9).

Experiments with coarse grained samples from the gneisses showed a plain square root in time relationship for the ec evolution.

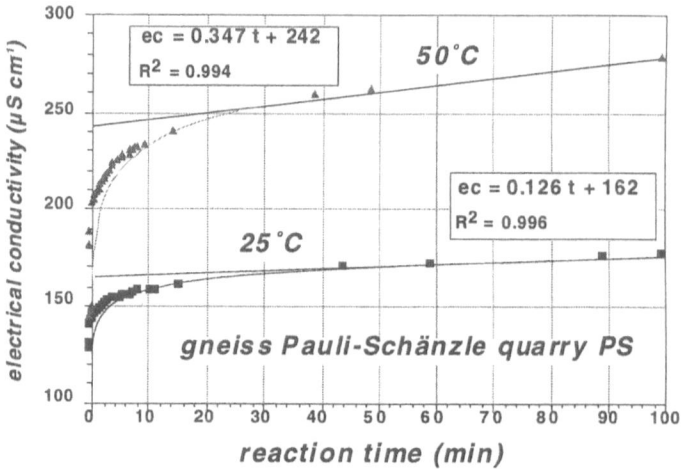

Figure 6: ec data from leaching experiments with Pauli Schänzle gneiss PS, fine powder at 25°C and 50°C.

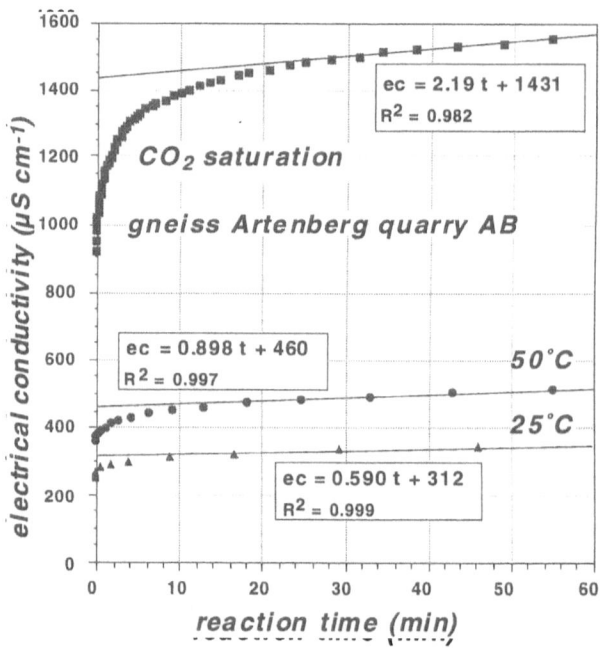

Figure 7: ec data from fine powder experiments of Artenberg gneiss, at 25°C, 50°C and with CO_2-saturation.

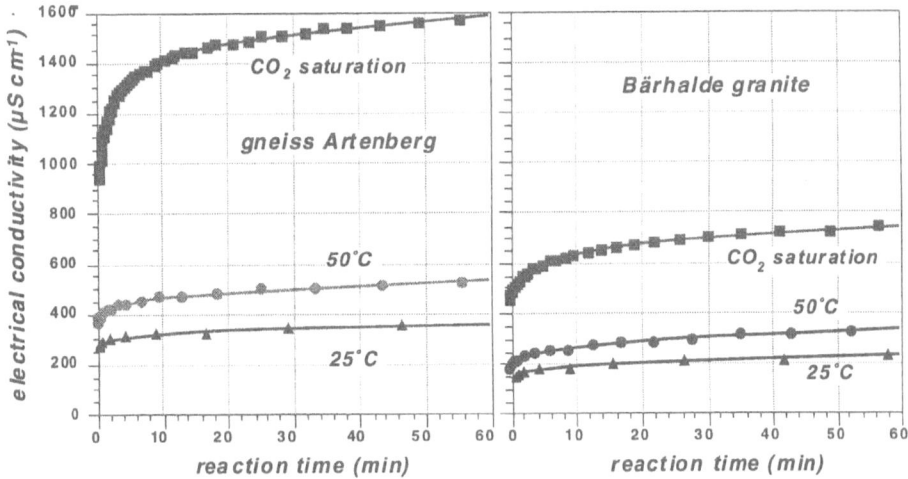

Figure 8: comparison of ec data from fine powder experiments of Artenberg gneiss and Bärhalde granite, at 25°C, 50°C and with CO_2-saturation.

The ec data of all long run experiments show that the kinetics of coarse grained material is fundamentally different from that of very fine powders. CO_2-saturated experiments resulted in high-TDS leachates for gneisses and, were consistently lower for granites. Also, all CO_2-free experiments with granite powders at 25° and 50°C released less solutes than the corresponding experiments with gneiss powders.

6.2. The evolution of pH

The progress of the dissolution process monitored by the electrical conductivity (ec) is accompanied by a systematic variation of pH during the experiments. As described above, during the very initial period of the experiment during which the fine powder is wetted by the reaction fluid (pure water or CO_2-water) the ongoing processes cannot be monitored directly. The first pH reading was typically possible after about 30 s reaction time. The pH evolution during the many experiments was extremely systematic and consistent. Three types of patterns can be distinguished: a) a power-law growth curve similar to the ec evolution pattern was observed during all CO_2-saturated experiments. The initial pH of the CO_2 water was close to 4. Very rapid reaction increases pH to about 6.1 in both gneiss experiments and to 5.6 in both experiments with granite. The final pH is reached after 3 - 4 min reaction time and then stays constant until the end of the experiment after 100 min. b) a similar power-law growth curve is observed with coarse grained powders, however, the first pH reading shows about 8.5 and it

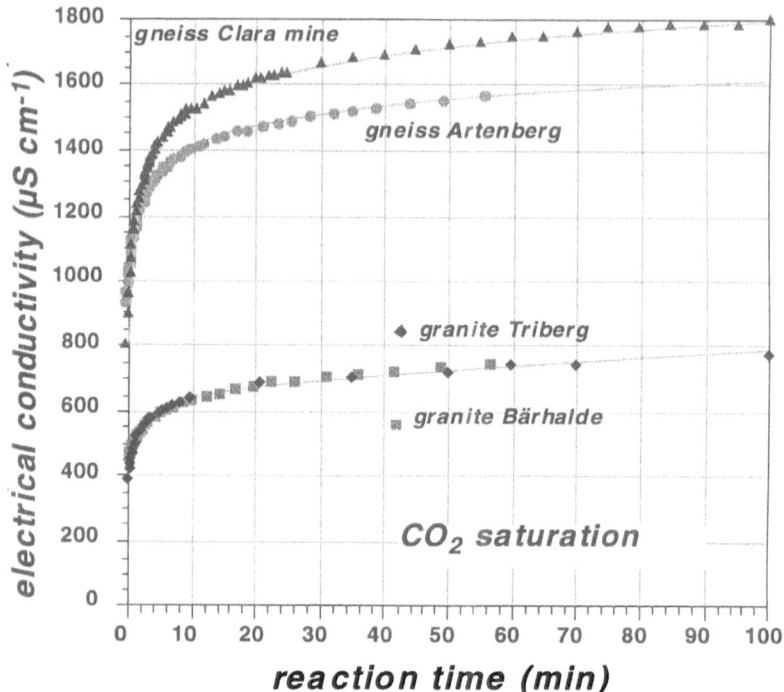

Figure 9: Fine powder leaching experiments with gneiss and granite at 25°C and under CO_2-saturated conditions.

increases to a pH close to 9.8 after 50 min reaction time. The pH increase is slower than in the CO_2-experiments. Fine powder experiments produced leachates with pH of about 8.5 - 9 for granites and pH of about 10 for gneisses. The pH of 50°C experiments was lower by about 0.5 units compared with 25°C runs. The pH of 50°C experiments reached the final value very rapidly (less than 30 s). An extreme behaviour was observed in the 25°C fine powder experiments. The first pH reading after 30 s was the highest pH (10 -10.5 for gneiss powders) in the experiment, the pH then decreased first rapidly then slower and finally assumed a constant value after about 20 min reaction time that was about 0.2 to 0.5 units lower then the initial pH. A similar behaviour observed in leaching experiments with sandstone of the Tensleep formation was attributed to the Kfs to Kao conversion by Shiraki and Dunn (2000).

The pH evolution pattern in the different experiments suggest contrasting governing processes. The 25°C data indicate that in the initial phase of the experiment H^+ of the water exchanges with cations attached to the reactive surface of the fine powder. The dominant cation released to the solution is K^+ followed by Na^+. After about 20 min an cation exchange equilibrium is established and pH remains constant. This process accounts for less than 10% of the total released K^+.

The main contribution to the observed high pH is most likely the closed system dissolution of secondary calcite in gneiss and granite. Cation exchange equilibrium is established in less than one minute in experiments at 50°C. In gneisses, pH is higher than in granites as a consequence of modal dominance of mica over feldspar in gneiss. Micas retain H^+ stronger than feldspar. In CO_2-saturated experiments, the starting solution had a high concentration of H^+ (low pH). Establishing an exchange equilibrium and a steady state pH takes a measurable amount of time (typically a few minutes).

Table 3: chemical composition of leachates from sample CM banded gneiss from the Clara mine

conditions	25 fg	25 fg	25 fg	25 fg	25 fg	25 fg C	25 fg C	25 fg C	25 fg C	25 cg	25 mg
time (min)	2	5	10	50	100	5	10	50	100	100	100
Na	14.19	14.38	14.35	17.45	19.31	24.55	25.02	30.99	34.97	1.15	1.65
K	40.47	44.98	45.64	50.93	54.20	84.15	82.14	89.33	98.17	1.97	1.58
Ca	7.62	7.95	7.86	7.14	6.54	139.45	148.07	148.46	120.76	4.10	4.05
Mg	5.75	6.32	6.44	6.45	6.30	103.42	101.24	127.67	143.89	0.49	0.44
Li	0.46	0.56	0.59	0.71	0.82	1.44	1.35	1.72	1.92	0.01	0.02
Rb	0.18	0.19	0.19	0.20	0.22	0.38	0.39	0.41	0.42		
Sr	0.04	0.04	0.03	0.03	0.03	0.41	0.43	0.47	0.45	0.02	0.03
Fe	0.08					0.06	0.07	0.08	0.20	0.07	0.11
Mn	0.00	0.00	0.00	0.00	0.00	2.26	2.60	2.69	3.04	0.00	0.00
Al	2.10	2.33	2.27	1.76	1.74	0.13	0.15	0.12	0.15	0.68	0.51
alkalinity	126.44	139.26	141.09	156.98	156.98	1172.74	1160.52	1331.54	1368.19	20.77	16.49
NO₃	0.31	0.22	0.30	0.28	0.29	0.42	0.40	0.36	0.69	0.20	0.15
SO₄	2.52	3.40	3.05	2.94	2.97	2.14	2.12	2.51	2.92	0.27	0.54
F	2.08	2.39	2.46	3.12	3.53	0.26	0.31	0.33	0.25	0.13	0.15
Cl	4.20	3.82	3.91	4.15	4.58	4.21	4.30	4.67	5.12	0.60	0.62
Br	0.05	0.05	0.05	0.05	0.06	0.06	0.07	0.07	0.08		
SiO₂	2.30	2.50	2.50	2.80	3.20	24.00	29.50	37.50	35.50	0.60	0.70
EC [µS/cm]	246	268	272	295	311	1395	1475	1688	1789	40	40
pH	9.78	9.70	9.68	9.60	9.58	6.09	6.06	6.12	6.12	9.75	9.70

conditions: 25 fg = 25°C fine grained (< 10µm); 50 fg = 50°C fine grained (< 10µm); 25 fg C = 25°C fine grained (< 10µm) & CO₂-saturation; 25 mg = 25°C medium grained (0.5 - 1 mm); 25 cg = 25°C coarse grained (1 - 2 mm); all concentrations in mg /l

Table 4: chemical composition of leachates from sample PS banded gneiss from the Pauli-Sch nzle quarry.

conditions	25 fg	25 fg	25 fg	25 fg	50 fg	50 fg	50 fg	50 fg	50 fg	25 cg C	25 cg C	25 cg C	25 cg
time (min)	5	10	50	100	2	5	10	50	100	10	50	100	50
Na	9.40	9.80	12.52	12.62	16.25	15.66	18.00	21.98	22.97	4.05	4.07	4.18	0.76
K	31.39	31.34	36.59	35.99	43.73	42.25	48.39	54.53	53.44	7.33	9.36	10.17	1.59
Ca	3.57	3.42	3.15	3.10	2.10	2.44	1.87	1.69	1.78	35.81	55.84	59.90	2.68
Mg	2.11	1.98	2.42	2.12	1.63	1.48	1.58	1.31	1.17	1.81	1.92	2.11	0.30
Li	0.15	0.14	0.18	0.18	0.26	0.23	0.31	0.39	0.49	0.28	0.31	0.32	0.08
Rb	0.09	0.09	0.10	0.10	0.17	0.15	0.20	0.22	0.21	0.02	0.03	0.03	0.01
Sr	0.00	0.01	0.00	0.01	0.00	0.00	0.00	0.00	0.00	0.05	0.06	0.07	0.00
Fe	0.73	0.96	1.28	0.77	1.20	0.99	1.24	1.23	1.07	2.32	8.12	12.16	0.29
Mn	0.01	0.02	0.02	0.01	0.02	0.01	0.02	0.02	0.02	1.21	2.21	2.70	0.01
Al	3.73	4.20	5.41	4.56	8.97	8.10	10.55	10.74	8.67	0.22	0.37	0.41	0.75
alkalinity	75.00	72.00	83.20	84.30	83.40	86.30	92.40	95.40	96.80	137.25	204.96	234.24	12.30
NO_3	0.27	0.17	0.54	0.13	0.19	0.36	0.13	0.16	0.13	0.00	0.64	0.00	0.11
SO_4	1.57	1.53	1.89	1.79	1.77	1.78	1.79	1.93	1.96	0.36	0.60	0.51	0.17
F	2.39	2.40	3.38	3.46	4.21	3.47	4.41	6.12	5.82	0.13	0.38	0.56	0.08
Cl	3.72	3.69	4.01	4.01	3.88	4.23	5.03	4.01	6.10	1.77	2.82	3.43	0.69
Br	0.03	0.04	0.04	0.03	0.04	0.04	0.04	0.04	0.04				0.00
J	0.15	0.19			0.18	0.20	0.22	0.69		0.55	0.65	0.32	0.00
P	0.00	0.00	0.00	0.00	0.00	0.09	0.09	0.18	0.15	0.00	0.00	0.00	0.00
SiO_2	2.08	2.76	3.21	3.17	3.94	5.63	4.82	6.18	6.46	1.01	1.80	2.40	0.51
EC [µS/cm]	166	169	196	194	215	211	244	279	277	213	312	373	26
pH	9.78	9.72	9.67	9.63	9.20	9.34	9.05	9.14	9.05	5.11	5.29	5.37	9.30

conditions: 25 fg = 25°C fine grained (< 10µm); 50 fg = 50°C fine grained (< 10µm); 25 fg C = 25°C fine grained (< 10µm) & CO_2-saturation; 25 mg = 25°C medium grained (0.5 - 1 mm); 25 cg = 25°C coarse grained (1 - 2 mm). all concentrations in mg /l

Table 5: chemical composition of leachates from sample AB, biotite gneiss from the Artenberg quarry.

conditions	25 fg	25 fg	25 fg	25 fg	25 fg	50 fg	50 fg	50 fg	50 fg	25 fg C	25 fg C	25 fg C	25 cg
time (min)	1	2	5	15	60	2	5	15	60	2	15	60	60
Na	13.11	13.58	14.75	15.80	18.39	20.48	20.60	24.83	28.73	23.50	25.53	35.27	1.27
K	64.10	65.19	66.98	63.06	72.22	98.20	91.74	100.24	117.87	135.89	151.37	168.81	2.22
Ca	3.35	3.07	3.30	3.27	2.95	2.20	2.28	2.12	2.02	110.79	137.67	141.67	8.23
Mg	2.33	2.33	2.47	2.21	1.73	1.39	1.21	1.37	1.35	39.81	50.97	62.73	0.31
Li	0.18	0.19	0.20	0.19	0.22	0.38	0.38	0.41	0.46	0.64	0.76	0.84	0.06
Rb	0.14	0.14	0.16	0.15	0.17	0.20	0.20	0.22	0.24	0.15	0.17	0.21	0.00
Sr	0.05	0.04	0.07	0.08	0.07	0.01	0.01	0.01	0.01	0.27	0.30	0.30	0.02
Fe	2.38	2.39	2.87	3.19	2.63	3.43	2.40	2.51	2.70	0.04	0.15	0.23	
Mn	0.05	0.04	0.06	0.07	0.05	0.06	0.04	0.04	0.04	1.51	2.68	3.22	
Al	9.48	9.27	9.56	8.13	7.83	14.00	12.12	11.75	12.05	0.08	0.12	0.98	0.01
Ba										0.34	0.67	0.83	0.22
alkalinity	150.00	145.00	148.00	141.00	161.00	196.00	187.00	205.00	241.00	807.03	971.73	1127.00	33.00
NO_3	0.06	0.08			0.07	0.09	0.09	0.14	0.13	0.09	0.10	0.12	0.64
SO_4	1.61	1.78	1.70	1.83	1.84	1.57	1.63	1.94	2.15	1.30	1.26	1.45	0.24
F	1.20	1.22	1.23	1.35	1.70	2.35	2.11	2.62	3.54	0.15	0.17	0.16	
Cl	3.28	3.33	3.82	4.22	5.72	3.84	3.63	5.22	6.82	3.41	3.10	4.02	0.95
J	0.06	0.09	0.05	0.07	0.08	0.08	0.16	0.13	0.12	0.03	0.04	0.03	0.15
SiO_2	4.00	4.00	4.20	4.20	5.40	5.30	6.10	6.60	6.90				0.30
EC [μS/cm]	280	299	305	281	342	398	390	433	507	1109	1402	1550	40
pH	10.18	10.29	10.20	10.03	10.03	10.03	9.97	9.97	9.96	6.13	6.02	6.06	8.44

conditions: 25 fg = 25°C fine grained (< 10μm); 50 fg = 50°C fine grained (< 10μm); 25 fg C = 25°C fine grained (< 10μm) & CO_2-saturation; 25 mg = 25°C medium grained (0.5 - 1 mm); 25 cg = 25°C coarse grained (1 - 2 mm). all concentrations in mg /l, Br and P below detection limit

Table 6: chemical composition of leachates from sample TB Triberg granite.

conditions	25 fg	25 fg	25 fg	25 fg	25 fg	50 fg	50 fg	50 fg	50 fg	25 fg C	25 fg C	25 fg C	25 cg
time (min)	2	5	10	50	100	5	10	50	100	10	50	100	50
Na	12.91	12.78	13.55	14.59	15.91	15.33	16.39	18.71	20.08	49.42	54.94	57.78	0.95
K	18.85	18.31	18.95	20.50	20.70	21.57	22.12	25.43	28.29	122.63	134.55	157.13	1.14
Ca	0.20	0.20	0.30	0.30	0.20	0.20	0.30	0.20	0.20	13.06	15.69	16.31	0.57
Mg	0.19	0.18	0.18	0.23	0.19	0.15	0.19	0.19	0.17	7.86	9.53	10.99	0.14
Li	0.14	0.14	0.15	0.15	0.15	0.22	0.16	0.15	0.13	0.64	0.71	0.69	0.00
Rb	0.07	0.07	0.07	0.08	0.08	0.13	0.09	0.11	0.12	0.48	0.50	0.55	0.01
Sr	0.00	0.00	0.00	0.00	0.00	0.00	0.00	0.00	0.00	0.04	0.05	0.06	0.00
Fe	0.29	0.41	0.25	0.44	0.36	1.10	0.22	0.48	0.52	0.07	0.09	0.29	0.07
Mn	0.01	0.01	0.01	0.02	0.01	0.03	0.01	0.02	0.02	0.54	0.74	0.92	0.00
Al	4.09	2.96	2.64	4.66	7.39	7.00	4.04	8.38	9.17	0.46	0.86	0.73	1.31
alkalinity	22.10	24.10	23.90	21.90	21.70	21.90	23.00	24.00	21.90	375.88	438.12	471.07	2.50
NO_3	0.26	0.36	0.26	0.44	0.31	0.25	0.20	0.23	0.31	0.53	0.46	0.44	0.14
SO_4	0.47	0.49	0.49	0.53	0.54	0.57	0.54	0.58	0.63	0.30	0.36	0.37	0.00
F	3.02	3.46	3.82	4.51	4.82	4.93	5.20	6.64	7.14	0.70	0.68	0.71	0.05
Cl	14.51	14.78	14.70	15.48	15.46	14.92	16.12	18.31	23.74	15.99	17.06	17.71	1.06
Br	0.12	0.12	0.12	0.12	0.13	0.13	0.13	0.14	0.15	0.15	0.19	0.17	0.00
J	0.17	0.13	0.17			0.21	0.50	0.73					0.00
P	0.00	0.00	0.00	0.09	0.15	0.13	0.15	0.25	0.25	0.00	0.00	0.00	0.00
SiO_2	2.74	2.93	3.38	4.96	5.41	6.76	7.85	12.03	13.14	26.11	34.24	38.95	0.21
EC [µS/cm]	117	119	123	135	139	137	145	170	188	644	721	764	9
pH	8.78	8.70	8.58	8.92	8.95	8.60	8.64	8.63	8.60	5.60	5.59	5.65	7.59

conditions: 25 fg = 25° C fine grained (< 10µm); 50 fg = 50° C fine grained (< 10µm); 25 fg C = 25° C fine grained (< 10µm) & CO_2-saturation; 25 mg = 25° C medium grained (0.5 - 1 mm); 25 cg = 25° C coarse grained (1 - 2 mm). all concentrations in mg /l

Table 7: chemical composition of leachates from sample BH Baerhalde granite

conditions time (min)	25 fg 1	25 fg 2	25 fg 5	25 fg 15	25 fg 60	50 fg 2	50 fg 5	50 fg 15	50 fg 60	25 fg C 2	25 fg C 15	25 fg C 60	25 cg 60
Na	17.03	19.06	21.26	20.65	24.95	29.93	30.44	31.02	35.49	47.97	50.71	76.35	0.59
K	23.23	25.57	32.40	25.97	33.71	42.88	42.08	42.48	47.76	104.79	90.70	131.63	1.80
Ca	0.29	0.24	0.21	0.24	0.16	0.28	0.12	0.23	0.33	4.04	7.42	8.00	0.23
Mg	0.04	0.05	0.06	0.05	3.05	0.09	0.05	0.10	0.10	1.24	2.09	1.66	
Li	0.51	0.56	0.62	0.60	0.66	1.00	1.00	0.94	0.90	1.46	2.19	2.10	
Rb	0.33	0.38	0.42	0.38	0.47	0.57	0.50	0.52	0.54	1.31	1.80	1.75	0.01
Sr	0.03	0.02	0.03	0.04	0.04					0.02	0.02	0.03	0.00
Fe	1.86	1.91	1.99	2.20	2.10	3.76	1.91	2.05	1.42	0.60	0.77	1.16	
Mn	0.05	0.05	0.05	0.05	0.05	0.05	0.02	0.03	0.02	0.52	0.55	0.92	
Al	9.19	10.74	9.46	9.17	9.78	21.43	14.81	15.88	11.30	3.84	3.93	4.06	0.12
alkalinity	31.72	38.43	34.16	31.72	40.72	43.31	46.42	40.14	45.75	301.34	304.00	432.49	1.70
NO$_3$	0.10	0.08	0.06	0.16	0.08	0.10	0.10	0.17	0.13	0.14	0.12	0.09	
SO$_4$	0.48	0.60	0.46	0.50	0.46	0.64	0.56	0.49	0.47	0.43	0.49	0.44	
F	14.17	15.98	15.94	16.74	18.71	24.87	24.80	24.58	27.19	4.90	6.25	5.83	0.13
Cl	9.51	9.79	18.70	10.34	13.36	12.73	15.34	16.09	23.46	9.79	11.52	12.38	1.84
Br	0.11	0.11	0.11	0.12	0.12	0.16	0.15	0.15	0.16	0.09	0.12	0.11	
J	0.66	0.56	0.67			0.75	0.92	0.81	0.95	0.03	0.51	0.51	0.03
P						1.39	1.69	1.55	1.76				
SiO$_2$	5.40	4.40	4.40	4.00	6.80	10.00	8.80	7.80	8.90				
EC [µS/cm]	138	159	195	180	217	247	265	275	321	487	346	728	10
pH	8.26	8.84	8.95	8.91	9.03	8.88	8.89	8.76	8.52	5.52	5.26	5.63	5.08

conditions: 25 fg = 25°C fine grained (< 10µm); 50 fg = 50°C fine grained (< 10µm); 25 fg C = 25°C fine grained (< 10µm) & CO$_2$-saturation; 25 mg = 25°C medium grained (0.5 - 1 mm); 25 cg = 25°C coarse grained (1 - 2 mm). All concentrations in mg /l

7. Water composition

All water composition data of all experiments are listed on Tables 3 through 7. The data permit to investigate the evolution of solute concentrations with increasing reaction time and also to explore the effect of different experimental conditions on the concentration patterns of the leachates. In this paper we will mostly concentrate on the latter aspect of the experiments.

7.1. General patterns

In all experiments the total of dissolved solids increased with time whereas the individual solutes normally also increased with time a few constituents first increased and then decreased with reaction progress. The TDS increase does not correlate with pH, there is no pH increase associated with the general TDS increase. The typical water characteristics emerge after very short reaction time. In fine powder experiments at 25°C and pure water of both gneiss and granite potassium is the dominant cation. In the CO_2-saturated experiments, there is more Ca and Mg extractable in two of the gneisses. The biotite-rich Artenberg gneiss still produces a very K-rich fluid. Both granites are K-dominated because the rocks contain plagioclase with extremely low An-content. Li is high in Bärhalde granite water.

Remarkable is the very high amount of fluorine that can be extracted particularly from Bärhalde granite. This granite contains F-biotite, F-apatite and topaz. However, also the other rock samples contain much extractable fluorine. Some of the leachates contain high concentrations of aluminum. Some of this may be colloidal Al or metastable concentrations. However, the Al content directly correlates with high pH and the fluorine concentration, this suggests that the observed high Al in water is present as Al-(OH) and Al-F complexes. The Bärhalde data (Tab. 7) further suggest that fluorapatite dissolves in 50°C experiments. This results in an increased phosphate concentration and an associated high-F water, which then together with high-pH also had a high Al concentration. Apatite does not dissolve at 25°C and in CO_2-saturated experiments where pH is low F and Al are low as well. Phosphorous is present in 50°C leachates only, its source is apatite. The source of fluorine is biotite, apatite and, in the Bärhalde granite topaz.

Cl and Br are predominantly released from fluid inclusions, however, some additional Cl is released mainly from dissolving apatite in the 50°C experiments. Silica increases significantly with reaction time in experiments with the Clara gneiss and the Triberg granite, particularly released in CO_2-saturated experiments. The waters become supersaturated with respect to quartz and chalcedony. Other experiments did not show much variation in the silica concentration. Sulfate is low in all leachates and it does not vary with reaction time. This suggests that during the experiments sulfide oxidation does not contribute solutes to the leachates.

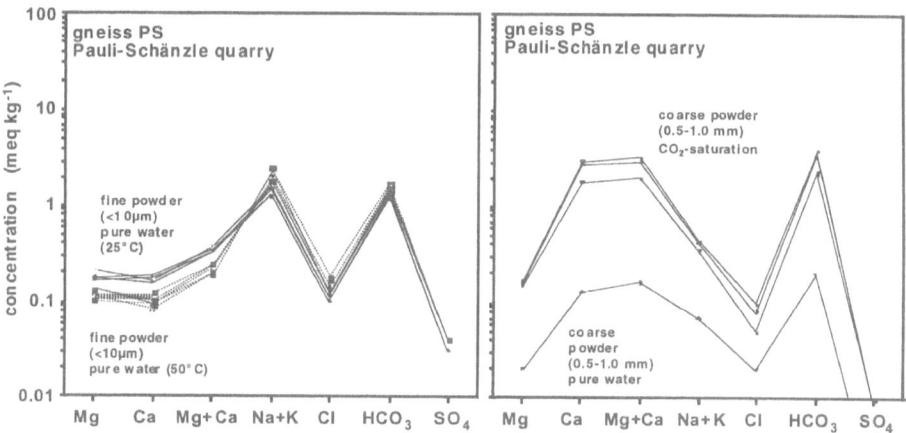

Figure 10: Composition of fine powder leachates at 25°C of CM gneiss and TB granite.

Significantly increased Mg, Li, and Rb concentrations in CO_2-saturated experiments suggest that in low-pH water biotite dissolution contributes significantly to the water composition. Sr is extremely low in both granite waters because of low-An plagioclase in these rocks, in the gneiss leachates Sr increases slightly in CO_2-saturated experiments that dissolve parts of the plagioclase. Mg concentration is highest in Clara gneiss waters, where it is released by biotite and chlorite dissolution (also Li is high). In the granites, biotite is more Fe-rich and consequently the biotite dissolution contributes less Mg to the water. The Mg content of the Triberg and Bärhalde leachates are directly correlated with the X_{Mg} of biotite in the rock. Clara: X_{Mg}^{bt}=0.45 Triberg: 0.22 Bärhalde: 0.05 and the corresponding Mg concentrations of the CO_2-leachates after 60 min are (mg/l): 144; 11; 2.

7.2. Major elements

The composition of the leachates shows characteristic patterns depending on rock type and experimental conditions. The composition patterns are remarkably consistent and reproducible. Three distinctly different patterns of major element compositions of the leachates of the gneiss CM from the Clara mine (Fig. 10) could be observed. At 25°C, the fine powder leachate is dominated by the cations Na and K and the anion HCO_3. The carbonate most likely has its origin in the rapid dissolution of secondary calcite in the gneiss. Ca is then exchanged with alkalis on the reactive surfaces, and $(K,Na)HCO_3$ is the net release to the leachate. A small amount of CO_2 is taken up from the air during the experiment in the open system. Similar leachates were observed by Bertrand et al. (1994) in experiments with Kfs-rich sandstone (Buntsandstein) at 150°C. Ca is lower than Mg, demonstrating the effectiveness of the Ca-removal by ion exchange. Chloride is significantly higher than SO_4. The amount of chloride corresponds to about 50 mg extractable Cl per kg of rock powder. All five leaching experiments of different reaction time show the same composition pattern. The experiment at CO_2-saturation shows a very different composition pattern (Fig. 10). Here the earth-alkalis greatly dominate over the alkalis. Still Mg is higher than Ca. TDS is higher by an order of magnitude. There is considerably more reaction progress under low-pH acidic conditions liberating 8 to 9 times more solutes and leading to a $MgCO_3$ dominated water. However, the CO_2-saturated extraction did not release more chloride than the pure water extraction. This suggests that chloride is exclusively contributed by opened fluid inclusions and not by dissolution of Cl-bearing minerals, such as biotite. The coarse grained powder produced a very distinct and different composition pattern (Fig. 10). In these leachates Ca is the most abundant cation, followed by alkalis and Mg. Cl is low as a result of less fluid inclusions opened in the grinding process. Ca released by calcite dissolution is incompletely retained by ion exchange in coarse grained material.

Three patterns can be distinguished also in leachates from the Triberg granite (Fig. 10). They are different from the patterns of the gneiss leachates. The most important difference is the very high Cl-concetration in the granite waters. Cl is the most abundant anion in the pure water experiments. More than 200 mg soluble chloride can be extracted from 1 kg rock powder. Alkalis remain the dominant cations also in the CO_2-saturated experiments, despite that the acid reaction increased the Ca and Mg content by 2 orders of magnitude. The granite bulk rock contains very little Ca in feldspar and Mg in biotite and much quartz with fluid inclusions. The patterns also suggest that the Cl is predominantly released from fluid inclusions. At 50°C an additional amount of Cl is released to the water with an equivalent amount of Na. This observation is difficult to explain. However, because no other solutes appear in the water, the additional NaCl may be released from fluid inclusions in small quartz grains dissolving in the warm water.

The patterns from the gneiss of the Pauli-Schänzli quarry show distinct patterns related to grain size, reaction temperature and CO_2-saturation (Fig. 11). The patterns are very similar to the ones from the Clara gneiss (Fig. 10). At 25°C the pure water experiment produces alkali-rich waters with low and equal concentrations of Ca and Mg. In 50°C experiments, both Mg and Ca further decrease, Na and K increase as well as chloride. The patterns are very consistent and the pattern characteristics develop in experiments of very short duration (1 min) longer experiments (100 min) maintain the pattern and displace it to higher concentrations. The experiment with the coarse powder produced a pattern identical to that of the Clara gneiss. The CO_2-saturation experiments were carried out with a coarse powder fraction of the Pauli-Schänzli gneiss. The water composition pattern is very similar to the coarse grained pure water experiment. It is evident that the pattern once established just evolves to similar patterns with higher concentrations.

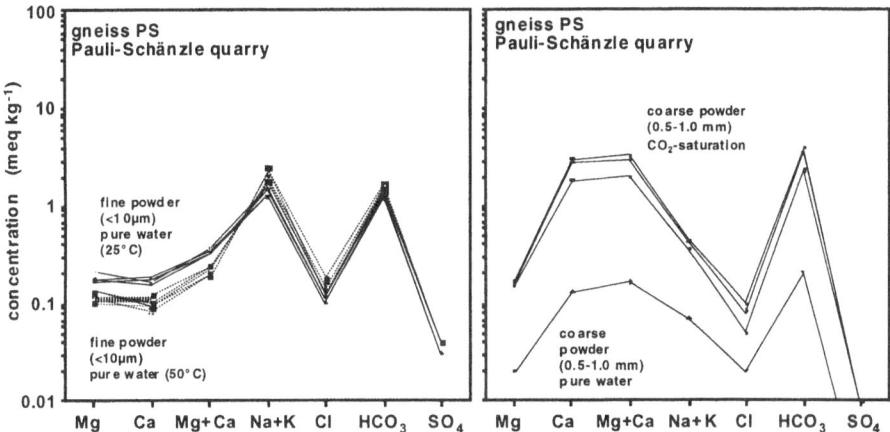

Figure 11: Composition of leachates of PS gneiss, fine powder experiments on the l.h.s, coarse grained material on the r.h.s.

The patterns from the Artenberg gneiss (AB) shown on Fig. 12 are very similar to the previous patterns from the other two gneisses. Relative to the 25°C pure water experiment Mg and Ca decreases at 50°C experiments. This is accompanied by an increase in alkalis and also, remarkably consistent, in chloride. In CO_2-saturated waters Mg and Ca increase dramatically (> 50 times) and become more abundant than the alkalis. However, the CO_2 experiments did not release additional chloride. The leachates of the coarse grained powder of the Artenberg gneiss show very similar patterns to the corresponding patterns of the other two gneisses.

All three gneiss patterns show very close relationships among each other and the data of the CG experiments are closely related to the compositional features of the Black Forest mineral waters (Fig. 1).

The experiments with the Bärhalde granite (BH) produced leachates with patterns that are very similar to the patterns from the Triberg granite (Fig. 13). The alkali metals dominate greatly over the earth alkalis and Ca over Mg. The resulting pattern has a characteristic step shape. Also in the Triberg granite a large amount of extractable chloride is stored in fluid inclusions. An additional amount of chloride can be mobilized in 50°C experiments but not in CO_2-saturated experiments at 25°C. The Bärhalde granite releases close to 300 mg Cl per 1 kg rock or 810 gr Cl per 1 m^3 of granite. This is 5 times the amount reported by Peters (1986) for the Böttstein granite in Northern Switzerland. The amount that can be extracted in coarse grained powders of the Bärhalde granite is about 10 times lower than in the < 10 μm powder. However, the characteristic step shaped pattern is also developed in the coarse powder experiments. The granite water patterns are closely related to the patterns of thermal waters from the Black Forest area (Fig. 1). It is therefore concluded that the mineral waters develop their characteristic pattern in gneiss and the thermal water derive predominantly from interaction with granite (see also Grasby et al., 2000).

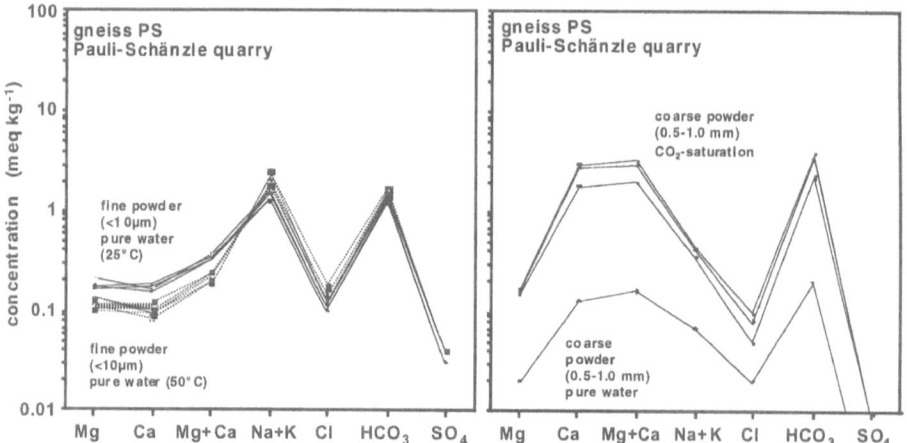

Figure 12: Composition of leachates from Artenberg gneiss, fine powder on the l.h.s., coarse grained material on the r.h.s.

The major element patterns are clearly related to the mineralogy of the samples. All gneisses contain plagioclase with a significant anorthite-component and the biotite or chlorite contains a high proportion of Mg-endmember equivalent. In contrast, both granites contain extremely Ca-poor plagioclase and almost pure annite as biotite. The contribution from fluid inclusions is low in gneisses and high in quartz-rich granite. Therefore, gneisses release much Mg and Ca to carbonate fluids and the granites produce predominantly alkali chloride fluids.

One consequence is also evident from molal X_{Na} numbers calculated for all leachates (Fig. 14). Gneiss leached by CO_2-saturated water produce Ca-rich fluids with X_{Na} centering around 0.3. This is close to the typical X_{Na} of about 0.3 - 0.4 measured in CO_2-rich mineral waters from the Black Forest area (Stober and Bucher, 1999a). Gneiss leached by pure water produces high-X_{Na} fluids. Granites always produce water with very low Ca and X_{Na} around 1. Even in CO_2-saturated water, the two granites used in the experiments produced waters with $X_{Na} \sim 0.9$. The X_{Na} relationship of the leachates of the three gneisses is in agreement with the composition of the average plagioclase composition in the three gneisses. The average plagioclase of the gneisses is An_{21}, that is a typical oligoclase, the maximum anorthite content measured was An_{32} in fresh plagioclase of the Clara and Artenberg gneiss respectively. Most of the Artenberg plagioclase was below An_{20}, however. One may conclude that the X_{Na} of the water is basically controlled by the average plagioclase composition in the aquifer matrix material and the P_{CO_2} in the water.

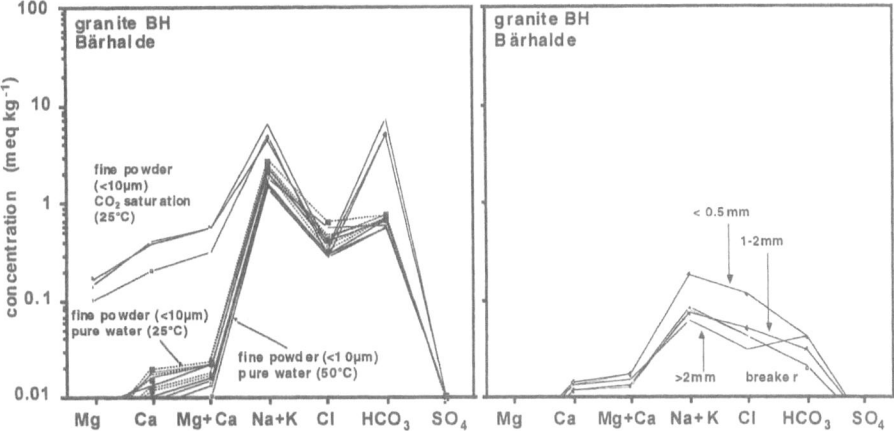

Figure 13: Composition of leachates from the Bärhalde granite, fine powder on the l.h.s., coarse grained material on the r.h.s.

7.3. Leachable halogens

Of particular interest and one major motivation to carry out the reported experiments was the halogen composition of the leachates. In many natural deep groundwaters in crystalline bedrock aquifers chloride is a major anion and its origin has been received considerable interest (Gascoyne et al., 1987; Edmunds et al., 1987; Kamineni, 1987; Bottomley et al., 1999; Stober and Bucher, 1999b). One major parameter that has been used to discriminate the origin of the chloride is the Cl/Br ratio. The range covers values from < 50 in residual brines to several thousand in waters from dissolving halite deposits. Average seawater has Cl/Br=288 (on a ppm basis). Also Cl/I ratios proved useful as source indicators for salinity (Böhlke and Irwin, 1992; Yardley et al., 1992). Finally, fluorine data of the Black Forest material is interesting because the granites and gneisses show minerals that are unusually rich in fluorine. The Bärhalde granite e.g. contains accessory F-apatite, fluorite and topaz.

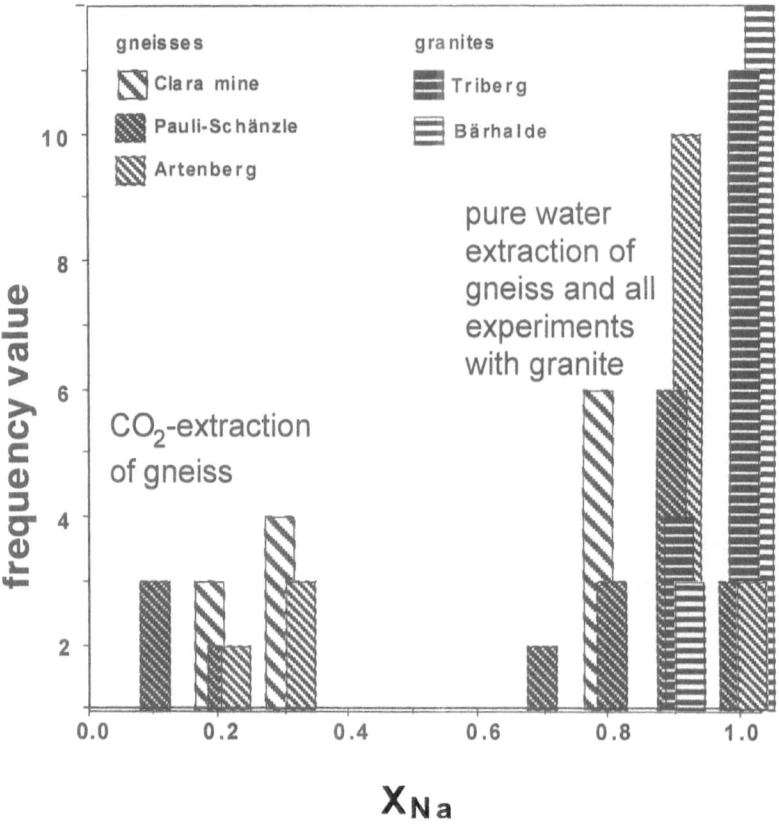

Figure 14: Molar ratios of Na/(Na+Ca) in leaching experiments.

The Cl/Br ratio in fluid inclusions of the Clara gneiss (CM) is 68 for CO_2-saturated and 78 for pure water experiments. Br data are available for the Clara gneiss only. The two granites had both higher Cl/Br ratios in fluid inclusion fluids, typically around 100 - 110. Both granite powders released some extra chloride during reaction at 50°C. This can also be seen on the Schøller patterns (Figs. 10 & 13). This indicates that after having dissolved the alkali halides from broken-up fluid inclusions a new chloride source is taped most likely structurally bound chloride in a mineral that dissolves easily in warm water. The prime candidate is apatite. This Cl-source is poor in bromide and the Cl/Br rapidly increases with reaction time. The maximum value reached after 100 min is Cl/Br = 160. Hence, the observed low Cl/Br of deep groudwater in gneiss and granite is direct evidence for formation of hydrate minerals from metastable high-T assemblages by retrograde low-T hydration reactions. The minerals incorporate Cl in the structure where it replaces OH-groups which has a similar ionic radius. The much larger Br and I ions are left behind and accumulate in the fluid phase. The observed high Br/Cl and I/Cl ratios (Fig. 15) of fluid inclusions in gneiss and granite can best be explained by preferential loss of Cl due to formation of minerals that can accommodate Cl in the structure. The initial fluid in the fluid inclusions may have been a mixture of meteoric and sea water and evolved along paths schematically indicated by the heavy arrows on Fig. 15.

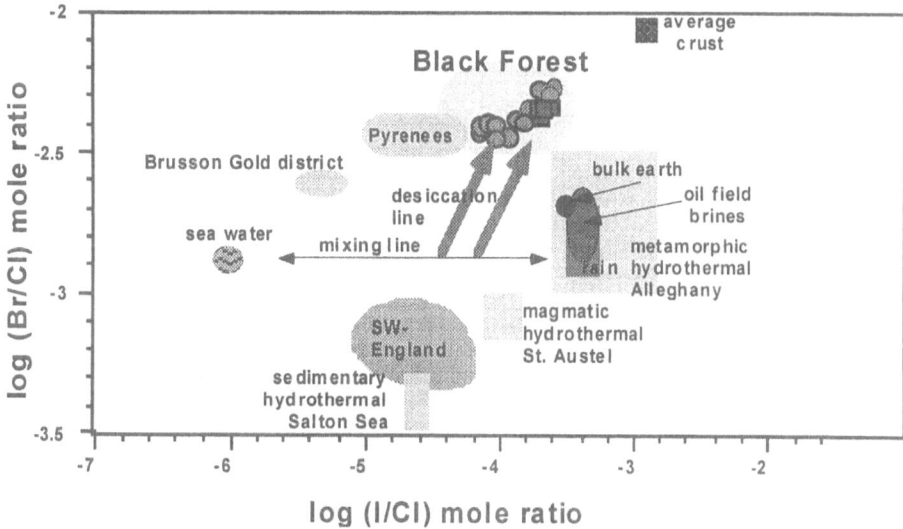

Figure 15: Halogen chemistry of leaching experiments, extractable Cl, Br and I in Black Forest granites (circles) and gneisses (squares) compared with halogen data from other sources. Data from Böhlke and Irwin (1992), Yardley et al. (1992) and this paper.

Acknowledgments

We are grateful to our responsible laboratory technician Sigrid Hirth-Walther who made the experimental and analytical work reported in this study possible. Much of the experimental work was carried out by Richard Liegl and Elvira Treß under the supervision of Sigrid Hirth-Walther and Erika Lutz. We also thank Erika Lutz for producing careful IC data. Reviews provided by Ryoji Shiraki (Davis) and Ian Hutcheon (Calgary) are gratefully acknowledged.

References

Azaroual, M., and Fouillac, Ch., 1997, Experimental study and modelling of granite-distilled water interactions at 180°C and 14 bar: Applied Geochemistry, 12, 55-73.

Banks, D., Reimann, C., Røyset, O., and others, 1995, Natural concentrations of major and trace elements in some Norwegian bedrock groundwaters: Applied Geochemistry, 10, 1-16.

Böhlke, J. K., and Irwin, J. J., 1992, Laser microprobe analyses of Cl, Br, I, and K in fluid inclusions: Implications for sources of salinity in some ancient hydrothermal fluids: Geochimica et Cosmochimica Acta, 56, 203-225.

Bottomley, D. J., Katz, A., Chan, L. H., Starinsky, A., Douglas, M., Clark, I. D. and Raven, K. G., 1999, The origin and evolution of Canadian Shield brines: evaporation or freezing of seawater? New lithium isotope and geochemical evidence from Slave craton. Chemical Geology 155, 295-322.

Bucher, K., and Stober, I., 2000, Hydrochemistry of water in the crystalline basement, In Stober, I, and Bucher, K., editors, Hydrogeology of Crystalline Rocks: Dordrecht, Kluwer Academic Publishers, 141-175.

Busenberg, E., and Clemency, C. V., 1976, The dissolution kinetics of feldspars at 25°C and 1 atm CO2 partial pressure: Geochimica et Cosmochimica Acta, 40, 41-49.

Edmunds, W. M., Kay, R. L. F., Miles, D. L., and others, 1987, The origin of saline groundwaters in the Carnmenellis granites, Cornwall (U.K.): Further evidence from minor and trace elements. In: Fritz, P., and Frape, S. K., ed, Saline Water And Gases In Crystalline Rocks: Ottawa, The Runge Press Limited, Geological Association of Canada Special Paper 33, 127-143.

Emmermann, R., 1977, A petrogenetic model for the origin and evolution of hercynian granite series of the Schwarzwald: Neues Jahrbuch Der Mineralogie, Abhandlungen, 128, 219-253.

Finger, F., Roberts, M. P., Haunschmid, B., and others, 1997, Variscan granitoids of central Europe: their typology, potential sources and tectonothermal relations: Mineralogy and Petrology, 61, 67-96.

Frape, S. K., and Fritz, P., 1987, Geochemical trends for groundwaters from the Canadian shield, In Fritz, P., and Frape, S. K., editors, Saline water and gases in crystalline rocks: Ottawa, The Runge Press Limited, Geological Association of Canada Special Paper 33, 19-38.

Garrels, R. M., 1967, Genesis of some ground waters from igneous rocks. Researches in geochemistry, V. 2. In: Researches in geochemistry, V. 2 Abelson, Philip H. ed.), 405-420. John Wiley and Sons, New York, NY, United States.

Garrels, R. M. and Howard, P. F., 1959, Reactions of feldspar and mica with water at low temperature and pressure. Clays and clay minerals. In: Clays and clay minerals Swineford, A. ed.), 68-88. International Series of Monographs on Earth Sciences Pergamon Press, Oxford, United Kingdom.

Garrels, R. M. and MacKenzie, F. T., 1975, Origin of the chemical compositions of some springs and lakes. Geochemistry of water. In: Geochemistry of water Kitano, Y. Ed.), 257-267. In the collection: Benchmark papers in geology Dowden, Hutchinson & Ross, Inc., Stroudsburg, Pa., United States.

Gascoyne, M., Davison, C. C., Ross, J. D., and others, 1987, Saline groundwaters and brines in plutons in the Canadian Shield. In: Fritz, P., and Frape, S. K., ed, Saline Water And

Gases In Crystalline Rocks: Ottawa, The Runge Press Limited, Geological Association of Canada Special Paper 33, 53-68.

Grasby, St. E., Hutcheon, I., and Krouse, H. R., 2000, The influence of water-rock interaction on the chemistry of thermal springs in western Canada: Applied Geochemistry, 15, 439-454.

Grimaud, D., Beaucaire, C. and Michard, G., 1990, Modelling of the evolution of ground waters in a granite system at low temperature: the Stripaground waters, Sweden. Applied Geochemistry 5, 515-525.

Hofmann, B., 1989, Genese, Alteration und rezentes Fliess-System der Uranlagerstätte Krunkelbach (Menzenschwand, Südschwarzwald): Nagra Technischer Bericht, 88-30, 195pp.

Illies, J. H., and Greiner, G., 1978, Rhine graben and the Alpine system: Geological Society of America Bulletin, 89, 770-782.

Kalt, A., Altherr, R., and Hanel, M., 1995, Contrasting P-T conditions recorded in ultramafic high-pressure rocks from the Variscan Schwarzwald (F.R.G.): Contributions to Mineralogy and Petrology, 121, 45-60.

Kamineni, Ch. D., 1987, Halogen-bearing minerals in plutonic rocks: A possible source of chlorine in saline groundwater in the Canadian Shield. In: Fritz, P., and Frape, S. K., ed, Saline Water And Gases In Crystalline Rocks: Ottawa, The Runge Press Limited, Geological Association of Canada Special Paper 33, 69-79.

Kretz, 1983, Symbols for rock - forming minerals: American Mineralogist, 68, 277-279.

Lahermo, P. W., and Lampén, P. H., 1987, Brackish and saline groundwaters in FinlandFritz, P., and Frape, S. K., ed, Saline Water And Gases In Crystalline Rocks: Ottawa, The Runge Press Limited, 103-109.

Lasaga, A. C., 1981, Rate laws of chemical reactions, In Lasaga, A. C., and Kirkpatrick, R. J., editors, Kinetics of Geochemical Processes: Washington, Mineralogical Society of America, Reviews in Mineralogy, 1-68.

Liegl, R., 1998, Eluier- und Laugungsversuche an Kristallingesteinen des Schwarzwaldes: Unpublished Diploma Thesis, University of Freiburg, 70pp.

Liegl, R., Stober, I., and Bucher, K., 1999, Experimental water-rock reaction of Black Forest gneiss and granite: Journal of Conference Abstracts, 4, 590.

Lippolt, H. J., and Kirsch, H., 1994, Isotopic Investigation of Post-Variscan Plagioclase Sericitation in the Schwarzwald Gneiss Massif: Chemie Der Erde, 54, 179-198.

Lodemann, M., Fritz, P., Wolf, M., and others, 1998, On the origin of saline fluids in the KTB: Applied Geochemistry, 13, 651-672.

Luce, R. W., Bartlett, R. W., and Olson, R. K., 1972, Dissolution kinetics of magnesium silicates: Geochimica et Cosmochimica Acta, 36, 35-50.

May, F., Hoernes, S. and Neugebauer, H. J., 1996, Genesis and distribution of mineral waters as a consequence of recent lithospheric dynamics: the Rhenish Massif, Central Europe. Geologische Rundschau 85, 782-797.

Mazurek, M. and Peters, T. 1992, Petrographie des kristallinen Grundgebirges der Nordschweiz und Systematik der herzynischen Granite. Schweizerische Mineralogische und Petrographische Mitteilungen 72, 11-35.

Möller, P., et al., 1997, Paleo- and recent fluids in the upper continental crust - Results from the German Continental deep drilling Program (KTB): Journal of Geophysical Research, 102, 18245-18256.

Möller, P., Stober, I., and Dulski, P., 1997, Seltenerd-Element-, Yttrium-Gehalte und Blei-Isotope in Thermal- und Mineralwässern des Schwarzwaldes: Grundwasser, 3, 118-132.

Nesbitt, H. W, Macrae, N. D. and Shotyk, W., 1991, Congruent and Incongruent Dissolution of Labradorite in Dilute, Acidic, Salt Solutions. Journal of Geology, 99, 429-442.

Pauwels, H., 1997, Geochemical results of a single-well hydraulic injection test in an experimental Hot Dry Rock geothermal reservoir Soultz-sous-Foréts Alsace, France: Applied Geochemistry, v. 12, p. 661-674.

Peters, T., 1986, Structurally incorporated and water extractable chlorine in the Boettstein granite (N. Switzerland): Contributions to Mineralogy and Petrology, 94, 272-273.

Savoye, S., Aranyossy, C., Beaucaire, M., Cathelineau, M., Louvat, D. and Michelot, J-L, 1998, Fluid inclusions in granites and their relationships with present-day groundwater chemistry. European Journal of Mineralogy 10, 1215-1226.

Schleicher, H., and Fritsche, R., 1978, Zur Petrologie des Triberger Granites (Mittlerer Schwarzwald): Jahreshefte Des Geologischen Landesamt Baden-Württemberg, 20, 15-41.

Shiraki, R., and Dunn, T. L., 2000, Experimental study on water-rock interactions during CO_2 flooding in the Tensleep Formation, Wyoming, USA: Applied Geochemistry, 15, 265-280.

Stenger, R., 1982, Petrology and Geochemistry of the Basement Rocks of the Research Drilling Projekt Urach 3, *in* Haenel, R., editor, The Urach Geothermal Project: Stuttgart, Schweizerbart'sche Verlagsbuchhandlung, 41-48.

Stober, I., 1996, Researchers Study Conductivity of Crystalline Rock in Proposed Radioactive Waste Site: EOS, Transactions of the American Geophysical Union, 77, 93-94.

Stober, I., and Bucher, K., 1999a, Deep groundwater in the crystalline basement of the Black Forest region: Applied Geochemistry, 14, 237-254.

Stober, I., and Bucher, K., 1999b, Origin of Salinity of Deep Groundwater in Crystalline Rocks: Terra Nova, 11, 181-185.

Stober, I., and Bucher, K., 2000, Hydraulic Properties of the upper Continental Crust: data from the Urach 3 geothermal well, *In* Stober, I., and Bucher, K., editors, Hydrogeology of Crystalline Rocks, Kluwer.

Stober, I., Richter, A., Brost, E., and Bucher, K. 1999, The Ohlsbach Plume: Natural Release of Deep Saline Water from the Crystalline Basement of the Black Forest : Hydrogeology Journal, 7, 273-283.

Thury, M, Gautschi, A, Mazurek, M, and others, 1994, Geology and Hydrogeology of the Crystalline Basement of Northern Switzerland. Synthesis of Regional Investigations

1981-1993 within the Nagra Radioactive Waste Disposal Programme: Nagra, Technical Report, 93-01, 1-347.

Treß, E., 1999, Laugungsversuche an Kristallingesteinen des Schwarzwaldes: Unpublished Diploma Thesis, University of Freiburg, 62pp.

Waldeck, H., 1970, Der Steinbruch am Artenberg bei Steinach i. K. (Schwarzwald) -eine petrographisch-mineralogische Betrachtung: (N.Jb.Min.), 76, 420-431.

White, A. F., Blum, A. E., Bullen, T. D., Vivit, D. V., Schulz, M. and Fitzpatrick, J., 1999, The effect of temperature on experimental and natural chemical weathering rates of granitoid rocks. Geochimica et Cosmochimica Acta 63, 3277-3292.

Wollast, R, 1967, Kinetics of the alteration of K-feldspar in buffered solutions at low temperature: Geochimica et Cosmochimica Acta, 31, 635-648.

Yardley, B. W. D., Banks, D. A., and Munz, I. A., 1992, Halogen compositions of fluid inclusions as tracers of crustal fluid behaviour. In: Kharaka, Yousif K., and Maest, Ann S., editors, Proceedings of the 7th international symposium on Water-rock interaction; Volume 2, Alberta Research Council, Edmonton, AB, 1137-1140.

The distribution of rare earth elements and yttrium in water-rock interactions: field observations and experiments.

P. MÖLLER
GeoForschungsZentrum Potsdam
Telegrafenberg, D-14473-Potsdam

Abstract

The partition of REE and Y abundances between the source rock and the aqueous phase has been studied in natural water and steam ranging from 10 to 400°C. Rare earth elements (REE) and Y abundances in aqueous fluids show a variability over 6 orders of magnitudes. For grouping of waters according to their source rocks, the patterns of REY/Ca ratio are much more appropriate than the conventional REY abundance patterns. Waters from mafic igneous rocks show higher REY/Ca ratios than those from granites. REY/Ca ratios in waters from metamorphites are lower than those of the igneous counterparts. Experimental leaching of the source rocks at pH values of about 3 yield patterns that differ from those of the natural fluids. Only natural acidic waters approach the results of the leaching experiments. The leaching results give insight into the distribution of REY in rocks and aids the interpretation of the anomalous behaviour of Eu, Ce, and Y. Anomalies inherited from the source rock can be distinguished from those acquired during fluid migration. Leachates only show inherited ones, whereas both types of anomalies occur in the natural waters. By comparing REY/Ca ratios of leachates and waters dimensionless retention factors of REY by rocks are defined, which are normally in the range of 10^2 to 10^4, which is mainly due to pH differences between natural and experimental leaching.

1. Introduction

The difficulties in studying the distribution of rare earth elements (REE) and yttrium (henceforth combined to REY) in water-rock interaction is that it is mostly impossible (i) to sample the crustal water uncontaminated by surface- or groundwater, and, if volatiles are present, under the in-situ PT conditions, and (ii) to sample the host rock the water interacted with. Most water sampling is done at 1 bar and outflow temperatures, i.e., conditions different from those at depth. For example, CO_2 degassing increases pH leading to carbonate precipitation. Due to cooling, silica, sulphides, and/or oxyhydroxides flocculate, and many trace elements are sorbed or co-precipitated. Furthermore, the changes in pressure and temperature conditions during the ascent of waters also influence the original composition at depth. The formation of metastable components, surface coatings, and ion exchange are a function of the rate of ascent. Additionally, most aquifers are petrologically inhomogeneous and the residence time of water is locally variable. Therefore, the REY abundances in waters will never represent equilibria *sensu stricto* with any sampled piece of rock suspected to represent the source rock. One can only attempt to make the best choice with respect to the reacted rocks. In spite of all these constraints, REY patterns of waters (natural leachates of rocks) and experimental leachates of rocks that are accepted to correspond to aquifer rocks will be compared in the following sections with the aim to study the systematics of REE and Y distribution

I. Stober and K. Bucher (eds.), Water-Rock Interaction, 97–123.

during water-rock interaction in the temperature range from 10 to 400°C. The group of REY are of particular interest in the study of the trace element fractionation processes during alteration of rocks and fluid transport because this series of elements yield information on fractionation mechanisms, the history of the altered rocks, and the temperature conditions of water-rock interaction.

REE in surface waters have been studied to some extend. REY distribution in seawater from all oceans (Klinkhammer et al., 1994; Elderfield, 1988; Möller et al. 1994a; Bau et al., 1996), in ground-, river- and lake waters have been investigated (Smedley, 1991; Johannesson et al., 2000; Banks et al., 1999; Fuganti et al., 1996; Bau et al., 1996). In this contribution, however, the REY abundances in river-, lake-, and seawater are not discussed.

The chemical composition of groundwater and geothermal water varies depending on the chemical and mineralogical composition and crystallisation history of the dominantly controlling source rocks (Garrels and Mackenzie, 1967; Humphris et al., 1978). The main alteration reactions in silicate-dominated rocks is the decomposition of plagioclase to kaolinite and smectite (Garrels, 1967; Drever, 1988), but these reactions seem not to be decisive for many trace element abundances in waters such as REE and Y, because high fractions of REY originate from the dissolution and alteration of accessory minerals (Möller et al., 1997a; Möller, 2000). This is only different in limestones and dolostones (Johannesson et al., 2000), where REE and Y are dominantly hosted by the rock-forming carbonates.

2. Sampling of water and aquifer rocks

2.1 MINERAL- AND THERMAL WATERS

Black Forest, Germany: The Ca-Na-HCO$_3$ (37°C) sparkling thermal water from Bad Wildbad, northern Black Forest, Germany, is considered to be derived from the Kegelbach granite (Stober, 1995) from which a drill core sample was available. The sparkling mineral water from Bad Peterstal (Na-Ca-HCO$_3$; 17°C) and the low-CO$_2$ water from nearby Hermersberg (Na-Ca-HCO$_3$; 17°C) originate from altered gneisses with granitic veinlets in the central Black Forest. Samples for leaching studies were available from drill cores. The thermal water from Bad Säckingen (Na-Ca-Cl; 25°C) is produced from a granitic aquifer in the southern Black Forest. The corresponding granite had to be sampled in an outcrop nearby. A detailed study of REE and Y from all these localities was published by Möller et al. (1997a).

Bohemia/Czech Republic: In Bohemia, Czech Republic, the sparkling mineral water (Na-Ca-HCO$_3$; 11-17°C) from Kyselka is produced from the altered Carboniferous Slavkovsky-les granite and the overlying Tertiary basalt of the Doupovske hory. The waters from wells in the local granite and the basalt with up to 2300 mg CO$_2$/kg were sampled separately at the water filling station of the Mattoni Company in Kyselka. From all these rocks drill core samples were available, which showed strong hematitisation and precipitation of calcite in fractures. At Jachymov, a thermal (34°C) Ca-Na-HCO$_3$ water was recovered from wells drilled from underground about 100 m into the muscovitised Krusne hory granite. This low-CO$_2$, thermal water is used for cures. A detailed study of water-rock interaction in this area was published by Möller et al. (1998).

Thermal springs and wells, Turkey: The Na-HCO₃ type geothermal water of Kula (56°C), east of the Gidez graben, West Anatolia, is obtained from bore holes into the basement rock of gneisses and mica schist, comparable with those at the geothermal power plant of Kizildere, Büyük Menderes graben. The water is carbonate-rich and travertine forms at natural outflows. This type of water is similar to that of a near-by springs with temperatures up to 100°C. In the Kizildere area, the thermal springs of Gölemezli, Tekkehamam, Babacik, and Yenicekent occur in the Neogene sediments and are located along faults of the northern and southern flanks of the Büyük Menderes graben. All spring waters are related to the mica schist of the area.

Nishiki-numa/Japan: The cold (8°C) Ca-Na-SO₄ water from the mildly acidic (pH 4) iron-rich spring at Nishiki-numa, Hokkaido/Japan, is compared with the basaltic andesite from which the water is derived (Bau et al., 1998). The pH value of the water is low because of the oxdising weathering of sulphide minerals in the andesite.

White smoker: The REE and Y data of dispersed flows with temperatures of about 30°C are from Teahitia, Society Islands, were sampled during the Cyana cruises (Michard et al., 1993). REE analyses of ocean island basalts (OIB) of Teahitia as well as Meahitia are reported by Hemond et al. 1994. The main problem with these waters is that the thermal regime is subject to much debate: the water may have derived (i) from interaction of basalt with limited amounts of low-temperature fluids containing only small fractions of seawater, (ii) from high-temperature interaction with basalt followed by significant mixing with seawater, or (iii) both processes (Michard et al., 1993).

2.2 HIGH-ENTHALPY GEOTHERMAL WATERS

Deep crustal fluid (KTB): The Continental Deep Drilling Project (KTB) produced about 270 m³ of a highly saline (70 g TDS/kg) Ca-Na-Cl brine from the open hole section between 3850-4000 m depth at temperatures of 129°C. The brine drained from fissures in the amphibolite into the borehole. Some mixing with brines originating from the overlying paragneisses cannot be excluded (Maiwald and Lodemann, 1994). This so-called "4000 m brine" was collected after a 3 months pumping test at about 30°C and 250 m below surface under an artificial N₂/CH₄ atmosphere to prevent access of oxygen. A detailed report is given by Möller et al. (1997b). The amphibolitic source rock was taken from drill core material.

Larderello geothermal field, Italy: One of the first industrialised, geothermal fields in the world was the region of Larderello, Tuscany/Italy, where nowadays 1000 to 3000 m deep wells produce superheated steam. The wells in the vicinity of Larderello with outflow temperatures of about 200°C and 15 bars have their sources either in phyllites (Radicondoli wells) or in the limestone-dolostone-anhydrite sequence of the Calcare Cavernoso (Carboli wells). The sampling procedure is described in detail by Möller et al. (subm.). The phyllitic rocks are from drill cores, whereas most of the limestones were collected from outcrops.

Monte Amiata geothermal field, Italy: This water-dominated geothermal field is located in southern Tuscany with bottom hole temperatures of 350°C at 3 km depth. At the well head the temperatures are still about 200°C at about 15 bars. Drill core chips of the Calcare Cavernoso were available. The Tertiary volcano of the Monte Amiata is considered as the heat source.

Geothermal field of Kizildere, Turkey: The CO_2-rich, Na-HCO_3-SO_4 waters from the geothermal pilot plant at Kizildere in the eastern Menderes graben, W-Anatolia, Turkey show temperatures and pressures at the drill heads of about 190°C and 13 atm, respectively. The recalculated bottom hole temperatures are about 220°C (Giese, 1997). The assumed aquifer rock is the underlying mica schist and gneisses that are intercalated by marble (Igdecik Formation; Giese, 1997). Because drill core material was only scarcely available, samples were collected from outcrops of the country rocks.

Black smoker fluids: The analyses of black smoker fluids (Na-Ca-Cl) from the East Pacific Rise (11°, 13°, 21°N) are taken from Klinkhammer et al. (1994). These ca. 400°C vent fluids were sampled by submersibles. The discussed mid-ocean-ridge basalt (MORB) has been dredged at the East Pacific Rise (21° S).

3. Leaching procedure

In order to study the REY distribution in rocks, a leaching procedure was applied (Möller and Giese, 1997). 1 g of powdered rock samples (grain size <100 μm) is weighted into a polyethylene bottle, where it is mixed with 4 g of ion exchange resin in H^+ form (BIORAD AG50W-X8) and 100 ml bidistilled water. The bottle is kept in a shaking water bath at 70°C. The pH is self-adjusting to values of about 3 and about 4 for felsic and basic rocks, respectively. After 20 hours, the ion exchange process is stopped by decanting the solution and separating the rock powder and resin by wet sieving. The resin is transferred into a chromatography column and the collected ions are eluted with 40 ml of 4 M HNO_3. In the eluate major and trace elements are determined by ICP-AES and ICP-MS, respectively.

Where possible, powdered drill core chips of the aquifer rock are chosen for the leaching experiments. Although the rocks are never the exact equivalents of the inhomogeneous aquifer rocks, they are assumed to represent the best material, to which the waters should be related. When comparing the results of experimental leachates with those of natural waters, one has to consider that in the experiments larger masses of minerals are dissolved than in the natural process according to pH values of 3 to 4; i.e., surface processes like sorption do not play that role as in nature. Fractionation of elements (including the trivalent elements Al and REE and Y) during experimental leaching is negligible as being proved by leaching the glassy MORB (Irber et al.,1996).

4. Analysis of waters, leachates and rocks

Because REY abundances in natural waters are in the pmol/Kg range, they have to be preconcentrated prior to analysis. The preconcentration procedure of REE and Y in waters and steam have been described in Bau and Dulski (1996) and Möller et al. (1998) and Möller et al. (submit.). Whole rocks were digested in a mixture of HF and HNO_3. The leaching of rocks followed the method described by Möller and Giese (1997). All chemical analyses of REE and Y were performed by ICP-MS. The necessary routine corrections of element contents are given by Dulski (1994). The major elements in waters have been determined by ICP-AES in samples adjusted to the range of measurement by dilution with ultra-poor water. The analytical results are compiled in Tables1 to 3.

5. Discussion

It is conventional to normalise the zigzag abundance distribution of REE in rocks and minerals by that of a reference material such as C1 chondrites (Coryell et al., 1963; Anders and Grevesse, 1989) or different local "shales" (McLennan, 1989). The normalisation generally leads to smooth patterns except for those samples in which individual elements (Ce, Eu, Y) behave in an anomalous manner. Because Y behaves like Ho in igneous rocks, Bau and Dulski (1995) suggested to include Y into the series of REE. This was also adopted by Douville et al. (1999) for waters from deep-sea hydrothermal systems. Grammaccioli et al. (1999) suggested to relate Y to Dy instead to Ho. Henceforth, when Y is included, the resulting patterns will be signified as REY patterns.

All waters and rocks and the leachates of the latter are additionally related to their corresponding Ca contents for two reasons: (i) REY are dominantly hosted by Ca minerals such as calcite, apatite, hydrous phosphates, fluorite, epidote and many more, and (ii) this normalisation eliminates the volume effect of the experimental leachates and enables the comparison of the leachates and waters. In these source-rock-normalised distribution patterns of REY/Ca ratios the *inherited* and/or *acquired* trends and anomalies can more easily be identified than by comparison of the individual chondrite-normalised REY patterns of waters and rocks. *Inherited* are features that are directly derived from the source rocks because fractions of Ce and Eu can be present in a form that is either more (Eu) or less (Ce) easily leached under natural conditions than the major part of them. *Acquired* features are gained either during fluid migration due to sorption processes at pore surfaces in the source rocks or selective coprecipitation and scavenging which also implies oxidation and reduction of Ce and Eu, respectively (Möller, 1998). Of particular interest is the fraction of Eu that is not bound like the other REE at high temperatures. This fraction is termed "excess Eu". All the *acquired* anomalies of Ce, Eu, and Y necessitate intensive interaction of water with minerals for which a high ratio of mineral surfaces to volume of water is a prerequisite as being given in pores and narrow fractures. In contrast, the interaction of water with wall rock minerals in fractures is too small to fractionate REY resulting in anomalous Eu and/or Y.

The source-rock and Ca normalised distribution patterns of Figures 1a to 3a, 4 and 5 also represent the patterns of an apparent distribution factors $^{app}K_d$ according to eq(1). Thus, these normalised patterns visualise the apparent partition of REY between aqueous and solid phases. This partition is termed apparent because for want of composition of the actual alteration minerals the whole rock composition is used:

$$^{app}K_d = \frac{(REY/Ca)_{water}}{(REY/Ca)_{source-rock}} \tag{1}$$

5.1 WATERS AND LEACHATES FROM MAFIC ROCKS

REY/Ca patterns of waters from the basic rocks show extreme scatter (Figure 1) and split into two groups, which differ by 3 orders of magnitude: (i) waters from basalts and andesite with high abundances of REY, and (ii) waters from amphibolite and OIB with low contents of REY (Table 1). The trends of REY/Ca patterns of the high-temperature vent fluids (MORB) in general decrease from La to Lu with an outstanding positive Eu

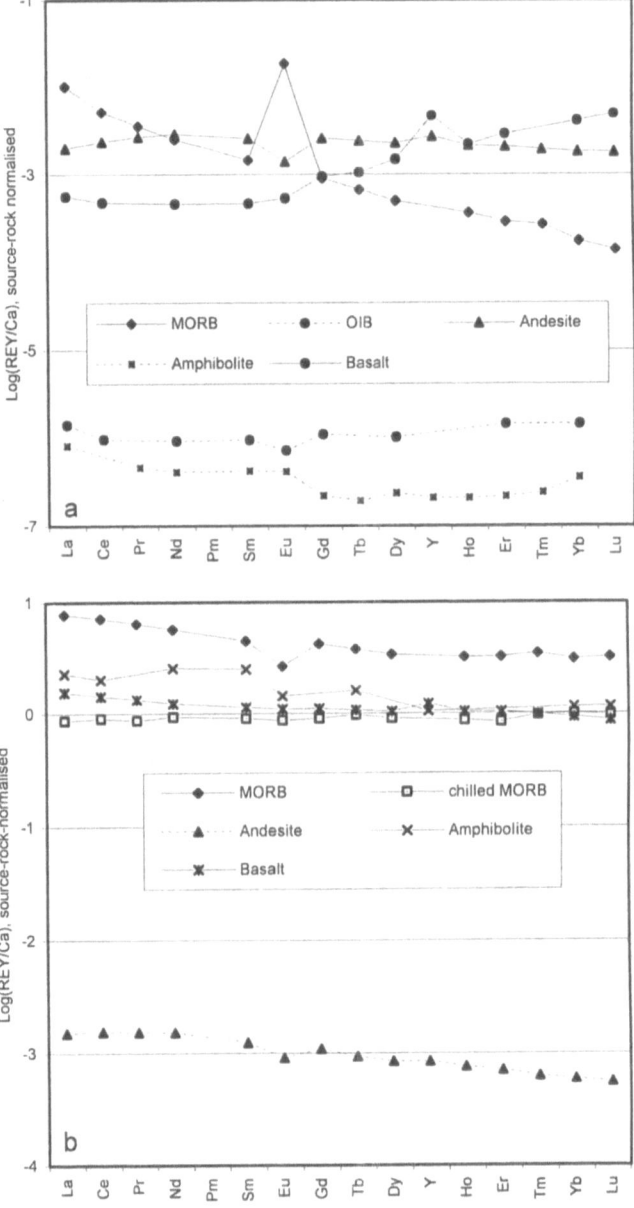

Fig. 1: Source-rock-normalised REY/Ca patterns of selected groundwaters and thermal waters from basic rocks (a) and corressponding results of leachates of the source rocks after 20 hours at 70°C (b). The igneous and metamorphic and altered source rocks are indicated by solid and broken lines, respectively. MORB: Mean of 22 vent fluids from the East Pacific rise (11°, 13°, 21°N) (Klinkhammer et al., 1994) normalised to MORB (Bau, unpubl.). OIB: Mean of 7 dispersed flows from Teahitia Seamount (Michard et al., 1993) normalised to OIB (Hemond et al., 1993). Andesite: Iron-spring, Nishiki-numa, Hokkaido, Japan normalised to local andesite (Bau et al., 1998). Amphibolite: Thermal water from the amphibolite of the Deep Drill Hole at 3850m, Oberpfalz, Germany (Möller et al., 1994). Basalt: Mean of two analyses of waters from two wells in the basalt of the Doupovske hory near Kyselka, Bohemia, Czech Republic (Möller et al., 1998).

anomaly. In contrast, the corresponding leachates of the MOR basalts show negative Eu anomalies (Figure 1b). The low-temperature waters from continental basalts increases from La to Lu (Figure 1a), but show no to slightly negative Eu anomalies, whereas the leachates show none. The low-temperature, dispersive flows of the Teahitia Seamount (Michard et al., 1993) show patterns that are very similar to leachates of the semi-crystalline MORB (Giese and Bau, 1994). In the semi-crystalline MORB and amphibolite REY are leached from solid phases that are more soluble than the plagioclase. Because of the positive Eu anomaly of plagioclase, waters and leachates exhibit a deficit of Eu (Irber et al., 1996). Rock-normalised leachates of the chilled margin of dredged pillows of MOR basalt are almost identical with the composition of the rocks Figure 1b. The source-rock-normalised REY pattern of the leachates of the glassy MORB are near one, indicating (i) homogeneous distribution of REY and Ca in the solid, and (ii) congruent bulk dissolution. Only the inner parts of the pillows, showing beginning of crystallisation, yield leachates that deviate significantly from the rock (Giese and Bau, 1994). The semi-crystalline MORB shows distinct distribution of REY in mineral phases that are easily soluble and Eu behaves significantly different from the other lanthanides (Irber et al., 1996). The accessibility of REY in the semi-crystalline MORB is high compared with the studied alkali basalt and amphibolite (Figure 1b).

Although the trend of the leachate of the amphibolite recovered from KTB is similar to that of the natural Ca-Cl brine from 4000 m depth the normalised REY/Ca ratios differ by 6 orders of magnitude (Figures 1a, b). Epidote is here the main alteration product that hosts REY. The absence of a significant Eu anomaly in the 4000 m fluid of KTB indicates that the metamorphic country rocks do not contain excess Eu.

The water from the sulphide-bearing andesite displays normalised REY/Ca abundances similar to those of the leachates (Figures 1a, b), which indicates that the experimental leaching yields results comparable with natural, acidic water. This finding demonstrates the usefulness of the experimental leaching procedure for studying the internal distribution and behaviour of REY in rocks. Furthermore, this correspondence proves the importance of pH activity in dissolution process (weathering and alteration). Leaching at about pH 3 in experiments and at about 6 to 7 under natural conditions results in about 1 to 4 orders difference in REE abundances. The higher the H^+ ion concentration is, the higher is the REY abundance in solution.

Detailed comparison of REY/Ca patterns of the spring waters and the corresponding leachates indicate that LREE are slightly more soluble. The REY/Ca trends of leachates of the amphibolite, MOR basalt, alkali basalt, and andesite resemble much more each other in composition (Figure 1b) than the corresponding waters (Figure 1a).

5.2 WATERS AND LEACHATES FROM FELSIC ROCKS

Waters and leachates from granites from various places show some features in common: the normalised waters show a very small scatter of HREE/Ca and Y/Ca ratios, whereas the light ones fan out over 2 orders of magnitude at maximum (Figure 2a). The corresponding leachates show the opposite: The light REE/Ca ratios show very similar trends, whereas the HREE/Ca and Y/Ca fan out over one order of magnitude (Figure 2b). This might be the result of leaching different suites of host minerals containing LREE and HREE including Y under natural and experimental conditions. The contrasting results of figures 2a and b could be due to the different pH of the fluid phases.

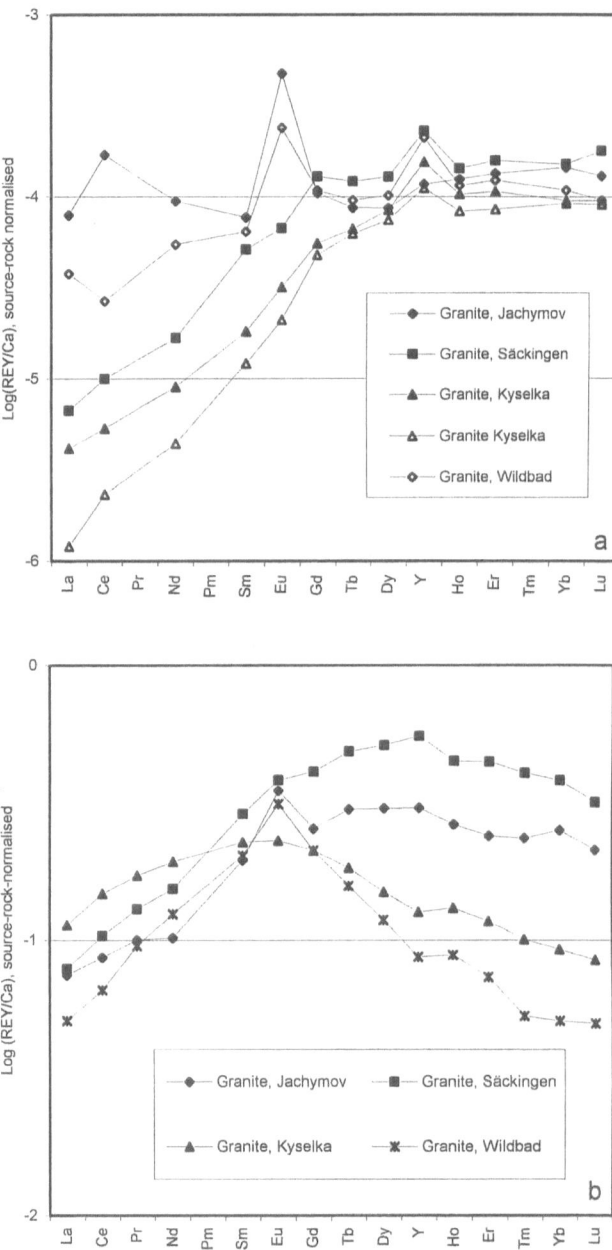

Fig. 2: Source-rock-normalised REY/Ca patterns of selected ground- and thermal waters from granites (a) and the leachates of the corresponding rocks (b). Jachymov and Kyselka are located in the Krusne hory and Slavkosky-les, respectively, Bohemia, Czech Republic (Möller et al., 1998). Bad Säckingen and Bad Wildbad are situated in the southern and northern part of the Black Forest, Germany (Möller et al., 1997a).

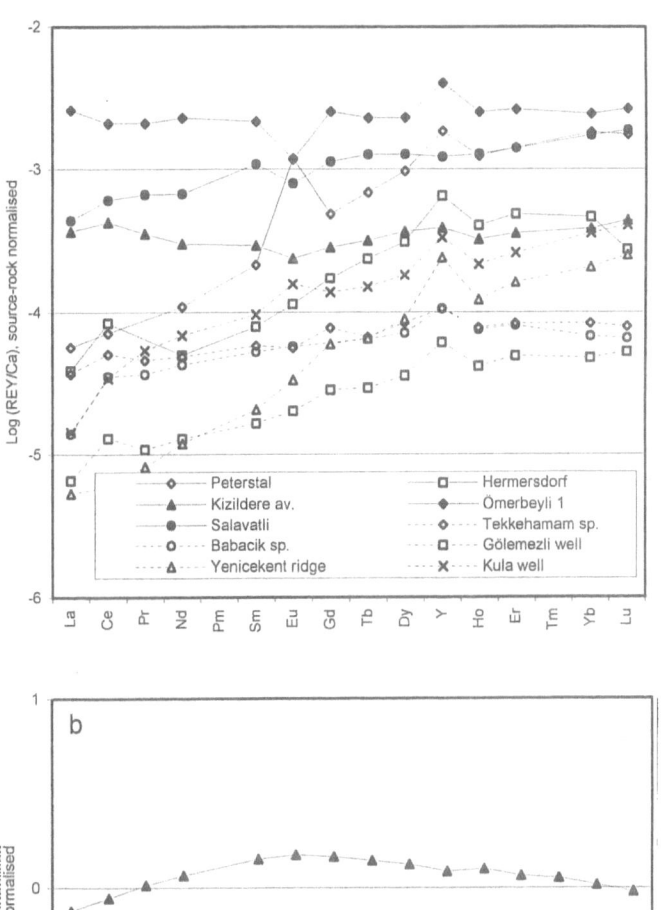

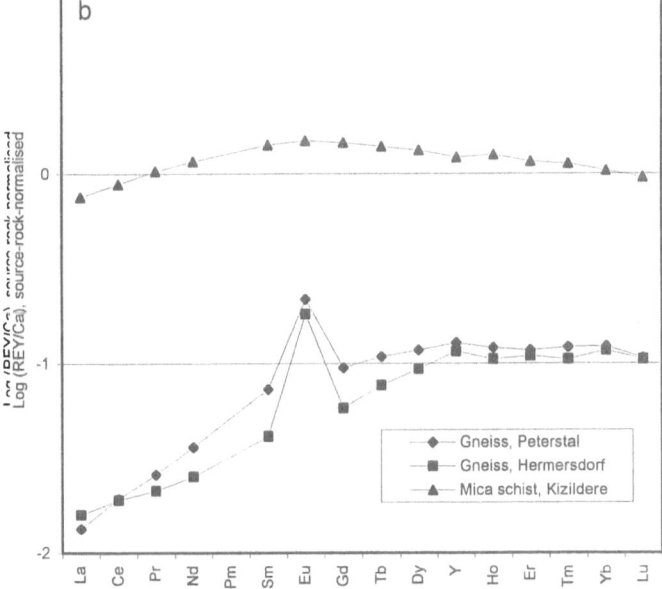

Fig. 3: Source-rock-normalised REY/Ca patterns of selected ground- and thermal waters from felsic meta-morhpites (a) and the leachates of the corresponding rocks (b). Peterstal and Hermersdorf are located in the central Black Forest, Germany. The other places represent localities in the Büyük Menderes graben/Turkey, excepting Kula, which is situated in the Gidez graben/Turkey. Kizildere represents the mean of 21 analyses of geothermal well waters of 145°C and 14 bars. The well waters in the surrounding of the Kizildere plant range between 50 and 100°C (Gölemezli, Tekkchamam, Babacik and Ycnicekent).

Comparing the behaviour of Eu in figures 2a and b it is obvious that the waters and the leachates from the granites of Kyselka and Bad Säckingen do not contain any excess Eu (at least not in the studied samples). In contrast, the natural waters and corresponding leachates of the granites from Bad Wildbad and Jachymov still leach minerals or surface coatings that are enriched in Eu relative to the bulk composition. A reason for the absence of Eu anomalies may be that the excess Eu has already been leached from the rocks. This may happen, if the process was ongoing over geological time. The experimental leachates of the Kegelbach (Bad Wildbad)- and the Kyselka granites show that the intermediate REY are enhanced leachable which is an effect of low pH.

The trends of REY/Ca patterns of waters from the felsic metamorphic rocks (Figure 3a) are similar to those of the granites but with enhanced REY/Ca ratios (Figure 2b). The leachates of the gneisses from Bad Peterstal and of the basement rocks of Kizildere (Figure 3b), however, are distinctly different from those of the granites (Figure 2b). The leachates of the gneisses exhibit strongly positive Eu anomalies, whereas those of the mica schists from Kizildere show none, which might be due to the dissolution of different types of feldspars. Remarkable is the positive Eu anomaly of the leachate of the gneisses from Peterstal and Hermersdorf. Because the water from Hermersberg (Figure 3a) shows no Eu anomaly but the leachate does, it is assumed that the latter tap Eu-enriched minerals such as K-feldspar, thereby indicating that the water-rock interactions are not the same in the two neighbouring localities. The flat trend of the source-rock-normalised REY/Ca ratios in the leachate of the mica schist from Kizildere resembles that of the corresponding mica schist.

Compared with the gneissic waters the enhanced LREE/Ca pattern of the Kizildere water (average of 19 analyses) might be due to the aragonite precipitation occurring during ascent in the casing and pipes. Under the high temperature and rapid mineral growth Ca is only insignificantly coprecipitated (Möller et al., submit.) The high-enthalpy water from Kizildere has a negligible tendency to negative Eu anomalies (Figure 3a) and the leachates of the mica schists indicate the absence of excess of Eu.

The rather flat REY/Ca patterns of the leachates of the mica schists are present also in the spring waters of the surroundings of Kizildere (Figure 3a). Compared with the thermal water at Kizildere, the thermal spring waters in the vicinity show a different trend of REY/Ca pattern. The difference between Kizildere geothermal well water and the spring water in the surroundings is that the former produce large amounts of steam since about 15 years now, whereas the springs are all of small productivity. Thus, the main conduits of the Kizildere system are significantly leached already, whereas the spring waters additionally interact with the Neogene sediments of the region, by which the REY patterns of the waters are changed. These pattern increase from La to Lu and show a marked Y anomaly.

5.3 WATERS FROM CARBONATE ROCKS

Oxygen-rich waters recovered from karsts and limestone are marked by negative Ce anomalies which are caused by sorption of Ce in precipitating oxyhydroxides during weathering (Figure 4). This anomaly is not restricted to waters from carbonate rocks but is a very common feature of all oxygen-rich waters. If sedimentary carbonates are deposited under oxidising conditions, they are free of authigenic sulphide mineralisations. All rocks containing the latter, are reducing and, therefore, unable to oxidise Ce. Marine

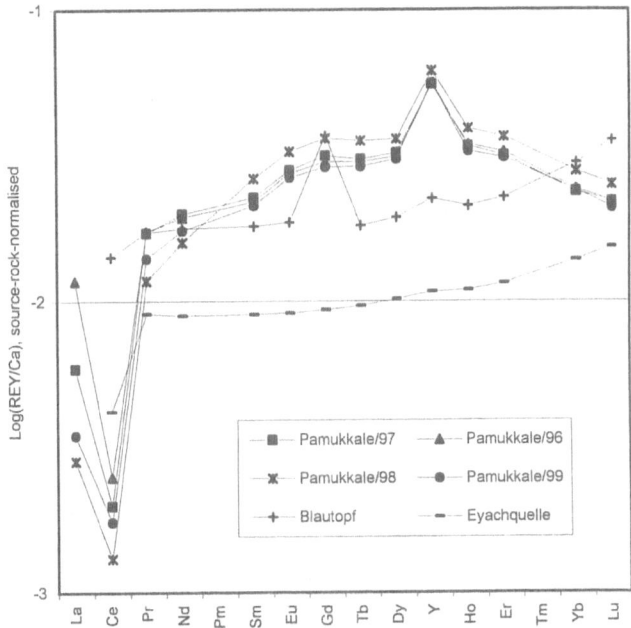

Fig. 4: Source-rock-normalised REY/Ca patterns of karst waters from Pamukkale/Turkey and Swabian Jura/Germany (Blautopf, Aachtopf, Eyachquelle). The Pamukkale water variable in LREE but constant in HREE over 4 years. The anomalous Gd in the Blautopf water is of anthropogenic origin.

carbonates also inherited a negative Ce anomaly from seawater, which can be lost during dolomitisation. This could be the reason why ground waters from limestones and dolostones show considerable and no Ce anomalies, respectively (Johannesson et al. 2000). The thermal water of Pamukkale/Turkey shows an extremely negative Ce anomaly, which might be attributed to the high Fe content in the aquifer rocks and long residence time of water in the aquifer compared to the karst waters from the Swabian Jura/Germany with their short residence times of only a few days.

5.4 SUPERHEATED STEAM

Normalising the superheated steam from Carboli and Radicondoli wells of the Larderello-Travale area to limestones and phyllites, respectively, and the Amiata wells of the Monte Amiata/Italy to limestone, the resulting patterns indicate a strong relationship between the source rocks and the steam (Figure 5). The rock-normalised REY/Ca ratios represent a rather flat pattern for the Radicondoli fluids, whereas some Carboli fluids show decreasing ratios of LREE/Ca (LREE: light REE). The Carboli steams indicate that LREE may be released more effectively than HREE and Y. The source-rock normalised REY/Ca patterns indicate that at least the lanthanides and Y are proportionally released according to their abundances in the rocks under the high TP conditions in thesource regions. Additionally, the source-rock-normalised Carboli REY/Ca ratios are higher than the Radicondoli ones. The REY/Ca ratios <1 of the steam from the Larderello wells (Radicondoli and Carboli) strongly point to the fact that REY are retained in

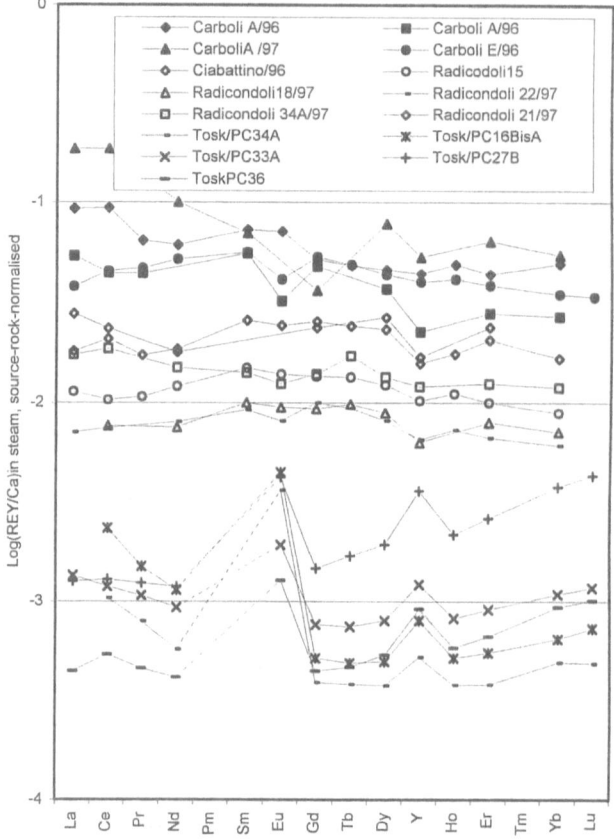

Fig. 5: Source-rock-normalised REY/Ca patterns of steam from wells in Tuscany/Italy. Carboli and Radicondoli wells tap the Calcare Cavernoso and phyllites, respectively. The PC wells from the Monte Amiata area
produce from the Calcare Cavernoso. The bottom hole temperature of the latter is >250°C. Sm in PC steams
could not be determined because of a contamination of the filters due to production engeneering.

the source region. The high-temperature steam from the Monte Amiata wells show
source-rock-normalised REY/Ca patterns much lower than those of the Larderello
wells, which means that the rock alteration is less proceeded. Due to temperature above
250°C Eu exhibits a positive anomaly. The presence of a positive Ho anomaly, which is
absent in the Larderello well fluids, indicates enhanced water-rock interaction. In general terms, the Monte Amiata aquifer is less equilibrated than the Larderello one.

5.5 ANOMALOUS BEHAVIOUR OF Ce, Eu AND Y

Ce anomalies are quite common in surface waters due to presence of free oxygen. For
instance, oxygen-rich seawater is characterised world-wide by a negative Ce anomaly
(Elderfield, 1988). It is assumed that this anomaly is caused by accumulation of Ce in

ferromanganese nodules and crusts (Elderfield et al., 1981; Hein et al., 1997, 2000). Bau (1999) and Kawabe et al. (1999) showed experimentally that Ce anomalies develop during aging of FeOOH precipitates. Starting with undifferentiating scavenging of all REE and Y, most of these elements are released during aging, only Ce is retained leading to the positive Ce anomaly in the final precipitate. In general, positive Ce anomalies form on Fe-Mn colloids and surface coatings (Braun et al., 1990) or by bacterial activity. All oxygen-rich surface waters show variable but negative Ce anomalies and so do all karst waters (Figure 4). The strongly negative Ce anomaly of the thermal water of Pamukkale with Eh values of about +400 mV might be due to long residence time of these waters in the aquifer. It is common to all of these waters that they are circumneutral, bicarbonate-rich, and high in Eh. It is assumed that the precipitation of FeOOH along the migration paths of the infiltrating meteoric waters also induce oxidation to Ce^{4+}, which is preferentially sorbed onto freshly precipitated oxyhydroxides. Because most igneous and metamorphic rocks yield waters with low Eh values at least at greater depth, negative Ce anomalies cannot develop in these waters.

Positive Ce anomalies are achieved, if such Ce-enriched FeOOH particles enter reducing regimes such as the high-salinity brines of the Tyro basin in the Mediterranean Sea, where the sedimenting particles are re-dissolved (Bau et al., 1996) or by carbonate complexation of Ce^{4+} in alkaline lakes (Möller and Bau, 1993).

Among the studied waters from felsic rocks, positive Ce anomalies have been found in the thermal water from Jachymov (Figure 2a) and the mineral water from Hermersberg (Figure 3a). This might be explained by the type of alteration in these areas. For instance, the granite of Jachymov has been altered along with the post-Variscan uranium mineralization in this area. Since U is only mobilised by oxidising fluids, it is plausible that also Ce was partly oxidised and fixed at mineral surfaces of the altered granite. The present-day water is chemically reducing and contains Fe^{2+}. If this type of water passes the Jachymov granite, some of the Ce^{4+} fixed at mineral surfaces is leached as Ce^{3+} and added to the Ce fraction from dissolving minerals.

None of the REY/Ca patterns of the experimental leachates (Figures 1b to 3b) show Ce anomalies, although they are quite common in the corresponding waters patterns (Figures 2a, 3a, and 4). This is due to the fact that the experimental leaching dissolves larger parts of the accessible minerals than natural leaching, which only interacts with the surfaces of minerals having obviously different chemistry than the bulk rock. The excess Ce in the waters from the Jachymov granite (Figure 2a) and the gneiss from Hermersberg (Figure 3a) must originate from minor solid phases such as surface coatings that only insignificantly contribute to the experimental leaching because their fraction is negligible.

The published enhanced Ce in the REY patterns of the KTB fluid is an artefact caused by drilling (Möller et al., 1994b) and is, therefore, not shown in Figures 1, 4, 5, and 6.

Strongly positive Eu anomalies are observed in the high temperature, acidic, vent fluids of the black smoker (Figure 1a), the steam of the Monte Amiata geothermal field (Figure 4), and some waters from granites (Figure 2) and gneisses (Figure 3). All the other waters and steams show no or negative Eu anomalies. Significantly positive anomalies always occur at temperatures above 250°C. This coincides with two features: (i) all high-temperature fluids are chemically reducing, and (ii) calculation (Sverjenski, 1984) and experiments (Bilal, 1991) showed that, depending on pH, an increasing amount of Eu is present in the divalent state at temperatures above about 250°C and fO_2 of the hematite/magnetite buffer (Bau and Möller, 1992). Thus, the earlier divalent fraction of

Eu behaves differently from the other REE at high temperatures. Due to the large ionic radius of Eu^{2+} (0.139^{VIII} nm compared to 0.121^{VIII} nm of Eu^{3+}; Shannon, 1976) it is not incorporated into Ca minerals at the same level as the other REE (Ca^{2+}: 0.126^{VIII}; Shannon, 1976). Because rocks rarely consist of large-ion-dominated minerals, into which Eu^{2+} would fit, Eu is enriched in residual fluids or is sorbed onto mineral surfaces. Its concentration in the range of pmol/kg is definitely too low to form an accessory Eu minerals. The fraction of "loosely-bound" Eu onto mineral surfaces often referred to as "excess Eu", is more easily leached from rocks by percolating water than the lattice-bound REE, although at low temperatures all leached Eu is trivalent. This excess Eu contributes to any *inherited* Eu anomaly.

Another contribution to an *inherited* anomaly is the alteration of feldspars, particularly plagioclase. Since most igneous feldspars show a positive Eu anomaly, their dissolution or alteration yields fluids with enhanced Eu. For instance, the high temperature alteration of plagioclase (>250°C) is considered as the source of the strongly positive Eu anomalies in black smoker fluids (Klinkhammer et al., 1994). In contrast to plagioclase, alkali feldspars are much more stable in contact with hydrothermal fluids and, therefore, their contribution of Eu is less than that of plagioclase. Separated biotite from felsic rocks, which hosts numerous tiny solid inclusions of accessory minerals, shows strongly negative Eu anomalies, which result from the many solid inclusions. Pure biotite has low REY abundances and negligible Eu anomalies (Bea et al., 1994). When biotite in felsic rocks is chloritised, the fluid inherits the high abundance levels of REY and the negative Eu anomaly from the tiny inclusions (Möller and Giese, 1997).

Because the excess Eu from the intergranular space of rocks is leached faster than REY and, thereafter, Eu is released from minerals with progress of leaching, the source-rock-normalised REY patterns of water loose the initially positive Eu anomaly with time. This may be the reason, why the thermal waters from Kyselka, Bad Säckingen, Hermersberg, and Kizildere only show insignificant Eu anomalies (Figures 2a and 3a). At high temperature, when Eu is partly divalent, sorption of the differently charged species could yield a fractionation effect. Considering Coulomb forces only, the divalent Eu is less strongly sorbed onto surfaces than the trivalent species, and as a result of this, the Eu^{2+} moves faster through pores than the trivalent REY (Möller and Holzbecher, 1998). Total Eu becomes enriched in the final fluid. For instance, at mid-oceanic ridges, the Eu anomaly is partly *inherited* from plagioclase alteration (Klinkhammer et al., 1994; Michard, 1989), but is also *acquired* by sorption processes during fluid migration. At temperatures below 200°C, an Eu anomaly can only be *inherited* from the source rocks (Bau and Möller, 1992). The superheated steam from Larderello (about 200°C) does not show a positive Eu anomaly, whereas the steam of the Monte Amiata geothermal field (about 300°C) does, which is the effect of higher temperature at depth of the latter.

Under igneous conditions Y behaves precisely like Ho (Bau and Dulski, 1965). In aqueous systems, Y displays a different behaviour, which may be related to small differences in solubility products of the hydroxides (Diakonov et al., 1998). During alteration processes, however, the small dissimilarity of sorption onto mineral surfaces, controlled by surface complexation (Bau et al., 1996; Diakonov et al., 1998) is enhanced in migrating fluids, and Y-Ho fractionation occurs. Different from anomalous Eu, the Y/Ho ratio is not principally dependent on temperature. In some groundwaters and thermal waters Y is enriched (see Y and Ho in Figure 1 and 3). This anomaly is only *acquired*, whilst the rock is not in a equilibrium with the migrating water. If such a process lasts long

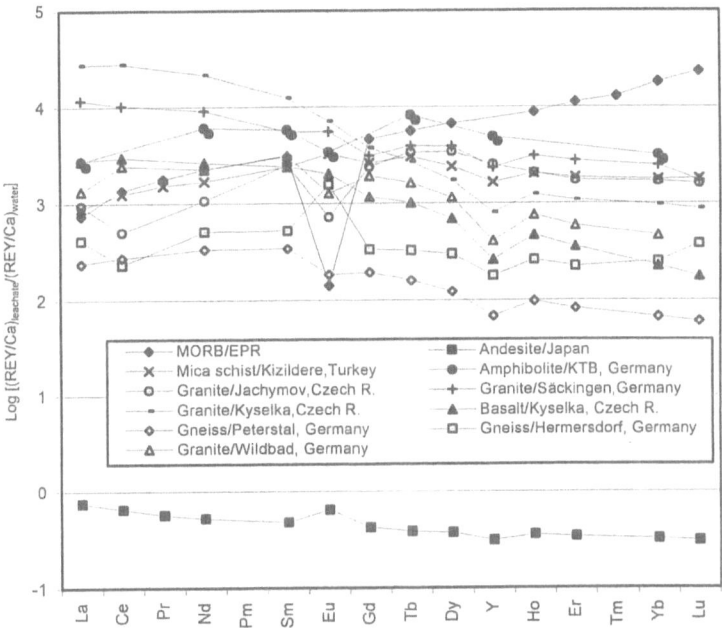

Fig. 6: Patterns of the retention of REY by the source rocks of the fluids. Only the acidic waters from an andesite yield results comparable with the leachates of this rock. In all other cases, the leachate are enriched by factors of 100 to 10^4.

enough, a steady state equilibrium may be reached and Y anomaly (like Eu anomalies) may vanish. Many waters exhibit *acquired* Y anomalies. Y in the steam of the long-time exploited Larderello geothermal field presents no anomaly, whereas in the young field of Monte Amiata Y is anomalous. It is suggested that the absence of Y/Ho anomalaous ratios indicate equilibrium conditions after long-time leaching of the source of Larderello (Figure 5). Similarly the vent fluids with temperatures above 400°C are (Figure 2a) and others are not anomalous in Y abundances (Douville et al., 1999).

5.6 COMPARISON OF REY/Ca PATTERNS IN NATURAL AND EXPERIMENTAL LEACHATES

The ratio of the REY/Ca of the waters and leachates (eq.2) yields insight into the tendency of the aquifer rocks to retain REE and Y (Figure 6).

$$R = \frac{(REY / Ca)_{leachate}}{(REY / Ca)_{water}} \qquad (2)$$

Although the pH and temperature in the natural leaching process are not comparable with those in the leaching experiments, the resulting retention patterns (R patterns) visualise that for most of the rocks the retention of REY is quite high (Figure 6). During natural leaching, Ca is more easily released than REY. Under such conditions, solubility of the major elements also control the solubility of the minor elements. If the REY form own minerals, their solubility would control the abundance in solutions that interacted with the crystalline rocks. Compared with the large spread of REY/Ca patterns of 6 orders of magnitude in natural waters, the less spread of R patterns in Figure 6 indicates that essentially the same processes act on REY-bearing compounds. The compounds may be not the same, but behave in a similar way.

The R values depend on the chemical complexation at mineral surfaces and in solution and the formation of secondary minerals during the natural alteration process. No systematic trend is recognisable for the effect of HCO_3^- in solution, which varies between less than 10 to more than 1000 mg/kg, and the pH values range from 5 to 8 at about 30°C (Möller, 2000). Tentatively, it might be assumed that the trends of R values are largely controlled by surface complexation as represented by hydrolysis reactions. Only the vent fluids with pH values between 3 and 4 show R values significantly increasing from La to Lu, thereby indicating that at these temperatures other processes control the REY distribution than those prevailing at lower temperatures and higher pH values.

Depending on the distinct conditions during crystallisation of igneous and metamorphic rocks the accessibility of REY from these rocks seems to be different. Solidification of igneous rocks proceeds by cooling, i.e. minerals form from melts at high temperatures. Many of them become unstable at low temperatures and, therefore, re-crystallise or are easily altered in the presence of fluids. The trace elements, initially mostly hosted by the major minerals, are, thereafter, partly hosted by hydrated accessory minerals. In contrast, during metamorphism the formation of new minerals follows increasing temperature and pressure, and the incompatible trace elements are rejected from major minerals. Thus, the distribution of REY is influenced by the chemical composition of the fluid and the crystallisation history of the rock. The studied felsic rocks are all coarse grained, whereas the basaltic and andesitic rocks are fine grained with variable amounts of glass. The minor phosphate contents of the less crystalline basic rocks, for instance, may give rise to the formation of more easily soluble accessory minerals than in the slowly crystallising phosphate-enriched felsic rocks.

The fraction of any solid dissolved by fluid-rock interaction, either naturally or experimentally, seldom represents the average composition of the rock. The REY in the solute are significantly controlled by the soluble tracer phases with REY patterns that may be totally different from those of the major mineral components. Since varying fractions of REY are bound to accessory minerals, detailed knowledge of the distribution of REY among the major minerals is not sufficient for deriving the REY patterns of waters that have interacted with the rocks. It is the solubility of the minor phases (particularly of the phosphates and carbonates; Irber, 1996) that control the behaviour of REY in water-rock interactions.

The quantities of experimentally leached REY obtained in this study (Table 3) mostly exceed by far those in natural waters (Table 1). This is because both processes affect different volumes of minerals and probably different solid phases as a function of pH and Eh values, temperature and mass dissolution. For instance, the experimental leachate completely dissolves carbonates, apatite, and hydrous phosphates within 20 hours, whereas the natural waters interact with the altered surfaces of minerals only. In natural leaching not only dissolution occurs but also precipitation of alteration minerals or amorphous coatings on surfaces, whereas in the experimental leaching dissolution prevails (Möller and Giese (1997). Because in batch-leaching experiments anomalies cannot be *acquired*, the absence of any significant Y-Ho fractionation during leaching is good evidence that the enhanced Y/Ho ratios in waters are due to sorption processes during migration.

The differences in REY/Ca patterns of the natural and experimental leachates are largely a function of pH, redox condition and temperature of water-rock interaction, with pH being the more important parameter. The dissolution depends on the H^+ activity in the fluids because H^+ ions are needed to break the bonds and start ion exchange at mineral surfaces. The fact that the retention is about unity for the iron-rich waters and andesite clearly points to pH as the most important parameter. At high temperatures the pH also increases leading to a similar effect (refer to steams of the Monte Amiata geothermal field, Figure 5).

Positive Eu and negative Ce anomalies in R patterns are due respectively to reduction and oxidation of these elements prior to leaching. During leaching, these elements are either more or less easily accessible than their trivalent fractions. Inherited anomalies of waters are largely eliminated by normalisation to the corresponding source rock. Acquired anomalies of waters yield negative anomalies in the R patterns. For example, the low retention of Eu in MORB is due to the enhanced leaching of Eu by black smoker fluids which is partly due to the decomposition of plagioclase under the acidic conditions in the smoker fluids and partly to less adsorption of Eu^{2+} during fluid migration. The decreased Y values of R patterns indicate that Y is often less strongly bound in minerals and their surfaces than its REE neighbours. This is not astonishing because Y is not a 4f element. In few cases also negative Ce (gneiss from Hermersberg) and Eu (granite from Jachymov) anomalies have been observed.

The high retention of Eu in the gneiss from Bad Peterstal is most probably due to leaching of excess Eu^{2+} at least in the sample subjected to leaching (Fig. 3b).

6. Conclusions

The waters clearly split into various groups characteristic for the types of rocks they interacted with. For comparison of natural waters with experimental leachates from the aquifer rocks it is appropriate to relate the REY/Ca ratios in the natural water to those in the experimental leachate. Source-rock- and Ca normalisation group the waters from the felsic and mafic igneous rocks and lead to much more coherent trends than the well-known REY patterns. *Inherited* anomalies are considerably eliminated in the REY/Ca retention patterns, whereas *acquired* ones are absent in the experimental leachates. Thus, naturally acquired anomalies can be distinguished.

Waters from felsic rocks inherit the negative Eu anomalies when present in the source, whereas waters from mafic rocks show none. The high-temperature (>250°C) steam show Eu anomalies which are similar to those of the black smoker fluids (about 400°C). Positive Eu anomalies are *inherited*, but at high temperatures they can also be acquired or those being inherited are enlarged.

All rocks studied show the chondritic Y/Ho molar ratio of about 50, those of the waters vary between 50 and 125. Where determined, many patterns show enhanced Y abundances in waters which indicate that these aquifer rocks are either in disequilibrium with the percolating water, or fresh rocks are constantly involved in the water-rock interaction due to progress of weathering or tectonic events.

Ce anomalies are created by alteration processes in rocks. Depending on the redox regime of the water, positive or negative Ce anomalies can be created under reducing or oxidising conditions, respectively.

The source-rock-normalised, 20-hours leachates show some similarities with the correspondingly normalised waters. The waters and leachates of the felsic and mafic magmatic rocks both form distinctive groups of source-rock-normalised REY/Ca patterns, whereas those of the metamorphic or strongly altered equivalents differ considerably in REY/Ca levels. For instance, REY/Ca patterns of leachates of gneisses are lower than those of granites, but the contrary is found for natural waters; those of the leachates of basalts are all in the same range, whereas the waters derived from the amphibolite separates from the basaltic water. Only the studied sulphide-bearing andesite behaves differently. The leached REY/Ca ratio from the glassy MORB is about unity, whereas it increases up to one order of magnitude after partial crystallisation.

When relating the leachates of rocks to the corresponding waters, a rather narrow band of retention patterns of REY is obtained covering two orders of magnitude with only one exception. Within this band different types of R patterns are observed. The retention of REY in the less altered granites and the basalt decrease from La to Lu, whereas the strongly altered rocks and most metamorphites show either horizontal or increasing patterns from La to Lu. Most of the pattern show negative Y and variable Eu anomalies. Thereby, it is indicated that the accessibility and/or chemical behaviour of these two elements is different from the other REE during weathering of rocks and migration of fluids.

Anomalies of Eu, Y and Ce reveal the history of the fluid-rock interaction. Eu is sensitive to temperature, whereas Y is not. Anomalous Eu is *inherited* and may be enhanced at temperatures above 250°C. Ce is sensitive to oxygen fugacity and pH. Y seems to be sensitive to pH and to ligands dominating REY complexation in solution and on surfaces. Some waters acquired enhanced Y/Ho ratios irrespective of their source rocks. Using the anomalies of REY patterns of waters, temperature-dependent reactions can be discussed. Waters from metamorphic rocks could have lost their excess Eu during intensive water-rock interaction. Such waters would not show Eu anomalies in source-rock-normalised patterns. Ce anomalies may still occur depending on the development of Eh with time. Although groundwater is geologically very young, it might be in a steady state equilibrium with the aquifer rocks. Under such conditions, an acquired positive Y and *inherited* Eu anomaly indicate temperatures of water-rock interaction below 200°C. Because in mildly acidic to neutral waters no or only positive Y anomalies are observed, this indicates that Y is released more easily from the rocks than REE and is less retained by sorption onto mineral surfaces.

In summary, REY represent a unique tool to study the behaviour of trace elements in water-rock interaction with time.

7. Acknowledgement

The assistance of G. Schettler, P. Dulski, M. Conrad, C. Wiesenberg, B. Zander and B. Richert in performing the analyses is greatly acknowledged.

8. References

Anders, E. and Grevesse, N. (1989) Abundance of elements: Meteoric and solar. *Geochim. Cosmochim. Acta* **53**, 197-214.

Banks D., Hall G., Reimann C., and Siewers U. (1999) Distribution of rare earth elements in crystalline bedrock groundwaters: Oslo and Bergen regions, Norway. *Appl. Geochem.* **14**, 27-39.

Bau M. (1999) Scavenging of dissolved yttrium and rare earths by precipitating iron oxyhydroxides: Experimental evidence for Ce oxidation, Y-Ho fractionation, and lanthanide tetrad effect. *Geochim. Cosmochim. Acta*, **63**, 67-77.

Bau, M. and Dulski, P. (1995) Comparative study of yttrium and rare earth element behaviour in fluorine-rich hydrothermal fluids. *Contrib. Mineral. Petrol.*, **119**, 213-223.

Bau M. and Dulski P. (1996): Anthropogenic origin of positive gadolinium anomalies in river waters. *Earth Planet. Sci. Lett.*, **143**, 245-255.

Bau, M. and Möller, P. (1992) Rare earth element fractionation in metamorphogenic hydrothermal calcite, magnesite and siderite. *Mineral. Petrol.*, **45**, 231-246.

Bau, M., Koschinsky, A., Dulski, P. and Hein, H.J. (1996) Comparison of the partitioning behaviours of yttrium, rare-earth elements, and titanium between hydrogenetic marine ferromanganese crusts and seawater. *Geochim. Cosmochim. Acta*, **60**, 1709-1725.

Bau M., Möller P., and Dulski P. (1996) Yttrium and lanthanides in eastern Mediterranean seawater and their fractionation during redox cycling. *Mar. Chem.*, **56**, 123-131.

Bau, M., Usui, A., Pracejus, B., Mita, N., Kanai, Y., Irber, W., and Dulski, P. (1998) Geochemistry of low-temperature water-rock interaction: Evidence from natural waters, andesites and Fe-oxyhydroxide precipitates at Nishiki-numa iron-spring, Kokkaido, Japan. *Chem. Geol*, **15**, 293-307.

Bea, F., Pereira, M.D., Corretge, L.G., and Fershitater, G.B. (1994) Differentiation of strongly peraluminous, perphosporous granites: The Pedrobernardo pluton, Central Spain. *Geochim. Cosmochim. Acta*, **58**, 2609-2627.

Bilal, B.A. (1991) Thermodynamic study of Eu^{3+}/Eu^{2+} redox reaction in aqueous solutions at elevated temperatures and pressures by means of cyclic voltammetry. *Z. Naturforsch.*, **46a**, 1108-1116.

Braun, J.J., Pagel, M., Muller, J.P., Bilong, P., Michard A., and Guillet, B. (1990) Cerium anomalies in lateritic profiles. *Geochim. Cosmochim. Acta*, **54**, 781-795.

Coryell, C.D., Chase, J.W., and Winchester, J.W. (1963) A procedure for geochemical interpretation of terrestrial rare earth abundance patterns, *J. Geophys. Res.*, **68**, 559-566.

Diakonov, I. I., Ragnarsdottir, K.V., and Tagirov, B. R. (1998) Standard thermodynamic properties and heat capacity equations of rare earth hydroxides: I. Ce(III)-, Pr-, Sm-, Eu(III)-, Gd-, Tb-, Dy-, Ho-, Er-, Tm-, Yb-, and Y-hydroxides. Comparison of thermochemical and solubility data. *Chem. Geol.*, **151**, 327-347.

Douville, E., Bienvenu, P., Charlou, J. L., Donval, J. P., Fouquet, Y., Appriou, P., and Gamo, T. (1999) Yttrium and rare earth elements in fluids from various deep-sea systems. *Geochim. Cosmochim. Acta*, **63**, 627-643.

Drever, J.I. (1988) The geochemistry of natural waters. 2nd edt. Prentice Hall, Englewood Cliffs, New Jersey, 437p.

Dulski, P. (1994) Interferences of oxide, hydroxide, and chloride analyte species in the determination of rare earth elements in geological samples by inductively coupled plasma-mass spectrometry. *Fresenius J..Anal. Chem.*, **304**, 193-203.

Elderfield, H., Hawkesworth, C.J., and Greaves, M.J. (1981) Rare earth element geochemistry of oceanic ferromanganese nodules and associated sediments. *Geochim. Cosmochim. Acta*, **45**, 513-528.

Elderfield, H. (1988) The oceanic chemistry of the rare earth elements. *Philos Trans. R. Soc. London*, Ser. A **325**, 105-126.

Fuganti, A., Möller, P., Morteani, G., and Dulski, P. (1996): Gadolinio ed altre terre rare usabili come traccianti per stabilire l`eta il movimento ed i rischi delle acque sotterranee: esempio dell`area di Trento. *Geol. Tec. Ambien.*, **4**, 13-18.

Garrels, R. M. (1967) Genesis of some ground waters from igneous rocks. In P. H. Abelson (ed.) *Researches in Geochemistry*, **2**, 405-420.

Garrels, R. M. and Mackenzie, F. T. (1967) Origin of the chemical compositions of some springs and lakes. In R. F. Gould (ed) Equilibrium concepts in natural water systems. *Am. Chem. Soc. Adv. Chem. Ser.*, **67**, 222-242.

Giese, L. (1997) Geotechnische und umweltgeologische Aspekte bei der Förderung und Reinjektion von Thermalfluiden zur Nutzung geothermischer Energie am Beispiel des Geothermalfeldes Kizildere und des Umfeldes, W-Anatolien/Türkei. PhD Thesis , Free University Berlin, 250p.

Giese, U. and Bau, M. (1994) Trace element accessibility in mid-ocean ridge and ocean island basalt: an experimental approach. *Min. Mag.*, **58A**, 329-330.

Grammaccioli, C.M., Diella, V.,and Demartin, F.(1999) The role of fluoride complexes in REE geochemistry and the importance of 4f electrons: some examples in minerals *Europ. J. Mineralogy*, 11, 983-992.

Hein, J.R., Koschinsky, A., Bau, M., Manheim, F.T., Kang, J.-K. andRoberts, L. (2000). Cobalt-rich ferromanganese crusts in the Pacific. In: D.S. Cronan, (ed.), Handbook of Marine Mineral Deposits. CRC Press, Boca Raton, Florida, 239-279.

Hein, J.R., Koschinsky, A., Halbach, P., Manheim, F.T., Bau, M., Kang,J.-K., and Lubick, N. (1997) Iron and manganese oxide mineralization in the Pacific. In: K. Nicholson, J.R. Hein, B. Bühn, B., and S. Dasgupta, (eds.) Manganese Mineralization: Geochemistry and Mineralogy of Terrestrial and Marine Deposits. Geological Society of London Special Publication No. 119, London, 123-138.

Hemond, C., Devey, C.W., and Chauvel, C. (1994) Source compositions and melting processes in the Society and Austral plumes (South Pacific Ocean): Element and isotope (Sr, Nd, Pb, Th) geochemistry. *Chem. Geol.*, **115**, 7-45.

Humphris, S.E., Morrison, M.A., and Thompson, R.N. (1978) Influence of rock crystallisation history upon subsequent lanthanide mobility during hydrothermal alteration of basalt. *Chem. Geol.*, **23**, 125-137.

Irber, W. (1996) Laugungsexperimente an peraluminischen Graniten als Sonde für Alterationsprozesse im finalen Stadium der Granitkristallisation mit Anwendung auf das Rb-Sr-Isotopensystem. PhD thesis, Free Univ. Berlin, 319p.

Irber, W., Bau, M. and Möller, P. (1996) Experimental leaching with cation exchange resin: a method to estimate element availabilities in geological samples. *J. Conf. Abstr.*, **1**, 280.

Johannesson, K.H., Zhou, X., Guo, C., Stetzenbach, K.J., and Hodge, V.F. (2000) Origin of rare earth element signatures in groundwaters of circumneutral pH from southern Nevada and eastern California. *Chem. Geol.*, **164**, 239-257.

Kawabe, I., Ohta, A., Ishii, S., Tokumura, M. and Miyauchi, K. (1999) REE partitioning between Fe-Mn oxyhydroxide precipitates and weakly acid NaCl solution: Convex tetrad effect and fractionation of Y and Sc from heavy lanthanides. *Geochem. J.*, **33**, 167-179.

Klinkhammer, G.P., Elderfield, H., Edmond, J.M., and Mitra, A. (1994) Geochemical implications of rare earth element patterns in hydrothermal fluids from mid-ocean ridges. *Geochim. Cosmochim. Acta*, **58**, 5105-5113.

Maiwald, U. and Lodemann, M. (1994) Continuing recordings of physicochemical and hydraulic parameters during the pumping test 1991 at KTB pilot borehole (KTB-VB) *Sci. Drill.*, **4**, 95-99.

McLennan, S.M. (1989) Rare earth elements in sedimentary rocks: Influence of provenance and sedimentary processes. In B.R. Lipin and G.A. McKray (eds) Geochemistry and mineralogy of rare earth elements. *Mineral. Soc.Amer.*, 169-200.

Michard, A. (1989) Rare earth systematics in hydrothermal fluids. *Geochim. Cosmochim. Acta* **53**, 745-750.

Michard, A., Michard, G., Stüben, D., Stoffers, P., Cheminee, J.-L., and Binard, N. (1993) Submarine thermal springs associated with young volcanoes: The Teahitia vents, Society Island, Pacific Ocean. *Geochim. Cosmochim. Acta*, **57**, 4977-4986.

Möller, P. (1988) The dependence of partition coefficients on differences on ionic volumes in crystal-melt systems. *Contrib. Mineral. Petrol.*, **99**, 62-69.

Möller, P. (1998): Rare earth elements and yttrium fractionation caused by fluid migration. In: M. Novak and J. Rosenbaum (eds.) Challenges to Chemical Geology. Czech Geol. Surv. Prague, 9-32.

Möller, P. (2000): Rare earth elements and yttrium as geochemical indicators of the source of mineral and thermal waters. In I. Stober and K Bucher Hydrology of crystalline rocks Kluwer Acad. Press, 227-246.

Möller, P. and Bau, M. (1993) Rare-earth patterns with positive cerium anomaly in alkaline waters from Lake Van, Turkey. *Earth Planet. Sci. Lett.*, **117**, 671-676.

Möller, P. and Giese, U. (1997) Determination of easily accessible metal fractions in rocks by batch leaching with acid cation-exchange resin. *Chem. Geol.*, **137**, 41-55.

Möller, P. and Holzbecher, E. (1998) Eu anomalies in hydrothermal fluids and minerals: A combined thermochemical and dynamic phenomenon. *Freib. Forsch.-H.*, **C475**, 73-84.

Möller, P., and 26 authors (1997b) Paleofluids and recent fluids in the upper continental crust: Results from the German Continental Deep Drilling program (KTB) *J. Geophys. Res.*, **102 B8**, 18233-18254.

Möller, P., Dulski, P., and Bau, M. (1994a) Rare earth element adsorption in a seawater profile above the East Pacific Rise. *Chem. Erde*, 54, 129-149.

Möller, P., Dulski, P., and Giese, U. (1994b) Rare earth elements in KTB-VB fluids. *Sci. Drill.* **4**, 113-122.

Möller, P., Dulski, P., Gerstenberger, H., Morteani, G., and Fuganti, A. (1998): Rare earth elements, yttrium and H, O, C, Sr, Nd and Pb isotope studies in mineral waters and corresponding rocks from NW Bohemia Czech Republic. *Appl. Geochem.*, 13, 975-994. Möller, P., Dulski, P. and Morteani G. (submit.a) Concentration of rare earth elements, yttrium and some major elements in liquid and vapour phases and their source rocks from the Larderello-Travale geothermal field (Tuscany, Italy). *Geothermics*

Möller, P., Dulski, P., Özgür, N., and Conrad, M. (subm.b) Distribution of some major and rare earth elements in fluids and scalings of the geothermal field of Kizildere, Menderes Graben, Turkey. Geofluids.

Möller, P., Stober, I., and Dulski, P. (1997a) Seltenerdelement-, Yittrium-Gehalte und Bleiisotope in Thermal- und Mineralwässern des Schwarzwaldes. *Grundwasser*, **2**, 118-132.

Shannon, R.D. (1976) Revised effective ionic radii and systematic studies of interatomic distances in halides and chalcogenides. *Acta Crystallogr.*, **A32**, 751-767.

Stober, I. (1995) Die Wasserführung des kristallinen Grundgebirges. Stuttgart, 191p.

Sverjensky, D.A. (1984) Europium redox equilibria in aqueous solution. *Earth Plant. Sci. Lett.*, **67**, 70-78.

Table 1: Composition of thermal fluids discussed in the text. References: 1: Möller et al., 1994; 2 Möller et al., 1997; 3 Möller et al., 1998; 4 Möller et al., subm.a; 5 Möller et al. subm.b; 6 Bau et al., 1998; 7 Klinkhammer et al., 1994; 8 Michard et al., 1993. Abbreviations: KTB: Continental Deep Drilling Project; ERP: East Pacific Rise; MORB: Mid-Ocean Ridge Basalt; OIB: Ocean Island Basalt; ## sample number (year/sample); (n) average of n samples; n.d.=not determined.

Locality	KTB	Säckingen	Wildbad	Peterstal	Hermersb.	Swab. Jura	Swab. Jura	Jachymov	Kyselka	Kyselka	Kysselka/HJ	Larderello	Larderello
Country	Germany	Germany	Germany	Germany	Germany	Germany	Germany	Czech Repl.	Czech Repl.	Czech Repl.	Czech Repl.	Italy	Italy
Source	4000 m	well	well	well	well	Blautopf	Eyach spr.	well	well BJ	well V5	well	Carb.A	Carb.A
Rock type	amphib.	granite	granite	gneiss	gneiss	limestone	limestone	alt. granite	granite	granite	basalt	limestone	limestone
##	(1)	(1)	(2)	(2)	(2)	98/40	98/45	(1)	(2)	(1)	(2)	96/51	97/48
Ref.	1	2	2	2	2	unpubl.	unpubl.	3	3	3	3	4	4
pH	8.4	6.8	7	6.2	8.2	7.7	7.8	7.5	5.8	5.7	6.2	7.1	7
T/°C	119	25	37	17	17	11	12	34	17	14	11	125	127
Eh/mV	n.d.	n.d.	n.d.	n.d.	n.d.	310	320	n.d.	n.d.	n.d.	n.d.	-21	-80
Na/mM	285	95.2	5.96	16.4	2.48	0.25	0.59	5.56	16.4	11.2	2.57	0.019	0.0042
K	5.6	3.74	0.213	0.74	0.15	0.024	0.019	0.21	1.64	0.08	0.667		
NH4	n.d.	n.d.	n.d.	n.d.	n.d.	n.d.	n.d.	n.d.	n.d.	n.d.	n.d.	n.d.	n.d.
Ca	393	7.43	0.9	10.2	0.35	3,150	3,050	0.45	1.93	1.28	5.45	0.0038	0.003
Mg	0.09	1	0.114	3.71	0.213	0.15	0.11	0.15	0.33	1	2.83	0.0052	0.00025
Cl	1240	102	3.75	0.96	0.073	3	0.75	0.11	0.366	32	0.068		
HCO3	0.74	8.41	6.62	32.1	3	3.97		6.46	21.6	16.7	21.6	3.86	12.2
F	0.2	0.089	0.045	0.11	0.021	0.05		0.22	0.16	0.19	0.005		
SO4	3.23					0.32	0.14	0.12	0.24	0.26	0.09		
La/pM	19	54	12	1032	19	45.9	19	14.7	77	15	483	1.2	1.24
Ce		156	20	2605	83	24.3	7.0	73.6	188	54	858	2.2	2.26
Pr	3.40					12.0	6.2					0.21	
Nd	13	102	19	1730	22	52	25	18.0	114	37	391	0.79	0.67
Sm	3.09	70	9.3	650	6.3	9.7	4.7	4.51	40	18	66	0.25	0.13
Eu	0.86	13	3.1	601	1.9	2.5	1.2	1.22	10	4.5	21	0.07	
Gd	1.62	145	17	1209	11	26.4	6.5	6.14	95	54	95	0.19	0.07
Tb	0.22	21	2.8	233	2.1	1.9	1.0	1.12	18	11	13	0.03	
Dy	1.65	110	16	1730	14	13.3	6.8	6.46	140	82	93	0.15	0.13
Y	15	1747	325	31052	272	422	197	88	2636	1252	2281	1.68	1.06
Ho	0.29	20	3.1	403	3.4	3.4	1.7	1.39	34	18	23	0.03	
Er	0.86	53	8.5	1316	11	11	5.4	3.76	104	55	73	0.07	0.05
Tm	0.13												
Yb	1.20	46	6.8	1472	8.9	27.5	11	3.38	95	61	75	66	0.09
Lu		7.4	0.73	230	0.8	3.9	2.0	0.38	14	8.83	13	9.97	

Table 1 continued.

	Larderello	Larderello	Larderello	Larderello	Larderello	Larderello	Mt.Amiata	Mt.Amiata	Mt.Amiata	Mt.Amiata	Mt.Amiata	Kizildere	Germencik
Locality													
Country	Italy	Italy	Italy	Italy	Italy	Italy	Italy	Italy	Italy	Italy	Italy	Turkey	Turkey
Source	Carb.E	Radic.15	Radic.18	Radic.22	Radic.34A	Radic.21	PC34A	PC16BisA	PC33A	PC27B	PC36	well KD	well OB1
Rock type	limestone	limestone	phyllite	phyllite	phyllite	phyllite	limestone	limestone	limestone	limestone	limestone	mica schist	mica schist
##	96/54	96/56	(2)	97/42	97/44	97/46	99/137	99/140	99/142	99/144	99/146	(19)	96/14
Ref.	4	4	4	4	4	4	unpubl.	unpubl.	unpubl.	unpubl.	unpubl.	5	5
pH	6.2	>6	7.2	6.2	6.2	6.1	n.d.	n.d.	n.d.	n.d.	n.d.	6.5	6.07
T/°C	>120	>130	>100	138	134	135	200	220	211	207	209	190	203
Eh/mV	-90	-70	-120	-80	-75	-70	H2S	H2S	H2S	H2S		6.5	
Na/mM	1.19	13.9	0.0014	77	0.37	0.0006	109	25.2	38.6	22.9	21.6	45.7	59.56
K			0.0013	1.22	0.0054		10.4	3.13	4.85	1.9	3.16	3	1.33
NH4	n.d.	n.d.	n.d.	n.d.	n.d.	n.d.	2.61	7.1	7.28	9.67	11.3	n.d.	n.d.
Ca	0.9	0.1	0.000	0.124	0.01	0.004	1.2	0.163	0.098	0.099	0.051	0.045	0.65
Mg	0.0032	0.0018	0.0011	0.0006	0.0015	0.0003						0.015	0.0708
Cl				3.17			125.9	33.6	47.2	31.7	32	2.99	44.03
HCO3	5.1	13.3	7.75	9.4	5.16	4.49	0.007	0.008	0.011	0.012	0.015	37.6	21.3
F			29.5				0.087	0.002	0.017	0.004	0.043	0.82	0.26
SO4							0.41	1.03	0.034	0.05	0.023	6.2	0.614
La/pM	92.0	11.3		8.15	2.35	1.29	4.36	14.1	12.1	11.34	4.00	2.06	235
Ce	201	20.6	3.75	17.2	5.05	2.19	17.1	38.5	19.6	21.2	8.93	4.64	371
Pr	28.7	2.53					1.80	3.39	2.43	2.81	1.05	0.45	44
Nd	128	9.87	1.52	7.58	1.67	0.69	5.22	10.4	8.47	10.79	3.77	1.30	161
Sm	37.1	2.30	0.38	1.65	0.30		15.7	11.7	12.6	13.9	12.0	0.22	27
Eu	7.6	0.49	0.082	0.53	0.06		2.50	3.09	1.33	2.92	0.88	0.04	3.1
Gd	38.4	1.95	0.33	1.67	0.27	0.16	1.04	1.38	2.04	3.90	1.19	0.21	30
Tb	5.3	0.28	0.051	0.23	0.05		0.15	0.20	0.30	0.68	0.19	0.04	4.2
Dy	27.2	1.59	0.29	1.22	0.24	0.16	0.87	1.15	1.85	4.49	1.24	0.24	24
Y	291	18	2.71	13.1	2.87	1.38	14.2	21.6	33.0	97.4	24.7	2.42	400
Ho	4.8	0.29		0.22			0.16	0.23	0.36	0.94	0.25	0.04	5.1
Er	11.7	0.77	0.15	0.59	0.13	0.086	0.43	0.63	1.03	2.97	0.75	0.12	15
Tm													
Yb	0.0	0.04	0.60	0.12	0.47	0.11	0.43	0.56	0.95	3.29	0.81	0.12	0.07
Lu							0.06	0.09	0.14	0.53	0.12	0.01	0.008

Table 1 continued

Locality	Salavatli	Tekkehamam	Babacik	Yenicekent	Kula	Boskoy	Pamukkale	Pamukkale	Pamukkale	Pamukkale	Hokkaido	Vent fluid	Teahitia
Country	Turkey	Turkey	Turkey	Turkey	Turkey	Turkey	Turkey	Turkey	Turkey	Turkey	Japan	EPR	Society Isl.
Source	well	pool	spring	ridge	well	well	hotel	hotel	Jandarma	Jandarma	iron sp.	smoker	Disp. flow
Rock type	mica schist	mica schist	mica schist	mica schist	mica schist	mica schist	limestone	limestone	limestone	limestone	andesite	MORB	OIB
##	96/43	(5)	97/03	96/39	99/03	96/12	97/28	96/42	98/26	99/29	(5)	(22)	(7)
Ref	5	unpubl.	unpubl.	unpubl.	unpubl.	unpubl.	unpubl.	unpubl.	unpubl.	unpubl.	6	7	8
pH	8.44	7.1	6.45	6.74	6.84	6.45	6.24	6.19	6.14	6.15	4	4	5-6
$T/°C$	50	73	60	26	57.4	61.6	37	23	32	33	8.1	400	30
Eh/mV		-88	-17		3	H2S	427		370		440		
Na/mM	45.8	35	22.3	23.8	41.35	62.82	1.7	1.7	1.74	1.74	1.15		
K	2.1	3.4	3.51	2.24	2.23	2.35	3.5	3.5	0.136	0.136	0.12		
NH_4	n.d.	n.d.	n.d.	n.d.	n.d.	n.d.	n.d.	n.d.	n.d.	n.d.	n.d.	n.d.	n.d.
Ca	0.82	1.95	6.1	7.5	2.37	1.843	11.8	11.8	11.500	11.500	2.19	25.2	10.48
Mg		0.58	3.5	1.844	3.86	0.536	4	4	3.83	3.83	0.97		
Cl	6.82	2.6	2.54	1.36	5.34	35.97	0.6	0.6	0.345	0.345	0.41		
HCO_3	46.7	5.7	7.5	30.2	49.6	36.11	9	9	18.48	18.48			
F		0.53	0.26	0.46	0.14	0.534	0.08	0.08	0.2	0.2	0.009		
SO_4	1.24	15.3	12.7	7.83	1.22	0.274	11.3	11.3	7.8	7.8	5.59		
La/pM	50.2	10.0	11.8	5.55	4.71	25.7	49.7	90.3	17.5	26.4	44100	2643	2.32
Ce	136	26.9	58.6	12.0	22.2	51.1	17.1	19.6	8.0	13.5	121000	4491	3.73
Pr	17.4	2.86	7.12	1.95	4.08	7.93	34.3	31.8	18.1	25.3	19400	585	
Nd	59.1	10.2	28.0	9.61	17.5	35.5	160	143	99.1	126	88400	2350	1.86
Sm	16.8	2.15	6.09	2.97	4.32	10.90	38.3	33.9	34.4	32.4	22200	478	0.39
Eu	2.6	0.45	1.43	1.03	1.52	3.64	10.6	9.5	9.5	9.0	4320	2194	0.09
Gd	16.6	2.72		8.10	5.90	19.39		52	52.8	48.7	25700	387	0.38
Tb	2.9	0.37	1.11	1.38	1.00	3.96	8.8	7.8	7.9	7.4	3950	53	
Dy	16.9	2.67	7.10	10.9	7.02	29.0	59.6	53	51.6	50.9	23400	262	0.23
Y	154	31.7	99.1	281	123	494	1379	1274	1190	1252	281000		
Ho	3.3	0.48	1.44	2.86	1.62	6.89	13.4	12.5	12.0	11.5	4780	41	
Er	9.8	1.38	4.18	10.3	5.30	20.8	38.7	36.4	34.8	34.1	13200	96	0.13
Tm											1870		
Yb	152	0.49	0.075	3.65	540	129	15.2	13.2	22.3	27.0	11200		
Lu	21.6	0.07	0.003	0.53	76	22.1	1.79	1.93	3.4	3.7	1700		0.09

Table 2: Composition of rocks being assumed to represent the typical aquifers rocks discussed in the text. References: 1: Möller et al., 1994; 2 Möller et al, 1997; 3 Möller et al., 1998; 4 Möller et al., subm.a; 5 Möller et al., subm.b; 6 Bau et al., 1998; 7 Hemond et al., 1994. Abbreviations: KTB: Continental Deep Drilling Project; ERP: East Pacific Rise; MORB: Mid-Ocean Ridge Basalt; OIB: Ocean Island Basalt.

Rock type	Amphibolite	Granite	Gneiss	Gneiss	Limestone	Granite	Granite	Granite	Alkalibasalt	Phyllite	Limestone
Locality	KTB, 3850 m	Black Forest Wildbad	Black Forest Peterstal	Black Forest Hermersberg	Swabian Jura	Krusne hory Jachymov	Black Forest Säckingen	Duopovske hory Kyselka	Duopovske hory Kyselka	Tuscany	Tuscany
Country	Germany	Germany	Germany	Germany	Germany	Czech Rep.	Germany	Czech Rep.	Czech Rep.	Italy	Italy
Ref.	1	2	2	2	unpubl.	3	3	3	3	4	4
Ca ppm	107000	2430	6285	7500	399000	3024	4000	913	108500	10000	350000
Y	42.70	9.12	22.80	19.90	5328	11.10	8.99	17.66	21.30	1.232	17.526
La	21.50	2.86	38.70	36.50	2339	4.32	15.08	30.19	59.04	1.100	9.10
Ce	50.60	7.08	78.50	74.30	761	10.08	29.30	58.21	124.24	2.217	16.8
Pr	6.93	0.95	9.64	9.26	312	1.24	3.35	6.53	15.38	0.265	2.327
Nd	30.50	3.45	35.20	33.70	1342	4.61	11.72	21.61	60.32	0.938	9.564
Sm	7.38	1.46	6.95	6.40	255	1.47	2.75	3.90	10.49	0.185	2.681
Eu	1.96	0.13	1.18	1.33	64	0.06	0.38	0.58	3.01	0.042	0.760
Gd	7.66	1.66	5.97	5.46	357	1.54	2.36	3.17	7.90	0.180	3.053
Tb	1.21	0.31	0.83	0.76	52	0.34	0.36	0.51	1.00	0.026	0.468
Dy	7.54	1.72	4.49	4.00	354	2.02	1.85	3.16	5.06	0.168	2.74
Ho	1.53	0.30	0.83	0.74	83	0.30	0.31	0.63	0.85	0.034	0.522
Er	4.43	0.77	2.38	1.94	258	0.78	0.75	1.91	2.11	0.103	1.383
Tm	0.61	0.11	0.33	0.29	127	0.11	0.11	0.30	0.26	0.015	0.182
Yb	3.91	0.73	2.14	1.77	201	0.67	0.70	2.02	1.57	0.093	1.099
Lu	0.58	0.09	0.35	0.28	31	0.08	0.10	0.30	0.22	0.018	0.156

Table 2 continued

	Limestone	Mica schist	Mica schist	Mica schist	Limestone	Andesite	MORB	OIB
Locality	Mt Amiata	Menderes Massif	Menderes Massif.	Menderes Massif	Pamukkale	Hokkaido	East Pacific Rise	Teahitia Sea-mount
	Piancastagnao	Kizildere	Kizildere	Kula		Nishiki-numa	East Pacific Ocean	Society Islands.
Country	Italy	Turkey	Turkey	Turkey	Turkey	Japan		
Ref.	unpubl.	5	5	unpubl.	unpubl.	6	Unpubl.	7
Ca ppm	390000	10000	9000	11000	390000	133	93000	69600
Y ppm	89	17.74	9.99	22.7	3042	13.9		
La	139	20.35	17.01	29.8	1610	4.66	3.25	37.5
Ce	140	40.46	39.22	60.8	1655	10.9	11.1	88.3
Pr	141	4.81	4.36	7.4	388	1.55	2.12	
Nd	144	17.18	15.66	28.2	1597	6.687	12.4	47.4
Sm	150	3.34	3.04	5.5	352	1.96	4.54	10
Eu	151	0.72	0.55	1.2	79	0.709	1.64	3.16
Gd	157	3.14	2.66	4.8	409	2.37	6.2	8.85
Tb	159	0.49	0.39	0.7	63	0.391	1.16	
Dy	163	3.11	2.21	4.2	416	2.53	7.82	6.08
Ho	165	0.71	0.43	0.8	90	0.558	1.7	
Er	167	1.50	1.18	2.4	284	1.61	5.1	2.47
Tm	169	0.86	0.18	0.3	99	0.248	0.69	
Yb	173	1.89	1.14	2.3	278	1.61	4.86	1.71
Lu	175	0.28	0.18	0.3	43	0.25	0.7	

(Rock type: Limestone, Mica schist, Mica schist, Mica schist, Limestone, Andesite, MORB, OIB)

Table 3: Composition of the leachates of the aquifer host rocks. References: 1 Möller et., 1997 ; 2 Möller et al., 1998 ; 3 Irber et al., 1996.
*=leached element in % of whole rock content.

Rock type	Amphibole	Granite	Granite	Gneiss	Gneiss	Granite	Granite	Granite	Alkalibasalt	Mica schist	Andesite	MORB, micro-cryst.	MORB, chilled margin
Locality	KTB 3850 m	Balck Forest Säckingen	Black Forest Wildbad	Black Forest Peterstal	Black Forest Hermersberg	Krusne h. Jachymov	Doupovske h. Kyselka	Doupovske h. Kyselka	Doupovske h. Kyselka	Menderes Massif Kizildere	Hokkaido	East Pacific Rise	East Pacific Rise
Country	Germany	Germany	Germany	Germany	Germany	Czech Rep.	Czech Rep.	Czech Rep.	Czech Rep.	Turkey	Japan	East Pacific Ocean	East Pacific Ocean
REF.	1	2	2	2	2	2	2	2	2	unpubl.	unpubl.	3	3
Ca ppm	10.2*	42	29	85	89	69	5.03	5.03	229	237.60	318.4	2.7*	3.2*
Y ppb	10.7*	46.20	9.50	39.20	27.20	85.40	12.30	12.30	55.30	99.09	28.1		
La	23.4*	4.46	1.75	7.05	6.92	20.42	18.84	18.84	195.35	86.12	16.7	21*	2.8*
Ce	20.6*	12.04	5.61	20.57	16.84	52.23	47.21	47.21	377.24	198.56	40.3	19.3*	2.9*
Pr		1.72	1.08	3.40	2.35	7.49	6.17	6.17	43.77	27.32	5.66	17.4*	2.8*
Nd	26.1*	6.49	5.15	17.31	10.21	31.06	22.92	22.92	157.72	106.78	24.4	15.4*	3*
Sm	25.6*	3.97	3.55	6.90	3.14	13.59	4.87	4.87	25.21	24.02	5.81	12.2*	2.9*
Eu	14.8*	0.31	0.49	3.45	2.86	2.48	0.73	0.73	6.98	5.47	1.54	7.2*	2.8*
Gd		5.41	4.21	7.63	3.76	16.60	3.69	3.69	18.55	24.49	6.21	11.5*	2.9*
Tb	16.5*	1.41	0.58	1.21	0.69	3.02	0.52	0.52	2.28	3.69	0.873	10.3*	3.1*
Dy		8.39	2.44	7.11	4.43	16.27	2.60	2.60	11.10	20.86	5.12	9.2*	2.9*
Ho		1.10	0.32	1.35	0.92	2.38	0.45	0.45	1.87	3.85	1.02	8.7*	2.8*
Er		2.56	0.68	3.75	2.51	5.75	1.23	1.23	4.59	9.80	2.73	8.8*	2.7*
Tm		0.35	0.07	0.54	0.36	0.74	0.16	0.16	0.54	1.35	0.375	9.4*	3.1*
Yb	11.7*	2.32	0.45	3.54	2.45	4.55	1.03	1.03	3.07	7.71	2.27	8.4*	3.2*
Lu	11.8*	0.25	0.05	0.50	0.35	0.52	0.14	0.14	0.41	1.04	0.333	8.7*	3.1*

A CASE STUDY OF GAS-WATER-ROCK INTERACTION IN A VOLCANIC AQUIFER: THE SOUTH-WESTERN FLANK OF MT. ETNA (SICILY)

A. AIUPPA[1], L. BRUSCA[1*], W. D'ALESSANDRO[2], S. GIAMMANCO[2], F. PARELLO[1]

1) *Dipartimento CFTA, Università di Palermo, via Archirafi 36, 90123 Palermo, Italy*
2) *Istituto Nazionale di Geofisica e Vulcanologia — Sezione di Palermo, via La Malfa 153, 90146 Palermo, Italy*

Abstract — New chemical and isotope data for 74 groundwater samples from the south-western slopes of Mt. Etna are presented. The processes responsible for the considerable chemical heterogeneity displayed by groundwaters were identified through factor analysis and by the use of mass balance calculations. A general hydrogeochemical model, concerning the interactions between the shallow volcanic aquifer, deep magma-derived fluids and the underlain sedimentary sequence, is also provided.

1. Introduction

The chemical features of groundwater are the result of several geochemical processes that are closely linked to the local natural environment and human perturbations. It is a matter of fact that underground water acquires salts from leaching of the host rock. This phenomenon is strictly dependent upon the residence time of water, that in turn is a consequence of water flow velocity and length of the flow path, the presence of reactive gas species, such as carbon dioxide, that enhance water aggressiveness towards the rock, and the chemical composition of host rocks. Also, knowledge of the depth of groundwater capture with wells is of the utmost importance in the case of stratified aquifers where each superimposed water body has its own peculiar chemistry.

In this work we present geochemical data on groundwaters from an area of the lower south and southwestern flanks of Mt. Etna (Sicily). This area was selected as a good case study because it is characterised by complex interactions between groundwaters, the host rocks, magma-derived gases, hydrothermal fluids, fluids from hydrocarbon reservoirs and possibly wastewaters, with marked lateral and vertical variations as a function of the different intensity of such processes.

In order to highlight the above processes, we first applied methods of statistical analysis to the gathered data. Several recent studies (e.g., Nicholson, 1993; Giammanco et al., 1998; Ceròn et al., 2000) showed the usefulness of statistical techniques to illustrate the main geochemical properties of an aquifer, to distinguish the many different sources of solutes and to track their diffusion paths. Once defined the main chemical processes responsible for water chemistry, we used a chemical mass balance approach in order to numerically reproduce the chemical evolution of the studied groundwaters and better constrain the reaction paths inside the local aquifers. The

---

* Present address: *Poseidon System, via Monti Rossi 12, 95030 Nicolosi (CT), Italy*

I. Stober and K. Bucher (eds.), Water-Rock Interaction, 125–145.

principles of mass balance calculations are widely discussed in Plummer and Back, (1980) and Plummer et al., (1983, 1990, 1994).

2. Study area and methods

2.1 GEOLOGICAL AND HYDROGEOLOGICAL SETTINGS

The study area is located on the lower portions of the south and southwestern flanks of Mt. Etna, covering a surface of about 150 km^2. The geological features of this area mainly consist of Etna s volcanic products in its central, northern and eastern parts, and of sedimentary rocks in the remaining parts. The volcanic rocks that outcrop in the study area are essentially represented by lava flows belonging to the Recent Mongibello Unit (< 5 ka, Romano, 1982) and subordinately to the Ancient Mongibello Unit (30 to 5 ka, Romano, 1982). The former unit is characterised mainly by hawaiites, whereas the latter is characterised by products ranging from trachyandesites to trachytes (Tanguy et al., 1997). Minor outcrops of basal subalkaline lavas are found just east of the town of Patern (Romano, 1982). These products are among the first erupted in the Etna area (age of 500 to 200 ka) and are made of tholeiitic basalts (Tanguy et al., 1997).

The sedimentary rocks that are found in the study area are for the most part bluish marly clays dated to lower and middle Pleistocene and fluvial terraced deposits dated to Holocene (Lentini, 1982). The formers outcrop mainly southeast of Patern . The latter are found mostly southwest of Patern and cover both the bluish marly clays and the volcanic products of Mt. Etna. Lastly, some minor outcrops of clays with sandy interbeddings (Tortonian age) are found in the southwest corner of the study area (Lentini, 1982). The presence of evaporitic rocks (gypsum, limestone and sometimes sulfur) deposited on the top of the sedimentary sequence beneath the lavas of the southern flank of Mt. Etna has also been hypothesised (Lentini, 1982).

From a hydrogeological viewpoint, the study area is characterised by two distinct types of rock permeability. In general, Etna s volcanics are characterised by high permeability values, 2.5×10^{-7} to 2.9×10^{-6} cm^2 (Aureli, 1973; Ferrara, 1975), while that of the sediments of Etna's basement is much lower, with an average value of 10^{-10} cm^2 (Schilir , 1988). Such a large permeability contrast, together with the remarkable amount of water and snow precipitations in the Mt. Etna area, implies that this volcano is an important reservoir of groundwater (about 7×10^8 m^3 of water available per year - Ogniben, 1966). The circulation of groundwater inside Mt. Etna is strongly influenced by the morphology of the sedimentary basement of the volcano. In fact, waters that percolate through the permeable volcanic rocks reach the impermeable basement and move along radial directions towards the outer boundaries of the volcanic edifice. However, because Etna's basement reaches its highest point (about 1200 m a.s.l.) slightly NW of the summit of the volcano and has a general slope towards SE, groundwaters tend to flow and accumulate in the southern and eastern flanks of the volcanic edifice (Ogniben, 1966), where most springs and water-wells are situated. Based on the underground water circulation, three main hydrogeological basins were recognised on Etna (Ogniben, 1966): the first roughly corresponds to the northern sector of the volcano, the second includes a large part of the western and southern flanks, and the third, the largest one, includes almost all of the eastern part of it (Fig. 1).

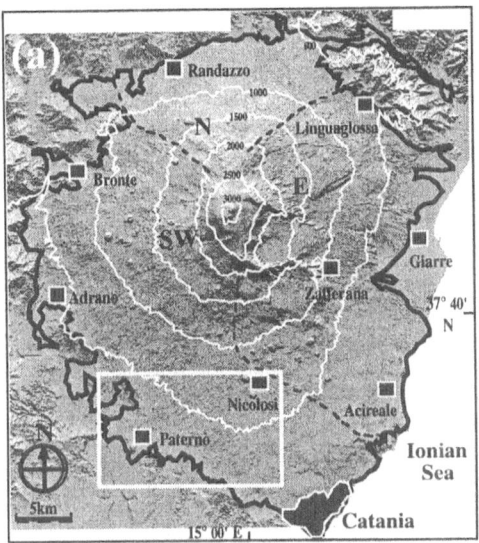

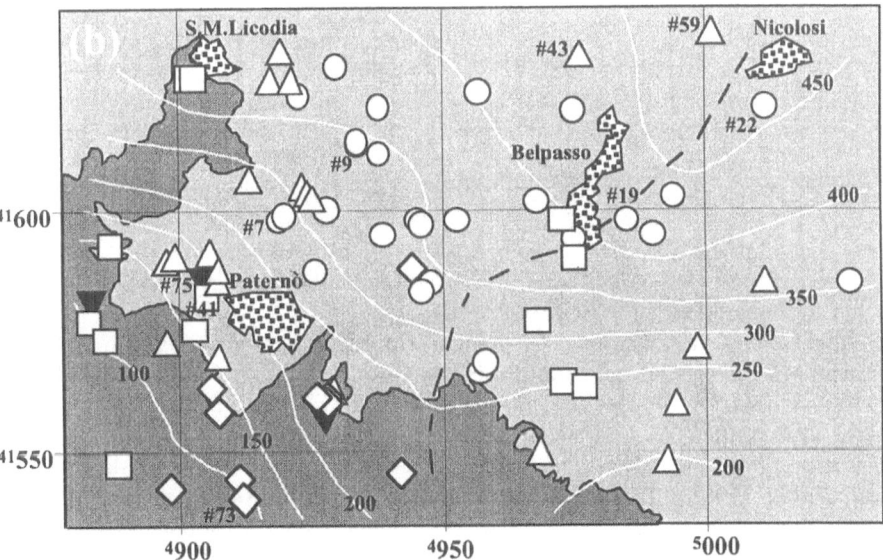

Figure 1 - Location of the study area and sampling sites. a) Mt. Etna area: the black solid line indicates the boundary between volcanic and sedimentary rocks, black dashed lines indicate the ground water divides of the three principal hydrogeologic basins (N, E and SW) and the white box delimits the investigated area; b) Patern area: pale grey = volcanites; dark grey = sedimentary rocks; white lines = water table contours (m a.s.l.); dashed line = local ground water divide; filled triangles = salinelle mud volcanoes. Water groups identified through factor analysis: I = circles; II = squares; III = triangles; IV = diamonds. UTM grid is displayed every 5 km (thin lines).

The study area lies in the second hydrogeological basins, and the contour lines for the altitude of the top of the aquifer indicate two main directions of groundwater flow, one towards southwest and the other roughly towards south (Fig. 1). Most water springs are located at the outer boundary of Etna s lavas, as a consequence of the above-mentioned marked difference in permeability between volcanic and sedimentary rocks. This difference is even more marked in the study area due to the widespread presence of impermeable clays in the local basement.

Previous studies assessed a high gas emission rate through the volcanic plume at Mt. Etna (Allard et al., 1991; D Alessandro et al., 1997a). Part of these gases, mainly composed of CO_2, is also released through the flanks of the volcano and dissolves in the aquifers. The distribution of this diffuse gas release is mainly controlled by the tectonic structure of the volcano (D Alessandro et al., 1999). Two main degassing zones, between Zafferana and Acireale to the east and between Nicolosi and Patern to the south-west, have been recognised based on high pCO_2 values either calculated (Anz et al., 1989; Brusca et al., 2001) or measured (D Alessandro et al., 1999). In the latter area, which corresponds to that investigated in the present paper, many CO_2-rich gas manifestations (bubbling waters, mud volcanoes) are present.

2.2 SAMPLING AND ANALYTICAL TECHNIQUES

In the studied area (Fig. 1), 74 groundwater samples were collected during a period of about two years (July 1997 — Oct. 1999). 50 of them were taken from wells, 22 from springs and 2 from drainage galleries. The wells were dug, often with large diameters (> 1 m), and have depth up to 350 m, although 56% have depth less then 100 m. Sampling sites were distributed almost uniformly over the area, with a higher density near the town of Patern (average density $\approx$ 0.5 samples/km^2). Samples were collected and stored in PE bottles. An aliquot for cation analysis was filtered (0.45 m MF Millipore filters) and acidified with HNO_3. Temperature, pH and Eh of water were measured directly in the field with portable instruments. All laboratory analyses were made at the Istituto Geochimica dei Fluidi, CNR of Palermo.

Total alkalinity was determined by titration with HCl 0.1 N, while Cl, NO_3, SO_4, Na, K, Ca and Mg contents were determined by ion chromatography. A spectrophotometric method was used for the determination of B and SiO_2, while Fe was determined by graphite furnace atomic absorption (GFAAS).

Dissolved gases were determined using the analytical method proposed by Capasso & Inguaggiato (1998). This method is based on the equilibrium partition of gases between a liquid and a gas phase after introducing a known amount of Ar in the sampling vial. Chemical analyses were carried out by a gas chromatograph using argon as carrier and equipped with a 4 m Carbosieve S II column. In order to evaluate the concentration of He, O_2, N_2, CO_2 a TCD detector was used while CH_4 concentration was determined by a FID detector.

$^{18}O/^{16}O$ ratio of water was determined by means of standard methodologies. Values are referred as $\delta^{18}O$ vs. V-SMOW standard with precision of -0.1 (1 σ).

All analytical results are reported in Table 1.

TABLE 1 - Chemical and isotopic analysis of the groundwaters

N.	sample	date	UTM	type	T°C	pH	Eh	Na	K	Mg	Ca	Alk	Cl	NO_3	SO_4	SiO_2	TDS	B	Fe	O_2	N_2	CO_2	He	CH_4	$\delta^{18}O$
1	S.Leonardo	09-10-97	4028 41585	115	16.4	6.35	212	119	19.6	108	64.7	811	118	n.d.	34.1	66	1340	1.90	n.a	6.21	13.0	282	n.d	0.05	-6.7
2	SAB	26-03-98	4075 41621	160	15.9	6.00	200	119	20.3	106	74.5	1030	39.7	1.2	25.4	90	1507	0.85	0.39	1.77	6.8	1510	1.26	0.52	-7.8
3	D'Urso	24-07-98	4022 41624	70	15.0	6.75	45	131	15.2	106	45.1	952	59.6	n.d	15.8	75	1400	0.44	n.a	8.81	23.8	159	n.a	3.27	n.a.
4	Ercolino	04-08-98	4053 41598	45	16.1	6.18	36	111	20.3	98.7	76.8	924	43.8	6.1	29.0	90	1399	1.00	0.55	7.58	22.9	1160	n.d	3.78	-7.8
5	Poggio Gurcio	24-09-98	4039 41595	42	18.5	6.16	185	115	21.9	103	91.0	930	47.9	13.9	49.1	91	1463	0.59	0.03	4.90	9.0	1410	n.a	0.67	-7.7
6	Piano Viti	05-10-98	4030 41630	115	16.0	6.31	113	115	18.4	75.5	47.5	735	50.8	2.4	11.2	85	1140	0.45	n.a	n.a.	n.a.	453	n.a.	n.a.	-8.5
7	S. Ardizzone	16-12-98	4019 41598	S	15.3	6.57	367	115	15.9	72.4	50.6	720	56.2	4.5	16.4	82	1132	0.53	n.a	2.69	4.1	196	n.d	0.08	-8.5
8	P. Ardizzone	16-12-98	4020 41599	15	15.4	6.48	260	114	18.6	71.8	48.6	738	56.8	5.2	16.6	82	1149	0.52	n.a	3.29	5.5	245	n.d	n.d.	-8.6
9	S. Vito 1	16-12-98	4034 41614	78	15.3	6.31	500	106	16.0	65.5	48.9	665	52.5	5.3	22.3	82	1063	0.49	n.d	4.47	7.8	339	n.d	0.05	n.a.
10	S. Vito 4	16-12-98	4036 41614	80	16.2	6.23	160	109	15.4	68.0	51.0	677	55.1	4.0	21.2	83	1083	0.55	n.a	n.a	n.a	401	n.a.	n.a.	n.a.
11	Guido	16-12-98	4038 41612	90	15.9	6.20	262	101	15.2	65.6	53.0	675	57.6	7.9	21.9	84	1082	0.61	n.a.	3.98	7.2	396	n.d	n.d.	-8.2
12	Raffo 1	16-12-98	4045 41598	48	16.3	6.07	201	110	16.9	95.2	82.0	924	39.7	5.0	38.6	88	1399	0.60	n.d	4.37	8.6	973	0.32	n.d.	-7.7
13	Raffo 2	16-12-98	4046 41597	48	16.1	6.12	179	108	17.9	95.2	80.7	912	40.8	5.5	39.0	87	1386	0.61	n.a.	4.13	7.5	867	n.a	n.d.	-7.5
14	Manganelli 1	18-12-98	4057 41566	G	16.2	6.72	145	114	22.9	126	86.0	1090	69.6	16.3	31.2	84	1642	1.07	n.d	4.50	7.4	191	n.d	n.d.	n.a.
15	Manganelli 2	18-12-98	4058 41568	110	19.6	6.60	77	120	26.6	122	85.9	982	93.1	34.1	60.0	83	1607	1.30	0.04	1.21	6.0	196	n.d	0.44	-7.7
16	Costanzo	25-02-99	4068 41603	90	16.7	6.11	130	125	21.5	260	85.2	1800	286	0.4	14.7	86	2752	6.27	1.01	1.41	8.5	2260	0.50	24.7	n.a.
17	Nicotra	25-05-99	4068 41602	70	16.4	5.90	90	152	27.9	144	123	1290	48.6	2.8	12.0	67	1816	1.41	0.34	0.34	1.13	1150	1.81	0.44	n.a.
18	Belpasso	26-07-99	4090 41595	70	17.5	6.03	15	193	24.2	220	159	1500	191	12.0	26.9	102	2324	3.31	0.75	4.61	15.2	1510	n.d	n.d.	-7.8
19	Motta	26-07-99	4085 41598	70	19.0	6.07	150	230	24.6	254	138	1940	281	0.4	8.0	98	2958	6.21	4.75	0.84	9.8	2260	n.d	10.1	-8.2
20	Bellia 2	13-09-99	4075 41594	60	17.6	6.25	39	206	16.4	284	273	1990	238	3.3	24.6	84	3032	4.47	n.d	2.88	11.3	1540	0.73	5.74	-7.9
21	Patellina	13-09-99	4026 41587	2	17.6	6.02	180	123	28.5	164	102	2050	73.7	3.7	30.5	84	2898	1.22	0.55	1.27	16.6	3140	3.01	21.8	-7.9
22	Serafica	14-09-99	4012 41622	230	17.9	6.13	257	139	15.6	102	36.5	857	89.0	1.2	41.6	78	1384	0.85	0.03	3.20	14.2	928	0.47	0.16	-7.3
23	Currune	20-09-99	4038 41622	95	14.6	6.28	245	218	14.9	59.4	141	677	42.2	6.7	18.7	72	1067	0.23	0.09	5.80	13.3	368	n.d	2.70	-8.7
24	Raffo	21-09-99	4048 41585	15	18.9	6.59	294	153	26.0	179	125	1740	74.8	1.6	41.1	n.a.	2441	1.25	n.d	1.63	13.1	462	0.30	n.d.	n.a.
25	Acquarossa	26-10-99	4046 41583	180	20.1	6.50	412	83.0	14.1	153	74.1	1480	67.0	0.9	8.0	84	2150	1.39	0.08	5.81	12.4	544	n.d	2.89	-7.7
26	A. Difesa sup.	26-10-99	4057 41625	210	16.2	6.08	380	166	41.8	84.0	153	784	35.8	6.4	43.1	77	1201	0.19	0.06	6.98	16.3	1720	n.d	32.9	-7.9
27	A. Difesa inf.	26-10-99	4057 41625	30	16.9	6.09	395	131	27.4	147	57.3	1490	56.0	2.0	49.0	85	2159	0.69	0.06	5.68	19.7	4270	n.d	n.d.	-8.6
28	La Mamma	26-10-99	4028 41600	30	16.3	6.51	260	94.3	28.2	74.8	36.9	732	50.7	9.4	43.2	76	1146	0.46	n.d	6.22	11.9	279	n.d	n.a.	n.a.
29	Macello Licodia	08-07-97	4002 41628	S	14.2	6.78	260	96.1	19.6	79.8	36.9	610	51.4	26.0	69.1	70	1059	0.28	n.a	n.a	n.a	139	n.a	n.a.	n.a.
30	Di Gennaro	10-12-97	4083 41577	79	17.4	6.94	288	211	22.7	131	103	1010	181	73.2	109	79	1922	1.19	n.a	5.36	11.1	78	n.d	0.44	n.a.
31	Canneto	13-02-98	4086 41573	S	14.9	7.50	160	103	25.4	74.5	51.3	750	103	11.5	51.4	52	1279	0.95	n.a	n.a	7.8	31	n.a	n.d.	-8.0
32	Finocchiaro	07-10-98	4075 41590	80	17.1	6.47	150	155	41.8	109	87.4	811	155	44.0	95.0	78	1552	1.70	n.a	4.24	7.0	267	n.d	0.06	-7.7
33	Ingheria	25-05-99	4068 41577	70	17.4	6.39	191	101	27.4	111	82.0	848	101	45.0	72.5	86	1494	1.26	n.d	3.75	7.0	195	n.d	n.d.	-7.6
34	Pracchio	25-05-99	4073 41564	70	16.9	6.60	140	112	27.4	112	77.8	903	81.5	32.3	59.5	75	1493	1.04	0.01	3.99	7.8	167	n.a	n.d.	n.a.
35	Ovile	26-05-99	4088 41547	S	18.1	7.59	110	216	32.1	123	70.1	881	216	75.0	136	72	1863	1.21	n.d	n.a	n.a	26	n.a	n.d.	n.a.
36	Cuscuna	28-07-99	4077 41563	70	17.3	6.59	143	117	27.2	117	84.1	900	117	32.6	49.7	70	1524	1.28	0.07	4.83	9.1	145	n.d	n.d.	-7.4
37	Rocca Patern	13-09-99	4002 41575	S	18.1	7.27	139	158	28.5	73.5	86.2	808	88.6	54.9	92.2	70	1495	0.81	0.06	4.59	9.7	76	n.d	0.02	-8.4
38	Olivine	13-09-99	4087 41593	70	18.2	7.85	135	189	29.3	109	57.1	894	112	55.2	69.1	70	1585	0.74	0.03	4.03	10.2	12	n.d	0.05	-8.2
39	Amore	15-09-99	4073 41598	80	17.9	6.43	202	674	61.4	139	74.9	1010	875	75.3	129	85	3119	1.26	0.02	2.74	13.1	514	0.34	0.67	-7.4
40	Cherubino	26-10-99	4003 41628	S	14.1	6.60	260	100	30.9	87.5	45.7	717	51.0	13.6	56.2	68	1170	0.16	0.01	6.24	12.8	104	n.d	n.d.	-8.8
41	Ossidi	26-10-99	4005 41582	S	19.7	6.63	195	508	56.7	145	88.6	1320	605	2.5	6.2	88	2821	4.10	1.37	0.54	12.5	324	n.d	4.29	-8.0

TABLE 1 - Continued

N.	sample	date	UTM	type	T;C	pH	Eh	Na	K	Mg	Ca	Alk	Cl	NO3	SO4	SiO2	TDS	B	Fe	O2	N2	CO2	He	CH4	δ18O
42	A.Sorrentine	12-01-98	998 41572	115	15.0	6.26	81	107	21.1	94.0	35.3	689	81.9	13.1	41.0	79	1162	1.35	n.a.	4.67	9.7	159	n.d.	0.05	-6.3
43	Ficominutilla	25-03-98	976 41633	300	22.6	6.35	50	263	25.4	198	189	2190	57.1	0.5	26.9	18	2967	2.17	11.6	0.14	4.7	1060	n.d.	6.93	-9.0
44	Barriera	29-06-98	917 41628	70	16.7	6.80	80	119	16.0	121	56.9	955	n.d.	n.d.	45.6	89	1468	0.33	n.a.	6.18	29.1	214	n.d.	0.31	-8.4
45	Camporotondo	23-07-98	011 41586	120	19.1	6.52	-78	150	14.9	245	55.3	1480	177	n.d.	25.4	105	2248	3.36	5.15	0.06	8.2	357	n.d.	3.69	-7.0
46	Rapisarda	24-07-98	919 41634	70	14.9	6.65	132	105	15.6	92.6	38.5	756	46.4	7.7	43.2	76	1182	0.29	0.02	4.27	8.9	145	n.d.	n.d.	-8.6
47	Monafia	04-08-98	905 41592	3	16.8	6.78	138	121	18.8	85.2	71.3	860	53.2	17.4	30.7	83	1341	0.63	n.a.	n.a.	n.a.	173	n.a.	n.a.	n.a.
48	Marano	07-10-98	968 41550	23	19.2	7.05	123	101	8.2	54.8	91.0	525	110	36.8	59.1	44	1029	0.59	0.01	1.51	8.4	48	n.d.	0.02	-6.2
49	Chiarenza	02-12-98	994 41560	90	15.6	7.37	150	106	18.4	78.9	47.9	631	104	23.9	43.7	71	1124	1.04	n.a.	3.10	5.6	44	n.d.	0.13	n.a.
50	Avana	17-12-99	897 41573	S	15.5	8.64	134	143	16.3	87.9	65.8	866	86.9	11.2	31.1	80	1388	0.83	n.a.	n.a.	n.a.	3	n.a.	n.a.	-8.2
51	Rocca Patern 2	27-01-99	907 41570	S	11.2	8.65	105	133	16.8	96.0	76.7	927	42.4	23.8	41.1	69	1425	0.80	n.a.	n.a.	n.a.	3	n.a.	n.a.	n.a.
52	Piano Tavola	26-04-99	992 41548	55	16.7	7.03	150	113	19.9	85.3	60.8	674	87.2	35.3	58.6	50	1184	0.89	0.02	3.93	7.2	46	n.d.	n.d.	-6.8
53	Porticello 3	26-05-99	929 41563	20	19.5	7.67	-171	392	13.7	53.1	65.7	467	562	3.3	57.1	33	1647	0.82	0.44	0.19	10.1	14	n.d.	n.d.	n.a.
54	Iuncio	26-07-99	897 41590	S	18.8	5.95	-178	248	18.3	127	120	1200	221	n.d.	24.3	92	2050	1.75	10.2	1.06	7.3	1720	n.d.	66.3	-8.4
55	Nocilla	26-07-99	899 41591	S	16.7	6.54	230	127	19.0	80.9	72.8	863	67.7	15.3	27.8	12	1285	0.62	0.09	3.32	6.5	104	n.a.	0.43	-8.4
56	Marbizzali	13-09-99	921 41628	70	n.m.	6.90	7	157	16.0	95.5	40.3	915	60.6	n.d.	7.9	38	1330	0.46	n.a.	0.46	n.a.	131	n.a.	n.a.	-9.0
57	Macello Patern	13-09-99	907 41586	S	18.9	6.90	118	160	20.7	106	92.8	1030	67.4	28.5	53.3	82	1639	0.91	0.05	4.25	7.7	93	n.d.	n.d.	n.a.
58	Piano Elisi	15-09-99	001 41638	323	19.0	6.40	166	113	28.9	83.6	46.5	598	96.8	3.2	80.2	75	1125	0.63	0.05	2.80	15.8	210	0.54	0.36	-7.6
59	Piano Elisi 2	15-09-99	001 41638	295	18.3	6.39	146	103	23.9	69.3	38.5	519	87.6	5.8	70.1	74	990	0.48	0.07	3.10	14.3	206	n.d.	n.d.	n.a.
60	Di Natale	25-10-99	923 41606	30	18.2	6.28	30	182	9.8	104	65.7	1150	36.2	0.4	4.0	95	1644	1.18	5.00	0.54	18.0	1170	3.60	5150	-8.6
61	Acqua Grassa	26-10-99	907 41588	S	19.2	6.10	73	175	27.8	134	126	1440	76.9	0.7	30.7	100	2111	1.03	9.88	1.35	9.3	3770	n.d.	35.5	-8.5
62	Romito	26-10-99	913 41607	S	15.5	6.72	-62	145	29.3	108	52.7	1020	79.1	n.d.	2.1	77	1511	0.44	4.44	0.07	13.5	171	0.48	56.7	-8.9
63	Liotta 1	26-10-99	924 41604	17	18.8	6.10	110	152	14.1	109	86.2	1090	82.2	2.2	10.6	91	1637	0.89	6.68	0.51	32.3	2380	3.89	149	n.a.
64	Liotta 2	26-10-99	925 41603	15	18.7	6.22	165	140	14.9	98.5	73.1	1000	68.8	3.2	16.0	88	1506	0.79	1.60	0.85	14.7	1040	1.13	8.48	n.a.
65	Petraro	08-07-97	944 41588	5	17.2	6.41	280	135	10.6	113	72.9	946	52.8	5.6	126	52	1514	0.98	n.a.	3.59	10.6	248	0.32	0.04	n.a.
66	Ciappe Bianche	10-12-97	906 41563	S	18.3	7.37	290	149	20.7	97.0	69.3	851	79.4	48.4	79.2	65	1460	0.87	n.a.	3.22	6.8	20	0.11	0.04	n.a.
67	Trefontane	10-12-97	907 41558	S	17.8	7.26	244	142	22.3	95.4	78.4	833	84.0	51.5	82.6	57	1446	1.02	n.d.	5.15	11.7	33	1.64	0.02	n.a.
68	Porticello 1	26-05-99	928 41561	10	n.m.	7.80	130	90.1	4.3	35.5	104	354	150	28.2	66.7	38	871	0.13	0.06	3.49	6.7	6.2	n.d.	n.d.	-5.5
69	Porticello 2	26-05-99	928 41560	S	n.m.	7.60	158	66.9	5.9	31.1	101	381	89.3	26.6	56.2	47	805	0.09	n.a.	1.99	n.a.	10	n.d.	n.a.	n.a.
70	Porticello 4	26-05-99	926 41561	S	18.0	8.18	90	148	8.6	50.7	98.0	397	227	30.4	76.8	42	1077	0.29	0.02	1.99	8.5	5.7	n.d.	0.07	-5.1
71	Cafaro	26-05-99	898 41542	S	18.4	7.56	110	197	8.0	94.3	68.9	839	94.3	59.6	105	75	1551	0.95	0.01	n.a.	n.a.	26	n.a.	n.a.	-7.4
72	Giosafat	26-07-99	911 41544	15	20.5	6.97	220	407	9.9	99.6	164	451	523	99.7	397	29	2180	1.03	0.04	2.92	7.0	33	n.d.	n.d.	-6.3
73	Mauta	26-10-99	912 41540	20	19.8	6.98	289	365	9.4	107	240	482	431	415	529	27	2604	0.71	0.01	5.42	13.7	42	n.d.	0.02	n.a.
74	Parco Zoo	26-10-99	942 41545	7	19.5	6.99	292	106	7.4	57.3	146	561	88.3	125	179	30	1300	0.19	0.03	2.72	14.7	59	n.d.	n.d.	-5.5
75	Stadio 4	17-01-97		MV	16.4	6.10	n.m.	25500	626	335	1150	1780	42200	6	27	84		229	n.m.						n.a.
averages	tot				17.3	6.69	160	162.2	21.2	110	85.2	953	126.9	27.3	58.8	71.7	1612	1.1	1.6	3.5	11.4	609	1.3	136	-7.7
	I				16.8	6.27	202	134.9	20.1	125	91.1	1111	86.3	6.3	29.5	81.2	1683	1.4	0.7	4.2	11.7	1043	1.1	6.9	-7.9
	II				17.0	6.91	183	220.2	35.3	108	72.7	882	210.6	41.6	76.5	74.1	1721	1.2	0.2	4.0	10.1	160	0.3	0.9	-7.9
	III				17.5	6.79	74	154.6	18.6	105	72.6	950	105.1	12.9	36.1	70.4	1521	1.0	3.5	2.2	12.2	577	1.9	391	-8.0
	IV				18.7	7.31	210	180.6	11.7	78	114	609	181.9	89.0	170	46.1	1481	0.6	0.0	3.6	10.0	48	0.9	0.1	-6.0

TABLE 1 - Chemical and isotopic analysis of the sampled groundwaters. UTM is the location of the samples in UTM coordinates. Type: S = spring; G = drainage galleries; MV = mud volcano; number = well depth. Eh is expressed in mV; ionic solutes, total dissolved solid (TDS) and SiO2 are expressed in mg/l, dissolved gases are expressed in ml STP per l of water (He and CH4 x 1000), δ18O is expressed in vs. V-SMOW standard, n.m. = not measured, n.a. = not analysed, n.d. = not detected. The samples are subdivided in four groups on the basis of factor analysis. Group I = samples 1-28; Group II = samples 29-41; Group III = samples 42-64; Group IV = samples 65-74. The averages for the whole data set and for each group are also reported.

TABLE 2 — Loadings of the orthogonally rotated solution of principal component analysis. Variables displaying a significant weight on each factor are evidenced in bold.

Factors	1	2	3	4
T	0.095	0.091	0.224	**0.658**
pH	**-0.677**	0.162	0.068	0.105
Eh	-0.032	-0.209	**-0.773**	0.120
Na	0.036	**0.839**	0.128	0.372
K	0.189	**0.700**	-0.174	-0.362
Mg	**0.831**	0.355	0.135	0.059
Ca	0.498	0.094	0.022	**0.762**
HCO_3	**0.903**	0.167	0.229	0.023
Cl	-0.076	**0.860**	0.075	0.352
SO_4	-0.302	0.240	-0.305	**0.747**
TDS	**0.678**	0.584	0.143	0.371
B	**0.655**	0.468	0.201	0.007
Fe	0.396	-0.024	**0.697**	0.084
O_2	-0.166	-0.186	**-0.762**	-0.143
CO_2	**0.785**	-0.181	0.144	0.162
CH_4	-0.021	-0.174	0.472	0.033
Expl. Var. %	25.3	18.1	14.0	13.7

Table 3 — Mass Balance Calculation with the *Netpath* code (Plummer et al., 1994)

runs	1a	1b	1c	3a	3b	3c	2a	2b	2c	4a	4b	4c
constraints	C, Ca, K, Fe, Mg, Na, Si			C, Ca, K, Fe, Mg, Na, Si, RS			C, Ca, Cl, Fe, Mg, Na, S, Si			C, Ca, Cl, K, Mg, Na, S, Si		
waters	# 22 → #19			#59 → #43			#75+#7→#41			#9 → #73		
mixing	no			no			yes			no		
Olivine	3.6	4.0	3.6	3.4	4.3	3.4	1.8	2.3	1.8	1.1	1.2	1.1
Cpx	1.8	1.8	1.8	0.6	0.6	0.6	-	-	-	-	-	-
Plagiocl.	5.1	5.1	5.1	11.6	11.6	11.6	6	6	6	1	1	1
CO_2 (g)	46.8	46.8	46.8	22.2	22.2	22.2	8.3	8.3	8.3	-11	-11	-11
CH_4 (g)	-	-	-	0.3	0.3	0.3	-	-	-	-	-	-
Goethite	-1.6	-1.7	-1.6	-	-	-	-	-	-	-	-	-
Kaolinite	-6.6	-	-	-15.2	-	-	-9.1	-	-	-1.6	-	-
Illite	-1	-1	-1	-1.9	-1.9	-1.9	-	-	-	-0.4	-0.4	-0.4
Mg-Mont.	-	-3.7	-	-	-8.5	-	-	-5.1	-	-	-0.9	-
Am. silica	-	-	-13.2	-	-	-30.4	-	-	-18.2	-	-	-5.7
Pyrite	-	-	-	-1.6	-1.6	-1.6	-0.05	-0.05	-0.05	-	-	-
Calcite	-	-	-	-	-	-	-1.3	-1.3	-1.3	-1.2	-1.2	-1.2
Gypsum	-	-	-	-	-	-	-	-	-	5.3	5.3	5.3
Halite	-	-	-	-	-	-	-	-	-	10.7	10.7	10.7

TABLE 3 - Results of MBC analysis carried out with the *Netpath* code (Plummer et al., 1994). Three solutions are reported for each water group (runs a-c). Constraints are chemical variables whose net changes from starting to final solutions (waters) are accounted for by the reaction model. Models 1 a-c, 3 a-c and 4 a-c describe groundwater evolution in terms of gas-water-rock interaction only. Mixing between two waters (1.2 % of #75 and 98.8% of #7) with further WRI is taken into account in models 2 a-c. Data are expressed in millimoles of each phase entering or leaving 1 litre of solution. Positive signs stand for phase dissolution, negative signs for phase precipitation.

2.3 STATISTICAL METHODS

Data from the 74 water samples were first processed by univariate analysis, the resulting mean values being listed in Table 1. Moreover, a multivariate method of analysis (factor analysis) was applied to a subset of 16 selected variables. Four factors, having eigenvalues higher than 1, were extracted and rotated according to the orthogonal Varimax procedure. The loadings of the rotated four-factor solution are shown in Table 2, along with the variance accounted for by each factor.

2.4 MASS BALANCE CALCULATIONS

Mass balance calculation (MBC) is a widely diffused approach applied to chemical data in order to model the reaction paths occurring along groundwater flow (Garrels and Mackenzie, 1967; Plummer and Back, 1980; Plummer et al., 1983, 1990, 1994). Basically, once an initial and a final water along a flow path are fixed, the mass balance calculation (MBC) ascribes the net changes in total concentration of aqueous solutes to dissolution and/or precipitation of a set of selected minerals and gases. MBC is actually an inverse method that attempts to evaluate the feasibility of a chemical process. In this sense, no thermodynamic constrains are taken into account during the stoichiometric calculations (Plummer et al., 1983), and several solutions are possible.

In this paper, the *Netpath* code (Plummer et al., 1994) was used for data handling and elaboration of MBC models. Results are summarised in Table 3.

3. Results

3.1 MAIN CHEMICAL FEATURES OF GROUNDWATER

At first glance, samples collected during the present survey display the general features of most Etnean groundwaters, which have been widely described elsewhere (Anz et al., 1989; Allard et al., 1997; Giammanco et al., 1998; Brusca et al., 2001). Patern groundwaters are cold to hypothermal fluids (T = 11.2-22.8 ¡C; mean 17.3 ¡C), ranging from slightly acidic to slightly basic (pH = 5.90-8.65; mean = 6.69). A wide range of redox conditions, from reduced to oxidised (-178 mV < Eh < 500 mV) is observed in the aquifer. Similarly, the total amount of dissolved salts (TDS) is highly variable, being comprised between 805 and 3119 mg/l. This significant heterogeneity of physico-chemical conditions is also reflected by the variable content of major ion species. To better put in light this feature, the triangular diagrams of Figure 2 were drawn. The representative points of seawater (SW) and of the few hypersaline thermal waters issuing at the base of Etna (Salinelle), all sharing a chloride alkaline composition, are also plotted for comparison. The diagrams highlight that, despite most Patern groundwaters display a magnesium bicarbonate composition, waters ranging from alkaline chloride to calcium chloride-sulphate are also observed.

Magnesium bicarbonate groundwaters form as a result of the dissolution of magmatic CO_2 into the aquifer, followed by CO_2 titration to bicarbonate throughout weathering of volcanic rocks (i.e., Anz et al., 1989). The high Mg content of the resulting solution depends both on the composition of the rocks (mainly alkali-basalts

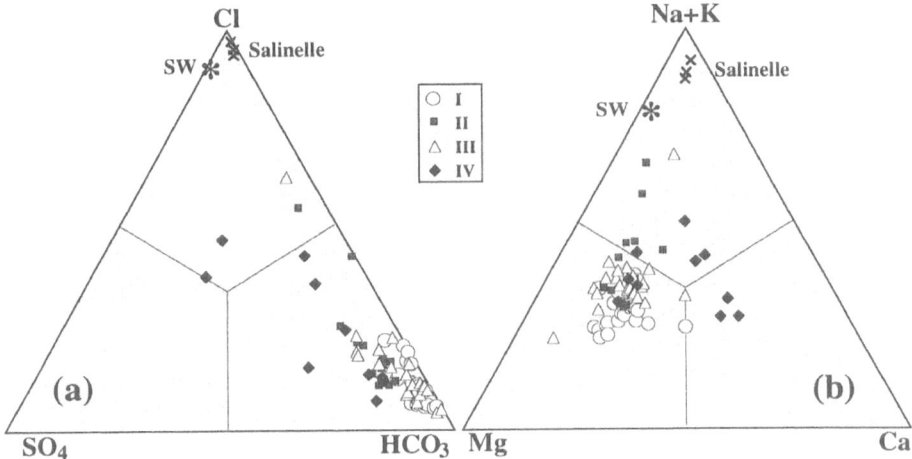

Figure 2 - Major anion (a) and cation (b) triangular plots. Water groups identified through factor analysis: I = empty circles; II = filled squares; III = empty triangles; IV = filled diamonds. Sea water (SW) and sedimentary thermal brines issuing in the Patern area (salinelle) are also plotted.

and hawaiites) hosting the aquifer and on the low pH and temperature of groundwaters, increasing the dissolution of ferro-magnesian minerals. The chemistry of groundwaters after dissolution of Etna s lavas displays chloride and sulfate concentrations rarely exceeding 100 mg/l, due to the quite low Cl and S content of the volcanites: 980-1680 ppm for Cl and 110-300 ppm for S (Metrich and Clocchiatti, 1989; Metrich et al., 1993). Therefore, the reason for the enhanced Cl and SO_4 concentrations measured in some samples from Patern , up to 875 and 529 mg/l, respectively, should be found in processes other than simple basalt leaching. Similarly, sodium content measured in some samples, up to 600 mg/l in samples 59 and 70, seems far too high to be only accounted for by the dissolution of volcanics.

The great variability of total dissolved content extends also to minor elements, such as B and Fe. Both elements, in fact, have coefficient variation (CV % = st.deviation/mean) exceeding 100%. Concerning iron, we may argue that the remarkable content measured in some groundwaters (up to 11.5 mg/l in # 43) is related to their strongly reduced conditions. In general, Fe concentration levels in normal oxidised groundwaters of Mt. Etna are significantly lower (less than 0.1 mg/l).

3.2 DISSOLVED GASES

All sampled waters are meteoric in origin, as suggested by oxygen and hydrogen isotopes (Anz et al., 1989, Allard et al., 1997). Accordingly, they contain atmospheric-derived gases, O_2 (from 0.06 to 8.81 ml/l, corresponding to pO_2 from 0.002 to 0.258 atm) and N_2 (from 4.12 to 32.3 ml/l; pN_2 from 0.240 to 1.997 atm). CO_2 is a major dissolved gas in most groundwaters, being often the prevalent species (from 3 to 4270 ml/l; pCO_2 from 0.003 to 4.62 atm). Nineteen of the sampled waters have pCO_2 values above 1 atmosphere and are characterised by a CO_2-rich free gas phase. Carbon isotope ratios revealed a magmatic origin for CO_2 in these manifestations (Allard et al., 1997; D Alessandro et al., 1997b). Only 16 samples displayed helium values higher than the

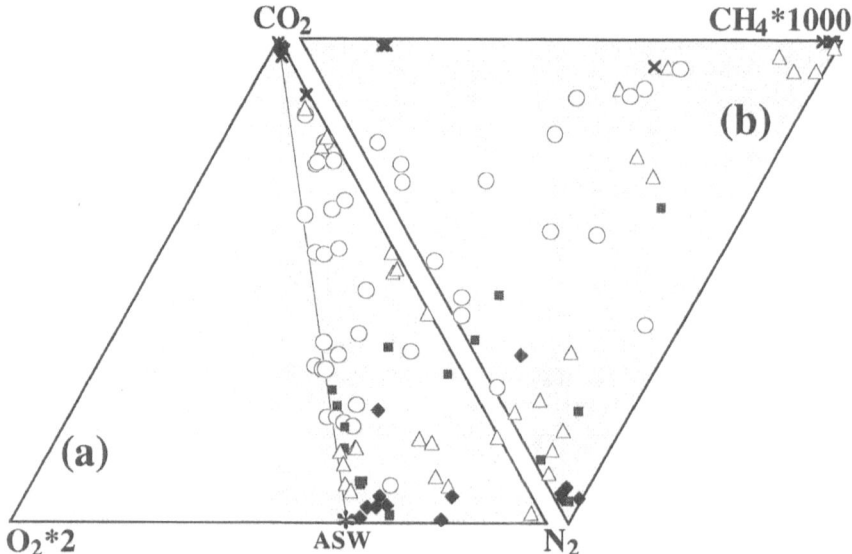

Figure 3 - CO_2-N_2-O_2 (a) and CO_2-N_2-CH_4 (b) triangular plots (symbols as in Figure 2). ASW = air-saturated water; crosses = free gases issuing in the same area.

detection limit (from 1.13×10^{-4} to 38.9×10^{-4} ml/l; pHe from 0.13×10^{-4} to 4.44×10^{-4} atm), but all of them were well above the values of air-saturated water ($\sim 0.5\times10^{-5}$ atm). All samples appeared to be enriched in He with respect to both the He/N_2 ratio of the atmosphere (6.6×10^{-6}) and the typical He/CO_2 ratio of the magmatic gases ($^-2\times10^{-5}$). Finally, 41 samples displayed detectable CH_4 values (from 0.15×10^{-4} to 5.15 ml/l; 0.39×10^{-6} to 0.141 atm).

The N_2, O_2, CO_2 triangular diagram (Figure 3a) shows that many samples plot along a mixing line between a component representative of air-saturated groundwaters (ASW) and a CO_2-rich end-member, similar in composition to the free gases emerging in the area. Other samples have N_2/O_2 ratios higher than ASW due to O_2 consumption during redox reactions in the aquifer. In some cases, this process leads to waters strongly depleted in O_2, with relatively high CH_4 contents and a clear H_2S smell. The enrichment in CH_4 is also evident in the N_2-CO_2-CH_4 triangular diagram (Figure 3b). The presence of a distinct gas component of thermogenic origin, characterized by CH_4 contents up to 35 %vol, was already revealed by D Alessandro et al. (1997b).

3.3 $^{18}O/^{16}O$ RATIOS

The analysed waters displayed $\delta^{18}O$ values in the range from —9.0 to —5.1 . On the basis of their isotopic signature they can be subdivided in three groups. A first group is formed by waters issuing southwest of Patern in the sedimentary part of the studied area (Figure 1). These waters have an average $\delta^{18}O$ value of —6.0 (range from —7.4 to —5.1) which, on the basis of the isotopic gradient reported by previous authors (Anz et al., 1989; Longo, 1999), correspond to an average recharge altitude of about 500 m. The remaining waters were collected upslope in the volcanic aquifer and are

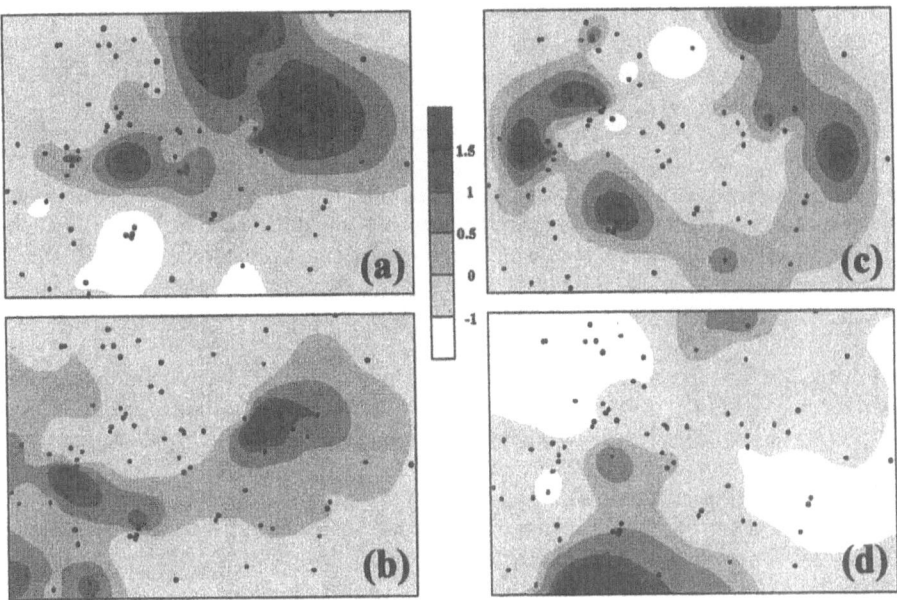

Figure 4 - Distribution map of the four factors obtained by Ordinary Kriging. (a) factor 1, (b) factor 2, (c) factor 3, and (d) factor 4.

separated by a groundwater divide (Fig. 1). The waters flowing in the eastern part display an average $\delta^{18}O$ value of —7.2 (range from —7.8 to —6.2) and those flowing in the western part a value of —8.2 (range from —9.0 to —7.4). These values correspond to average recharge altitudes of about 900 and 1200 m, respectively.

3.4 FACTOR ANALYSIS

Factor analysis attributed 71% of the total variance in the chemistry of groundwaters to four factors (Table 2). Factor 1, which explains 25% of the total variance, clusters pH, Mg, HCO_3, CO_2 (g), TDS and B. The significant high positive loadings of HCO_3 and CO_2 (g) and the high negative loading of pH indicate that this variable association is related to the feeding of groundwaters with deep CO_2, which is gradually converted to bicarbonate under open CO_2 conditions. The marked correlation between alkalinity and total dissolved solids (TDS) is an effect of the higher extent of rock leaching promoted by gas-charged acidic waters, while the positive loading of Mg on factor 1 can be explained by the high rate of dissolution of ferro-magnesian minerals (Aiuppa et al., 2000).

The association of Na, Cl and K is reflected on factor 2, which explains 18% of the total variance. Both B and TDS display significant loadings on Factors 1 and 2, which likely indicates two independent sources for these variables. Variables clustering on Factor 2 may be indicative of a contribution to the aquifer from seawater-like brines rising from the sedimentary basement. The occurrence of this process (Brusca et al., 2001) is further suggested by the presence near Patern of mud volcanoes, known as *salinelle* (Figure 1). These manifestations, interpreted as the surface outflow of an hydrothermal system hosted in the carbonate or quartz-arenite layers of the sedimentary sequence below Etna, release connate sedimentary fluids (CH_4, higher hydrocarbons, and Na-Cl brines) along with magmatic gases (mainly CO_2 and He) (Chiodini et al., 1996; D Alessandro et al., 1997b).

Factor 3 accounts for 14% of the total variance. It is characterised by high loadings from Fe, O_2, Eh and, to a lesser extent, CH_4 and is clearly related to redox reactions. The high negative loadings of Eh and O_2 are indicative of oxygen consumption under reduced redox conditions, prevailing in the deepest parts of the aquifer. Low Eh values are likely related to the rise of reduced gases (mainly light hydrocarbons and H_2S) from the sedimentary basement (Fig. 3b). These reduced redox conditions are in turn responsible for the high aqueous mobility of iron, which displays a loading of 0.7 on factor 3.

Finally, the last factor is entirely dominated by Ca and SO_4 and, subordinately, by water temperature. This association is likely to reflect the interaction of groundwaters with evaporite deposits, which are known to be present in the sedimentary sequence lying below the Etnean volcanics (Romano et al., 1979; Lentini, 1982).

In order to better constrain the chemical processes revealed by PCA, we performed surface mapping of the factor scores of the four-factor solution by Ordinary Kriging (Swan and Sandilands, 1995). Maps (Fig. 4) were utilised to evaluate the spatial distribution of the extracted factors.

The map of Factor 1 (Fig. 4a) reveals the presence of maximum values in the northern sector of the investigated area, between the villages of Belpasso and Patern . As factor 1 is defined by the association of HCO_3, CO_2 $_{(g)}$, pH and Mg, Fig. 4a enables us to identify a well-defined area of ascent of slightly acidic gas-rich groundwaters. Samples 16, 18, 19, 20, 21 and 27, which are indeed among the gas-richest waters (Table 1), display the highest factor scores.

The spatial distribution of Factor 2 (Fig. 4b), furthermore, confirms that the relevant association of variables (Na, Cl, K and TDS) is related to mixing between groundwater and *salinelle-type* brines. In fact, the maximum factor scores are observed near the mud volcanoes of Simeto (#29, 31, 35, 38) and Stadio (#37, 41). The only exception is represented by sample 39, whose high TDS contents and Na-Cl composition should be ascribed to human pollution. This is also inferred by its Br/Cl ratio (0.0006), which is very different from those of seawater (0.0035) and salinelle brines (0.0027-0.003) markers. The Br/Cl ratio measured in sample 39 is consistent with pollution of groundwaters from salts used in the industrial manufacturing of preserves.

Factor 3 (Fig. 4c), ascribed to Fe, O_2, Eh and CH_4, displays a sort of ring-shaped distribution pattern of its highest values, that are located basically along the margins of the central CO_2 degassing area (Fig. 4a). Maximum values of this factor are observed at sites where very reduced conditions prevail, such as in samples 43, 45, 53 and 54. More important, most of the samples with high scores on Factor 3 correspond to springs or to very deep wells. This suggests that the particularly reduced conditions measured at these sites result from sampling of the deepest horizons of the aquifer, which are likely to be the most depleted in oxygen due to both enhanced water-rock interaction and to the dissolution of methane and other hydrocarbons rising from the underlying sedimentary sequence. A longer and slower water circulation is also suggested by the very negative $\delta^{18}O$ values, indicating a higher average recharge area, and by the very low tritium values (< 1 TU for # 61 — P. Allard and C. Federico, pers. comm.).

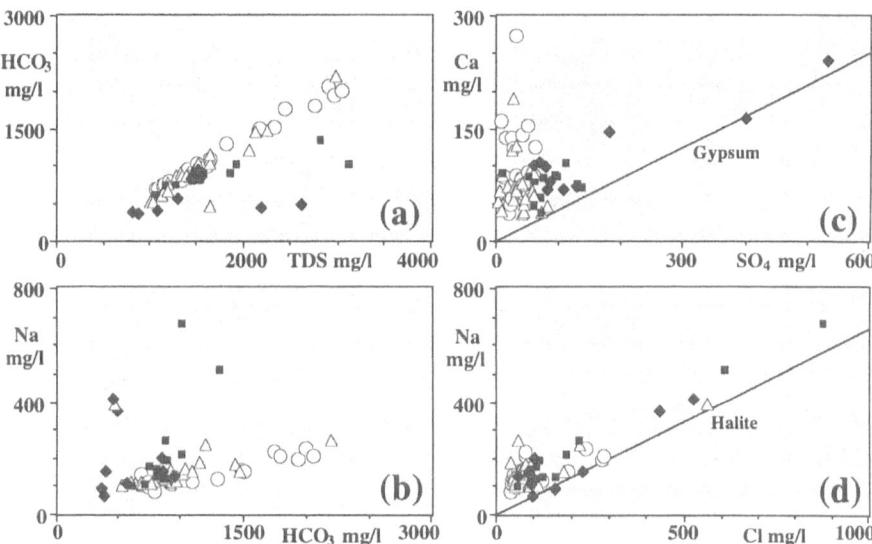

Figure 5 - Correlation diagrams of the sampled waters (symbols as in Figure 2); (a) HCO_3^- vs. TDS; (b) Na^+ vs. HCO_3^-; (c) Ca^{2+} vs. SO_4^{2-}; (d) Na^+ vs. Cl^-. The stoichiometric dissolution lines for gypsum and halite are also reported in c and d .

Finally, the highest values for Factor 4 are observed in the southernmost sector of the aquifer (Fig. 4d) at sites 68, 69, 70, 71, 72, 73 and 74. Given the Ca-SO$_4$ association that defines Factor 4, we may argue that interaction of groundwater with evaporite deposits is prevalent in this area. The chemistry of gases dissolved in these samples indicates a composition very close to air-saturated waters with no or very little CO_2 or CH_4 addition (Fig. 3). $\delta^{18}O$ values of these samples are close to the meteoric values measured in the same area (Longo, 1999), indicating that most of these waters have a very short circulation path.

4. Discussion

The previous considerations have emphasised that, despite the rather limited spatial extent of the investigated area (about 150 km^2), a significant heterogeneity in the chemistry of groundwaters exists. This suggests that a combination of geochemical processes is likely to control the physical and chemical variables of local groundwater. Such an assumption is supported by multivariate analysis on the acquired data, which highlighted the following chemical processes: (i) leaching of the host volcanic rocks, driven by high contents of dissolved CO_2 in the aquifer; (ii) mixing with Na-Cl brines that rise from the sedimentary basement of Mt. Etna and are represented by the saline fluids discharged by the *Salinelle* ; (iii) redox transformations, caused by interaction between the pristine O_2-rich meteoric waters and reduced gases — mainly hydrocarbons and H_2S — rising from depth; (iv) dissolution of evaporitic rocks of the sedimentary basement.

Process (i) is likely to be prevalent or at least to be active throughout the whole studied area. Fig. 5a is a HCO_3^- vs. TDS scatter diagram where the samples from the

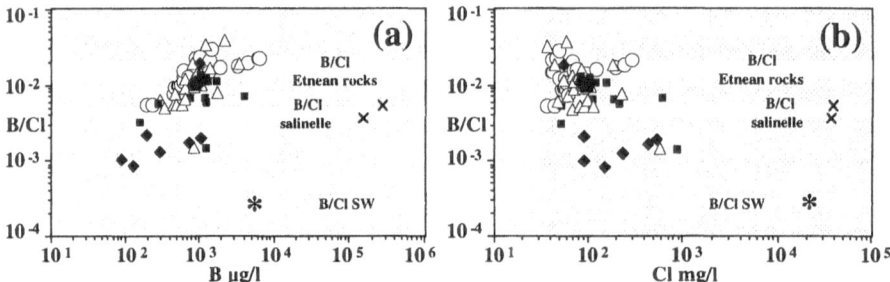

Figure 6 - B/Cl vs. B (a) and Cl⁻ (b) correlation diagrams (symbols as in Figure 2). Crosses indicate the salinelle composition, asterisk indicates seawater composition and white bands indicate B/Cl ratios in Etnean volcanites, salinelle and seawater, respectively.

Patern area were divided into 4 groups based on the relative weights (i.e., the factor scores) exercised on each of the four factors extracted by factor analysis. In other words, this means that a sample is ascribed to group 1 if it shows the highest factor score on Factor 1 (with respect to factor scores on Factors 2, 3 and 4), and so on. This procedure is based on the basic statistical concept stating that a factor score corresponds to the actual value of individual cases (water samples) on each extracted factor.

As expected, a good correlation between TDS and bicarbonate is observed for samples belonging to group 1. The same holds true for samples of group 3, except sample 53, which shows a prevalent Na-Cl composition. The increasing bicarbonate content in these samples is related to the different CO_2 supply in the various sectors of the aquifer and the different extent of CO_2 conversion to HCO_3^- through water-rock interaction. Moreover, the variable depth of the wells used for groundwater exploitation seems also to affect water chemistry, as several evidences suggest a chemical stratification of the aquifer. For instance, samples 26 and 27, collected in the same well at two different water levels (respectively 180 and 210 m), show temperature, TDS, bicarbonate and major cations contents that increase with depth (Table 1). Similar conclusions can be drawn from the analysis of data from samples 58 and 59, both collected in the Piano Elisi well near Nicolosi village (Table 1).

By contrast, Fig. 5a points out that many samples belonging to groups 2 and 4 depart from the general trend, since they are enriched in Na, Ca, Cl and SO_4, as shown in Figs. 5 b-d.

In order to better constrain the geochemical processes responsible for the high salt contents of groups 2 and 4 water samples, the relations between two conservative elements such as boron and chlorine are examined in detail in Figs. 6a-b. In both plots, it is evident that samples belonging to groups 1 and 3 (again except sample 53) cluster close to the B/Cl ratio typical of Etnean volcanics (Pennisi et al., 2000), confirming their nature of pure volcanic groundwaters. Some samples from groups 2 and 4, instead, display considerably lower B/Cl ratios, as typical for waters related to or interacting with sedimentary fluids (Pennisi et al., 2000). In particular, group 2 waters are shifted toward the composition field of the *salinelle* brines, which are then the most likely responsible for the peculiar characteristics of these waters. Group 4 groundwater samples are characterised by even lower B/Cl ratios, close to the compositional field of seawater. This further suggests their interaction with marine sediments, and, again, their high Ca and SO_4 contents (Figs. 5 e-f) probably reflect leaching of gypsum- and halite-bearing deposits (i.e., Messinian evaporites of the Gela Nappe).

4.1 GEOCHEMICAL MODELLING

Table 3 lists the results of chemical mass balance calculations carried out using the *Netpath* code (Plummer et al., 1994). Four separate computations were performed in order to distinguish the distinct processes that determine the chemical evolution towards the four envisaged groups of waters. We first discuss the genesis of pure volcanic groundwaters (groups 1 and 3) and then their interaction with sedimentary fluids and/or evaporite rocks.

4.1.1 Group 1 waters

As previously stated, these waters acquire their composition through dissolution of magma-derived CO_2 into the aquifer and consequent intense basalt leaching. Thus, we selected a low-salinity well (#22) as representative of the initial stage of the process and a downgradient gas-rich well (#19) as representative of its final stage. Due to the complexity of the studied chemical system, modelling was carried out on a limited set of constraining chemical parameters (C, Ca, K, Fe, Mg, Na, Si). The hydrological system was assumed to be open to gaseous carbon dioxide along the whole water path. Furthermore, the main rock-forming minerals of Etna hawaiites (Tanguy et al., 1997) were used as possible solid reactants, i.e.: olivine (Fo 0.8), clinopyroxene (En 0.42, Fs 0.11, Wo 0.47) groundmass plagioclase (An 0.3, Ab 0.6, Or 0.1). All these mineral phases are known to be slightly to strongly subsaturated in Etnean groundwaters, and are thus the most susceptible to water dissolution. By contrast, plagioclase phenocrysts and Fe-Ti oxides are generally supersaturated in Etnean groundwaters (Aiuppa et al., 2000), and are considered not to take part in the weathering process.

Identification of possible secondary minerals is a more difficult task. Goethite and amorphous silica are generally supersaturated in Paternò groundwaters (Figs. 7 a-b). Calcite supersaturation is observed in the neutral to slightly basic waters, mainly belonging to groups 3 and 4, whereas most of group 1 and 3 samples display various degrees of undersaturation. At some sites, carbon dioxide exsolution following travertine precipitation is likely to occur upon water discharge and may reasonably explain the low Ca content in some of the studied waters (Fig. 7c). However, calcite is unlikely to be massively deposited during water-rock interaction at depth. Among clay minerals, Mg-montmorillonite, kaolinite and illite, all being characteristic phases of the low-temperature weathering of basaltic rocks (Kristmannsdottir, 1978, 1982; Douglas, · 1987), are commonly supersaturated in Etnean groundwaters (Aiuppa et al., 2000).

In the absence of more detailed data on alteration assemblages at Mt. Etna, we may note that more than one set of geologically plausible phases is able to produce the composition of sample 19. Table 3 shows the three possible solutions of MBC (models 1 a-c), which differ exclusively for the secondary minerals that are assumed to be the final product of weathering. A general reaction path describing the chemical evolution of group 1 groundwaters can be schematised by the following equation:

$$\text{Initial water (\#22)} + 46.8CO_{2\ (g)} + 5.1plg_{30} + 1.8px_{42} + aOl_{80} \Rightarrow$$
$$\Rightarrow \text{Final water (\#19)} + 1.0\text{Illite} + b\text{Goeth} + cX$$

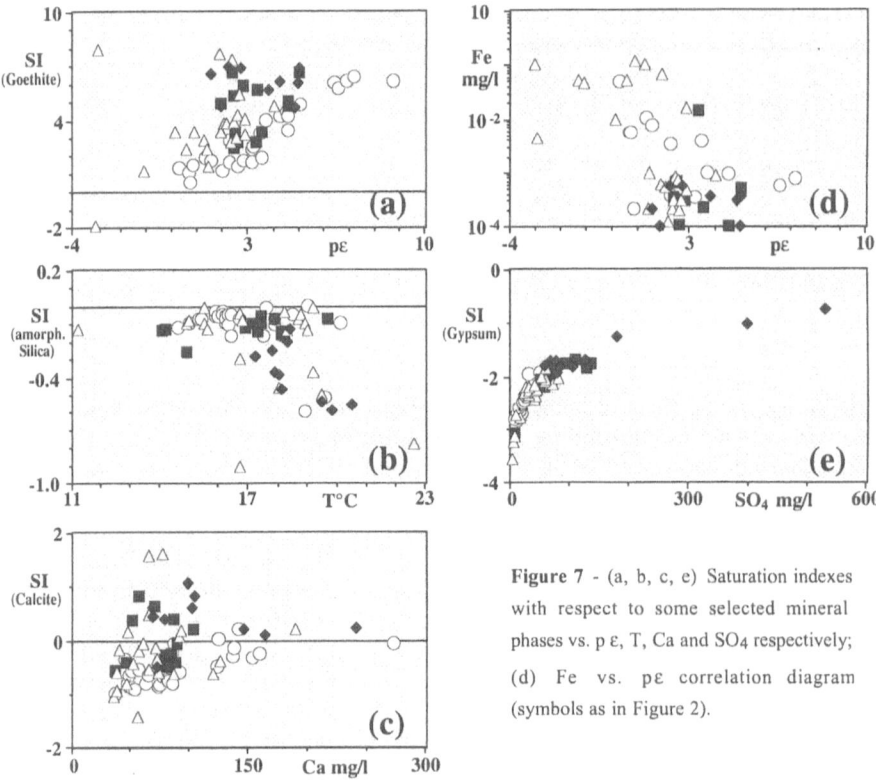

Figure 7 - (a, b, c, e) Saturation indexes with respect to some selected mineral phases vs. p ε, T, Ca and SO₄ respectively; (d) Fe vs. pε correlation diagram (symbols as in Figure 2).

where a, b and c are reaction coefficients and X is a variable secondary mineral — alternatively, kaolinite (1a), Mg-Montmorillonite (1b) or amorphous silica (1c).

4.1.2 Group 3 waters

Many waters of this group display the lowest dissolved O_2 contents and the highest dissolved CH_4 contents (Fig. 3 a-b). Also, the fact that these groundwaters evolve under reduced conditions is suggested by the low Eh values measured at fluid discharge and by their high Fe contents (Fig. 7d).

In order to provide a reaction path for these waters, redox transformations have to be taken into account. Thus, the redox state coefficient (RS; Plummer et al., 1994) was included among the constraints of the model. The presence of a reducing agent — leading to atmospheric oxygen consumption — was considered by assuming that the system is open to a mixed CO_2 + CH_4 gas phase. In other words, we ascribed total carbon net changes upon reactions that follow dissolution of both CO_2(g) and CH_4(g) and — eventually - to calcite precipitation. It should be pointed out that other electron donor species are likely to play a major role in redox transformations and oxygen consumption. In fact, H_2S and light hydrocarbons can be released from oil and gas reservoirs hosted in the sedimentary sequence below Etna and eventually rise towards the surface along with methane (D Alessandro et al. 1997b). However, these compounds were not considered in modelling due to the lack of reliable data on their concentrations both in Etna s groundwaters and in the interacting gas phase.

Mass balance calculations were carried out to model the reaction path from an oxygen-rich meteoric sample (# 59) to a downgradient O_2-depleted, CH_4-Fe-rich one (# 43). The mineral phases which are supposed to eventually enter or leave the aquatic system are the same as in the previous case (models 1 a-c), except for the fact that pyrite is now considered among the possible secondary minerals. The main results of the computation, listed in Table 3 as models 3a-c, are the following: i) water evolution is consistent with the dissolution into the aquifer of a carbon-rich gas phase having a CO_2/CH_4 ratio of about 70, that is similar to the ratio measured in gas manifestations from the Patern area (values in the range 2 - 180; D Alessandro et al. 1997b) ; ii) the high iron content in the final water is related to goethite undersaturation, while iron sulphides may represent the only Fe sink; iii) calcite is not included in any of the possible reaction paths, confirming it is a minor or negligible product of weathering; iv) plagioclase, olivine and clinopyroxene are the main solid reactants that account for Ca, Na, Mg and Fe net changes in group 3 groundwaters; v) as in the previous case, a multiple solution is obtained (models 3a—c).

4.1.3 Group 2 waters

The MBC approach was also attempted to check if the composition of group 2 groundwaters is consistent with the previously hypothesised mixing process between *shallow volcanic waters* (equivalent to group 1 waters) and *salinelle-type sedimentary brines* . In order to perform such a computation, we considered water sample 41 as the final evolutionary step of this process, which is indeed suggested by its high TDS and high sodium and chloride contents. Notably, this water sample was collected only a few hundred meters far from and downgradient to a zone where salinelle-like fluids reach the surface (Fig. 1), which further strengthen our hypothesis. Thus, we tested the possibility of accomplishing for the composition of sample 41 by mixing a salinelle-type brine (Stadio 4; Table 1) and a dilute water sample (# 7 of group 1), collected upgradient to # 41. The tested model (permitted by a special option in the *Netpath* code) first determines the mixing proportions of the two initial waters by using a single mass balance equation for a conservative species such as chloride or boron. The calculated mixing fractions, which do not display appreciable variation whatever we choose Cl or B, indicate that 98.8% of the final water is from the shallow volcanic end-member, and only 1.2% is accounted for by the saline sedimentary end-member. Based on these proportions, the model calculates the composition of the final solution if only mixing occurs, without further mineral-gas-water reactions. Then, any discrepancy between the computed and the observed values for the selected chemical parameters is ascribed to mass transfer of mineral and gas phases. The main reactions appear to be dissolution of olivine and plagioclase, titration of $CO_{2(g)}$, and precipitation of calcite, pyrite, and either a clay mineral (kaolinite, Mg-montmorillonite) or amorphous silica (Table 3, models 2 a-c). The reliability of this process in confirmed by the fact that calcite, pyrite and amorphous silica are actually saturated in the final water represented by sample 41.

4.1.4 Group 4 waters

Finally, the reaction paths describing the chemical evolution of shallow volcanic groundwaters upon interaction with the evaporite deposits of the sedimentary basement are investigated. In this case, chemical modelling was carried out by selecting starting

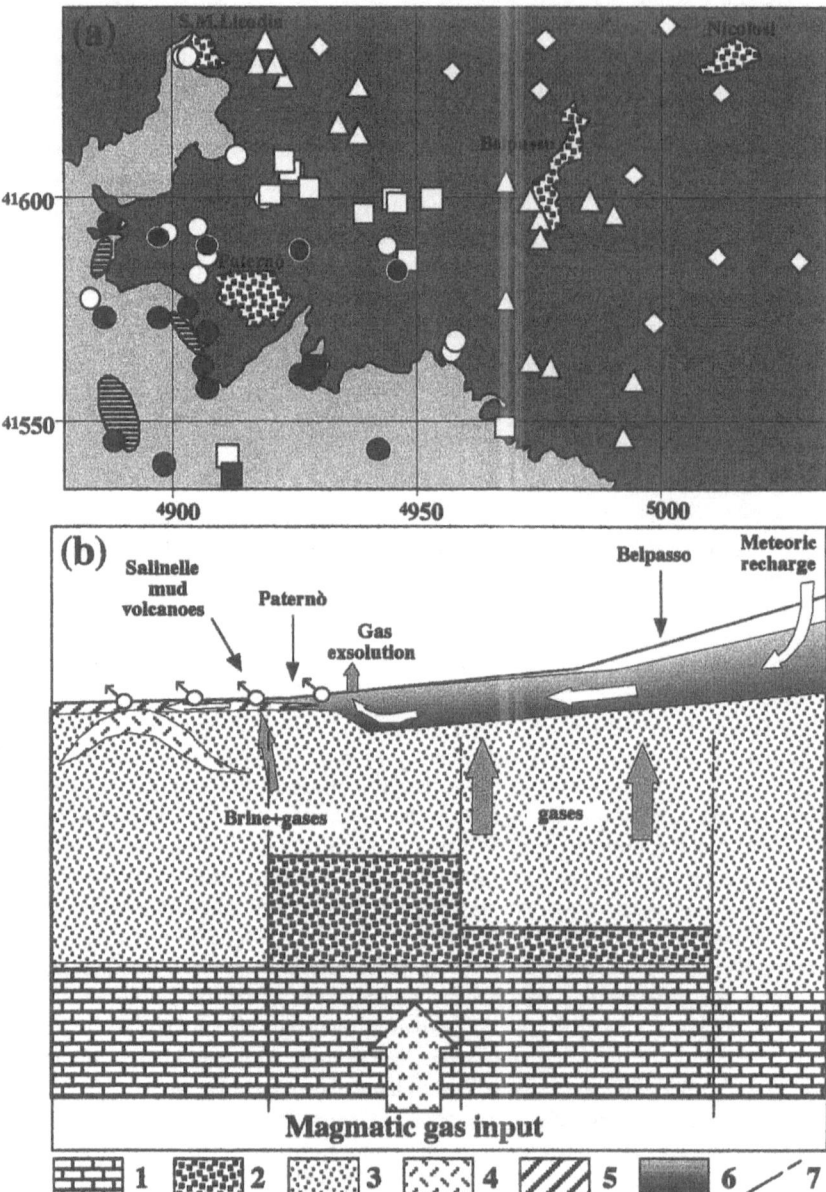

Figure 8 - Geochemical model of the studied area. a) Map of the area; Dark grey = volcanites; pale grey = sedimentary rocks; dashed areas = travertine deposits; circles = springs and very shallow wells (< 10m); squares = shallow wells (< 50 m); triangles = deep wells (< 100m); diamonds = very deep wells (> 100m). Black circles indicates the presence of a bubbling CO_2-rich gas phase while grey circles indicate waters with $SI_{calcite} > 0.2$. b) Schematic cross section of the area: (1) permeable fractured rocks (carbonate or quartzarenite); (2) geothermal reservoir hosted in (1) containing a gas-rich (hydrocarbons and magmatic CO_2) brine with estimated temperatures of about 150 ¡C (Chiodini et al., 1996); (3) Impermeable sedimentary sequence; (4) Evaporite rocks hosted in (3); (5) low permeability fluvial terraced deposits; (6) stratified volcanic aquifer; (7) faults.

(# 9) and final (# 73) waters and a set of constraints (C, Ca, Cl, K, Mg, Na, S, Si), thus fixing the chemical system. The results of MBC, listed in Table 3, clearly indicate that gypsum and halite dissolution can easily explain the net S and Cl changes observed from the starting to the final solution. CO_2 degassing and calcite precipitation, both suggested by models 4 a-c, are consistent with the low carbon content, high pH, and $SI_{calcite} \geq 0$ displayed by group 4 groundwaters.

5. Conclusions

The geochemical processes occurring in a complex groundwater system can be successfully investigated by coupling multivariate statistical methods and mass balance calculations. The interpretation of the collected data shows that groundwater chemistry in the Patern area is controlled by a complex interaction between several geochemical processes — namely, dissolution in the aquifer of magmatic and/or crustal gases, leaching of volcanic and/or sedimentary rocks, mixing with saline brines of natural or anthropogenic origin.

Based on geological and geophysical data, a hydrogeological cross section of the studied area is drawn, accounting for the variable chemical composition of the aquifers (Fig. 8). This schematic figure shows that Etna volcanics are underlain by a thick sedimentary sequence, mainly composed of a deep fractured carbonate or quartzarenite basement and of an impermeable clayey cover. In the studied area, the rigid substrate forms a structural high, as pointed out by the existence of a positive gravimetric anomaly (Loddo et al., 1989). This dome-shaped structure allows for the formation of a hydrothermal reservoir, fed by magmatic (CO_2, He) and crustal (CH_4) gases (Chiodini et al., 1996). This gas phase rises from the reservoir along tectonic discontinuities that cut the impermeable strata, and represents a continuous supply of volatiles to the volcanic aquifer. This is reflected by the acid reducing conditions acquired by the originally meteoric groundwaters along their flow path. Such conditions in turn enhance the aggressiveness of groundwater towards the host rocks that cause dissolution of rock-forming minerals (olivine, pyroxene and plagioclase). The distribution pattern displayed by group 1 and 3 waters (Fig. 4 a-c) highlights the location and extent of the area mostly affected by deep gas supply (Belpasso-Patern sector).

Upon flowing further toward south-southwest, groundwaters are forced to uprise and eventually discharge at the surface, as suggested by the occurrence of several springs exsolving a free CO_2-rich gas phase near Patern village (Fig. 8). This process is likely due to the reduced thickness of the aquifer and to the permeability contrast between fractured lavas and fluvial terraced deposits. From there onward, the main chemical processes affecting groundwater composition are the following: a) further carbon dioxide degassing and calcite precipitation, leading to the formation of travertine deposits (Fig. 8); b) mixing with salinelle thermal brines (less than 2%), giving rise to the formation of group 2 waters; c) interaction with evaporite deposits present within the sedimentary sequence. Both processes b) and c) lead to water compositions characterised by enrichments in Na, Cl, SO_4 and Ca with respect to pure volcanic waters .

Acknowledgements — We whish to thank S. Bellomo and M. Longo for technical assistance in the field and in the laboratory. Careful and quick review by L. Marini helped to improve the quality of this paper, and the precious editing work by I. Stober and K. Bucher was greatly appreciated. Work supported by Italian National Group for Volcanology (G.N.V.) and IGF-CNR, Italy.

6. References

Aiuppa, A., Allard, P., D'Alessandro, W., Michel, A., Parello, F., Treuil, M. and Valenza, M., Mobility and fluxes of major, minor and trace metals during basalt weathering and groundwater transport at Mt. Etna volcano (Sicily). Geochim. Cosmochim. Acta, 64 (11), 2000, 1827-1841.

Allard, P., Carbonnelle, J., Dajlevic, D., Le Bronec, J., Morel, P., Robe, M. C., Maurenas, J.M., Faivre-Pierret, R., Martin, D., Sabroux, J.C. and Zettwoog, P., Eruptive and diffuse emissions of CO_2 from Mount Etna. Nature 351, 1991, 387-391.

Allard, P., Jean-Baptist,e P., D'Alessandro, W., Parello, F., Parisi, B. and Flehoc, C., Mantle-derived helium and carbon in groundwaters and gases of Mount Etna, Italy. Earth Planet. Sci. Letters 148, 1997, 501-516.

Anz , S., Dongarr , G., Giammanco, S., Gottini, V. Hauser, S. and Valenza, M., Geochimica dei fluidi dell'Etna. Miner. Petrogr. Acta 32, 1989, 231-251.

Aureli, A., Idrogeologia del fianco occidentale etneo. Proc. 2nd Int.l Congress on Underground Waters, Palermo, Italy, 1973, 425-487.

Brusca, L., Aiuppa, A., D Alessandro, W., Parello, F., Allard, P. and Michel, A., Geochemical mapping of magmatic gas-water-rock interactions in the aquifer of Mount Etna volcano. J. Volcanol. Geotherm. Res., special issue Magma degassing through volcanoes (in press).

Capasso, G. and Inguaggiato, S., A simple method for the determination of dissolved gases in natural waters. An application to thermal waters from Vulcano Island. Appl. Geochem. 13, 1998, 631-642

Cer n, J.C., Jim nez-Espinosa, R. and Pulido-Bosch, A., Numerical analysis of hydrogeochemical data: a case study (Alto Guadalent n, southeast Spain). Appl. Geochem. 15, 2000, 1053-1067.

Chiodini, G., D'Alessandro, W. and Parello, F., Geochemistry of the gases and of the waters discharged by the mud volcanoes of Patern , Mt.Etna (Italy). Bull. Volcanol. 58, 1996, 51-58.

Condomines, M. and Tanguy, J.C., Age de l Etna determine par la m thode du d s quilibre radioactif ^{230}Th/^{238}U. C. R. Acad. Sci., Paris, 282D, 1976, 1661-1664.

Continisio, R., Ferrucci, F., Gaudiosi, G., Lo Bascio, D. and Ventura, G., Malta escarpment and Mt. Etna: early stages of an asymmetric rifting process? Evidences from geophysical and geological data, Acta Vulcanol., 9 (1/2), 1997, 45-53.

D'Alessandro, W., Giammanco, S., Parello, F. and Valenza, M., CO_2 output and $\delta^{13}C(CO_2)$ from Mount Etna as indicators of degassing of shallow asthenosphere. Bull. Volcanol. 58, 1997a, 455-458.

D'Alessandro, W., De Gregorio, S., Dongarr , G., Gurrieri, S., Parello, F. and Parisi, B., Chemical and isotopic characterization of the gases of Mount Etna (Italy). J. Volcanol. Geotherm. Res. 78, 1997b, 65-76.

D Alessandro, W., Inguaggiato, S., Federico, C. and Parello, F., Chemical composition of dissolved gases in groundwaters from Mt. Etna, Eastern Sicily. Proceedings of the 5th Int. Symp. Of the Earth s Surface, Reykjavik, Iceland, 16-20 Aug. 1999, Balkema, 1999, 491-494.

Douglas, G.R., Manganese-rich rock coatings from Iceland. Earth Surface and Landforms 12, 1987, 301-310.

Ferrara, V., Idrogeologia del versante orientale dell'Etna. Proc. 3^{d} Intl. Congress on Underground Waters, Palermo, Italy, 1975, 91-144.

Garrels, R.M. and Mackenzie, F.T., Origin of the chemical compositions of some springs and lakes. In: Equilibrium concepts in natural water systems: Advances in chemical Series, 67, Am. Chem. Soc., Washington, 1967, 222-242.

Giammanco, S., Ottaviani, M., Valenza, M., Veschetti, E., Principio, E., Giammanco, G. and Pignato, S., Major and trace elements geochemistry in the ground waters of a volcanic area: Mount Etna (Sicily). Wat. Res. 32, 1998, 19-30.

Kristmanndottir, H., Alteration of basaltic rocks by hydrothermal activity at 100-300¡C. In International Clay Conference 1978 (eds M. Mortland and V.C. Farmer), 358-367, Elsevier, Amsterdam, 1978.

Kristmanndottir, H., Alteration in the IRDP drill hole compared with other drill holes in Iceland. J. Geophy. Res. 87, 1982, 6525-6531.

Lentini, F., The geology of the Mt. Etna basement. Mem. Soc. Geol. It. 23, 1982, 7-25.

Loddo, M., Patella, D., Quarto, R., Ruina, G., Tramacere, A. and Zito, G., Application of gravity and deep dipole geoelectrics in the volcanic area of Mt. Etna (Sicily). J. Volcanol. Geotherm. Res. 39, 1989, 17-39.

Longo, M., Studio chimico ed isotopico delle precipitazioni nell area etnea. unpublished degree thesis, Science Faculty, Palermo University.

Nicholson, K., Geothermal fluids. Springer-Verlag, Berlin, 1993.

M trich, N. and Clocchiatti, R., Melt inclusion investigation of the volatile behaviour in historic alkaline magmas of Etna. Bull. Volcanol. 51, 1989, 185-198.

M trich, N., Clocchiatti, R., Mosbah, M. and Chaussidon, M., The 1989-1990 activity of Etna magma mingling and ascent of H_2O-Cl-S-rich basaltic magma. Evidence from melt inclusions. J. Volcanol. Geotherm. Res. 59, 1993, 131-144.

Ogniben, L., Lineamenti idrogeologici dell'Etna. Rivista Mineraria Siciliana 100-102, 1966, 1-24.

Pennisi, M., Leeman, W.P., Tonarini, S., Pennisi, A. and Nabelek, P., Boron, Sr, O, and H isotope geochemistry of groundwaters from Mt. Etna (Sicily) - hydrologic implications, Geochim. Cosmochim. Acta 64/6, 2000, 961-974

Plummer, L.N. and Back, W., The mass balance approach: Applications to interpreting the chemical evolution of hydrologic systems. Am. J. Sci. 280, 1980, 130-142.

Plummer, L.N., Parkhurst, D.L. and Thorstenson, D.C., Development of reaction models for groundwater systems. Geochim. Cosmochim. Acta 47, 1983, 665-685.

Plummer, L.N., Busby, J.F., Lee, R.W. and Hanshaw, B.B., Geochemical modelling of the Madison aquifer in parts of Montana, Wyoming ans South Dakota. Water Resource Res. 26, 1990, 1981-2014.

Plummer, L.N., Prestemon, E.C. and Parkhurst, D.L., An interactive code (NETPATH) for modelling NET geochemical reactions along a flow PATH, version 2.0. U.S.G.S Water-Resources Investigations Report 94-4169, 1994.

Romano, R., Succession of the volcanic activity in the Etnean area. Boll. Soc. Geol. It. 23, 1982, 27-48.

Schilir , F., Proposta metodologica per una zonazione geologico-tecnica del centro abitato di Maletto. Geologia Tecnica 3/88, 1988, 32-53.

Swan, A. R. H. and Sandilands, M., Introduction to geological data analysis. Blackwell Science, Oxford, 1995.

Tanguy, J.C., Condomines, M. and Kieffer, G., Evolution of the Mount Etna magma: Constraints on the present feeding system and eruptive mechanism. J. Volcanol. Geotherm. Res. 75, 1997, 221-250.

MODELING CHEMICAL BRINE-ROCK INTERACTION IN GEOTHERMAL RESERVOIRS

M. KÜHN[1], J. BARTELS[2], H. PAPE[2], W. SCHNEIDER[1] & C. CLAUSER[2]

[1] Technische Universität Hamburg-Harburg, Wasserwirtschaft & Wasserversorgung, Schwarzenbergstr. 95, D-21073 Hamburg, Germany

[2] University of Technology Aachen (RWTH), Applied Geophysics, Lochnerstr. 4-20, D-52056 Aachen, Germany

Abstract

Application of the Debye-Hückel theory for chemical reaction modeling of geothermal brines does not yield sufficiently accurate results. Thus, for the development of a new chemical reaction module for the numerical simulation model SHEMAT (Clauser and Villinger, 1990), the Pitzer formalism (Pitzer, 1991) is used to calculate aqueous speciation and mineral solubilities. It is based on an extended code of PHRQPITZ (Plummer et al., 1988). Using temperature dependent Pitzer coefficients, the system Na-K-Mg-Ca-Ba-Sr-Si-H-Cl-SO_4-OH-(HCO_3-CO_3-CO_2)-H_2O can be modeled with sufficient accuracy for temperatures from 0° to 150°C. The incorporated carbonic acid system (set in parentheses in the list above) is valid for temperatures from 0 to 90°C, only. Flow, heat transfer, species transport, and geochemical reactions are mutually coupled for modeling reactive flow. Changes in porosity and permeability influence the flow and transport properties of the reservoir. These changes are taken into account by a relation derived from a fractal model of the pore space structure (Pape et al., 1999).

A conceptual case study of the injection behavior of a geothermal installation focuses on the immediate vicinity of the well. The injection of cold water has a great influence firstly on the hydraulic conductivity of the aquifer indicated by continuous head pressure increase at the well and secondly on the equilibria between the minerals of the formation and the geothermal fluid. Reservoir changes are studied for the two cases of temperature dependence of solubility, prograde (i.e. barite) and retrograde (i.e. anhydrite). Dissolution of anhydrite induced by cooling down increases the permeability of the formation in a growing region around the borehole and precipitation at the temperature front decreases it. During the initial period of reinjection considered in this study, the negative effect on the injectivity by the colder water is partially compensated by the dissolution of anhydrite. Precipitation of barite around the borehole does not alter the permeability of the formation significantly because the volume of relocated mineral is too small.

I. Stober and K. Bucher (eds.), Water-Rock Interaction, 147–169.
© 2002 Kluwer Academic Publishers.

1. Introduction

In north Germany, low enthalpy geothermal heat is mainly produced from deep sandstone aquifers, usually by doublet installations. The hot water is tapped by means of a production drilling from a suitable aquifer and cooled down by heat exchangers to provide the secondary district heating water cycle. In order to protect environment and to maintain the hydraulic pressure in the aquifer the cooled water is generally reinjected. This process permits to optimize the use of energy (Darwis et al., 1995). In their majority the used thermal waters are high saline brines (at least 100 g/L total dissolved solids) which often additionally contain toxic constituents like boron, arsenic or heavy metals (Weres, 1988). Therefore the cooled brines would not be discharged at the surface. Experience from the petroleum and geothermal energy industry indicated, that reinjection of the cooled, highly saline solutions into a sandstone aquifer often leads to a reduction of the permeability of the formation rock (Ungemach, 1983). In this case the reinjection pressure has to be increased in order to maintain the injection rate. Since the reinjection pressure cannot be increased arbitrarily, the productivity of the heat mining system is reduced on the long run.

This permeability reduction can be attributed to two principal mechanisms. On the one hand pore throats can be plugged by particles either introduced into the formation or eroded by the injected water (Vernoux and Ochi, 1994; Kühn et al., 1998). On the other hand the hydraulic path can be restricted by precipitation of minerals due to mixing of injected water and formation water or temperature effects. The precipitation products of some minerals are colloids, e.g. iron hydroxides (Potter et al., 1981), silica (Brown et al., 1995), and heavy metal sulfides (Akaku, 1990), other minerals form well shaped crystals, e.g. anhydrite (Akaku, 1990), calcite (Sanyal et al., 1985), and barite (Vinchon et al., 1993).

Under in situ conditions, geothermal reservoir fluids are in equilibrium with co-existing mineral phases or close to saturation. In the north German sedimentary basin, this is the case for earth alkaline sulfates and carbonates. When cool water is reinjected into the hot reservoir the equilibrium between fluid and reactive minerals is disturbed. This may cause temperature induced dissolution or precipitation. Predicting the permeability changes caused by chemical reactions in these aquifers over the entire period of operation one requires understanding of the complex interactions of the involved processes, namely flow, heat transfer, transport of dissolved species, and chemical reactions. This can be achieved by numerical simulation only. Therefore, both a new chemical reaction model based on the Pitzer formalism (Pitzer, 1991) and a new relationship between porosity and permeability (Pape at al., 1999) were incorporated into the numerical code SHEMAT for simulating coupled flow, heat and mass transport in porous media (Clauser and Villinger, 1990).

This numerical simulation tool was applied for two case studies investigating the change of formation properties around an injection well of a geothermal doublet installation due to possible dissolution and precipitation of the mineral phases anhydrite and barite. In the investigated range of temperature, anhydrite and barite are minerals with retrograde and prograde solubility respectively, which means that the solubility of anhydrite increases with temperature whereas the opposite is true for barite. In some of the hydrothermal reservoirs in northern Germany diagenetic anhydrite cementation caused a substantial reduction of permeability compared with uncemented sandstone of the same formation. Kühn *et al.* (1997) showed, that precipitation of barite during the brine reinjection is possible from the thermodynamic point of view with reaction kinetics as the only restricting factor.

2. Numerical model

2.1 CHEMICAL MODULE INCORPORATING PITZER'S EQUATIONS

An ideal model for the calculation of chemical reactions in geochemical or engineering numerical codes would have the following properties: consistency with the laws of thermodynamics; compact mathematical form; high accuracy over wide ranges of temperature, pressure, and concentration; applicability to systems including most of the elements of the periodic table. No model currently exists which satisfies all of these requirements.

For calculating species activities in solutions of higher ionic strength an ion interaction model was developed by Kenneth Pitzer and his associates in the 1970s (Pitzer, 1973, 1975; Pitzer and Mayorga, 1973, 1974; Pitzer and Kim, 1974), because the Debye-Hückel theory does not yield sufficiently accurate results. Due to its internal consistency and the possible accuracy of the calculations the Pitzer equations currently represent the most promising approach for calculating mineral solubilities in brines (Wolery and Jackson, 1990).

The numerical model SHEMAT contains a chemical reaction model which is a modification of the geochemical simulation code PHRQPITZ (Plummer *et al.*, 1988). The PHRQPITZ model implements on the equations and Pitzer interaction parameters presented by Harvie and Weare (1980) and Harvie *et al.* (1984). The code comes with a partially validated data base, which is valid for 25°C and includes the system Na-K-Mg-Ca-H-Cl-SO$_4$-OH-HCO$_3$-CO$_3$-CO$_2$-H$_2$O.

For geothermal applications a chemical module for calculating thermodynamic equilibria needs to be valid for temperatures far beyond 25°C. In the specific case of north Germany the temperature range of 0 to 150°C is of interest. An extensive literature study provided data for varying elements and temperatures between –60 and 350°C, mainly 25 to 250°C (Table 1). With this data the PHRQPITZ code is extended so that temperature dependent Pitzer coefficients can be calculated.

In order to describe the temperature dependence of the Pitzer ion interaction parameters, equation (1) is widely used in the literature quoted above. Temperature dependent equations for the various model parameters were derived from a fit of the isothermal parameter values to the temperature.

$$P(T)=a_1 +a_2 T +\frac{a_3}{T}+a_4 \ln(T)+\frac{a_5}{(T-263)}+a_6 T^2 +\frac{a_7}{(680-T)}+\frac{a_8}{(T-227)} \quad (1)$$

Tab. 1:   Pitzer interaction parameters for elevated temperatures. The table denotes the authors and the covered systems.

Authors	Ref.	System
Greenberg & Moller (1989)	1	Na-K-Ca-Cl-SO$_4$ (0 to 250°C)
Spencer et al. (1990)	2	Mg-Cl-SO$_4$ (-60 to <25°C)
Pabalan & Pitzer (1987a)	3	Mg-Cl-SO$_4$ (>25 to 250°C)
Pabalan & Pitzer (1987b)	4	Na-OH (0 to 350°C)
Holmes et al. (1987)	5	H-Cl (to 350°C)
Holmes & Mesmer (1992)	6	H-HSO$_4$-SO$_4$ (25 to 200°C)
Harvie et al. (1984)	7	HCO$_3$-CO$_3$-SO$_4$/Na/K/Mg/Ca/H (25°C)
He & Morse (1993)	8	Na-K-Ca-Mg-Cl-SO$_4$-H$_2$CO$_3$ (0-90°C)
Azaroural et al. (1997)	9	Na-K-Ca-Mg-Cl-SO$_4$-HCO$_3$-SiO$_2$-H$_2$O (<250°)
Monnin C. (1999)	10	Na-K-Ca-Mg-Ba-Sr-Cl-SO$_4$-H$_2$O (25-200°C)

Here, $P(T)$ denotes a Pitzer ion interaction parameter, and a_1 to a_8 are the temperature coefficients, while T is the absolute temperature. In PHRQPITZ, those terms representing critical and supercooling temperature regions outside our range of temperature are dropped (Plummer et al., 1988). Equation (2) shows the resulting fit for the temperature dependence $P(T)$ in terms of the temperature coefficients c_1 to c_5. T_R is the reference temperature of 298.15°K. Coefficient c_1 is the value of the Pitzer ion interaction parameter $P(T)$ at T_R.

$$P(T) = c_1 + c_2(T - T_R) + c_3\left(\frac{1}{T} - \frac{1}{T_R}\right) + c_4 \ln\left(\frac{T}{T_R}\right) + c_5\left(T^2 - T_R^2\right) \quad (2)$$

In Table 2 the Pitzer interaction parameters and their coefficients of temperature dependence (c_1 to c_5) are listed as they are implemented in the chemical module of SHEMAT. The differences between the temperature dependent Pitzer interaction parameters given in the literature by equation 1 and the ones refitted by equation 2 are stated with the value σ defined in equation 3.

$$\sigma = \sqrt{\sum_{i=1}^{n} \frac{\left(P(i)_{literature} - P(i)_{refit}\right)^2}{n}} \quad (3)$$

The temperature dependent functions of the ion interaction parameters (equation 1 and 2) coincide very well even though the refit was performed by a reduced number of temperature coefficients. Table 3 contains additional Pitzer interaction parameters which are implemented without temperature dependence, according to He and Morse (1993).

Tab. 2: The Pitzer ion interaction parameters and the coefficients of the temperature dependence (c_1 to c_5) as adapted from literature values (compare column "Ref." with Table 1 for original source) and applied in the numerical code. The value c_1 contains the interaction parameter at the reference temperature of 25°C. Column "σ" depicts the difference between the literature values based on Equation 1 and their refit to Equation 2 (σ is defined in Equation 3).

Parameters	Ref.	c_1 [25°C]	c_2	c_3	c_4	c_5	σ
$\beta^{(0)}$ Na-Cl	1	7.53591E-02	-2.36974E+03	-1.79464E+01	4.67000E-02	-2.08219E-05	8.02E-05
$\beta^{(0)}$ K-Cl	1	4.80800E-02	-7.58485E+02	-4.70624E+00	1.00721E-02	-3.75994E-06	3.94E-10
$\beta^{(0)}$ Mg-Cl	3	3.51088E-01	2.38419E-06	4.65661E-09	-9.31654E-04	5.93915E-07	7.12E-10
$\beta^{(0)}$ Ca-Cl	1	3.05320E-01	-2.17620E+04	-1.85146E+02	5.23401E-01	-2.45985E-04	2.51E-03
$\beta^{(0)}$ Sr-Cl	10	0.283440975	-512.2149976	2.07801E-08	-0.01095575	9.44275E-06	4.27E-12
$\beta^{(0)}$ Ba-Cl	10	0.290748207	-1336.530013	-5.302131123	0.0006375	4.60872E-06	4.76E-10
$\beta^{(0)}$ Na-SO$_4$	1	1.86933E-02	-1.97239E+04	-1.60348E+02	4.37051E-01	-1.98804E-04	1.82E-03
$\beta^{(0)}$ K-SO$_4$	1	5.55358E-02	-1.41843E+03	-6.74729E+00	8.26907E-03	-2.52243E-13	5.70E-10
$\beta^{(0)}$ Mg-SO$_4$	3	2.15092E-01	-5.46923E+03	-4.22887E+01	1.06604E-01	-4.29019E-05	2.95E-04
$\beta^{(0)}$ Na-OH	4	9.19001E-02	6.59191E+03	6.14228E+01	-1.86361E-01	9.20267E-05	3.05E-03
$\beta^{(0)}$ Na-HCO$_3$	8	2.80021E-02	6.82886E+02	6.89959E+00	-1.44593E-02	2.64233E-13	2.64E-11
$\beta^{(0)}$ K-HCO$_3$	8	-1.07023E-02	-7.02620E-04	-4.70504E-06	1.00000E-03	-1.06581E-14	8.61E-10
$\beta^{(0)}$ Mg-HCO$_3$	8	-9.31256E-03	-2.73406E+05	-2.60712E+03	8.25084E+00	-4.34000E-03	1.49E-08
$\beta^{(0)}$ Ca-HCO$_3$	8	1.82545E-01	-5.76521E+05	-5.66112E+03	1.84473E+01	-9.98900E-03	3.66E-08
$\beta^{(0)}$ Na-CO$_3$	8	3.62048E-02	1.10838E+03	1.11986E+01	-2.33017E-02	4.27658E-13	9.69E-11
$\beta^{(0)}$ K-CO$_3$	8	1.28800E-01	1.40667E-05	3.72529E-09	1.10000E-03	-7.10543E-15	2.80E-09
$\beta^{(1)}$ Na-Cl	1	2.77031E-01	-4.80698E+03	-3.92387E+01	1.06571E-01	-4.71391E-05	1.97E-04
$\beta^{(1)}$ K-Cl	1	2.18025E-01	-6.59453E+03	-5.39339E+01	1.47670E-01	-6.61920E-05	2.76E-04
$\beta^{(1)}$ Mg-Cl	3	1.65119E+00	-2.28882E-05	1.49012E-08	-1.09438E-02	2.60169E-05	5.08E-09
$\beta^{(1)}$ Ca-Cl	1	1.70813E+00	-1.71661E-05	2.98023E-08	-1.54170E-02	3.17910E-05	4.95E-09
$\beta^{(1)}$ Sr-Cl	10	1.625584216	-1064.593008	6.70552E-08	-0.0336065	4.24357E-05	7.42E-10
$\beta^{(1)}$ Ba-Cl	10	1.250020317	4374.11002	15.87517039	0.003224999	-6.77403E-06	1.32E-08
$\beta^{(1)}$ H-Cl	5	0.2945	0	0	1.419E-04	0	0.0
$\beta^{(1)}$ Na-SO$_4$	1	1.09940E+00	-1.52214E+04	-1.12685E+02	2.87135E-01	-1.24691E-04	8.38E-04
$\beta^{(1)}$ K-SO$_4$	1	7.96385E-01	2.06713E+03	-5.96046E-08	2.35793E-02	-9.94760E-14	3.15E-09
$\beta^{(1)}$ Mg-SO$_4$	3	3.36625E+00	-5.77887E+03	3.42727E-07	-1.47980E-01	1.57607E-04	1.73E-08
$\beta^{(1)}$ Na-OH	4	2.53001E-01	-1.02942E+04	-8.59606E+01	2.39060E-01	-1.07959E-04	5.44E-09
$\beta^{(1)}$ Na-HCO$_3$	8	4.40052E-02	1.12939E+03	1.14109E+01	-2.44673E-02	4.38760E-13	1.79E-09
$\beta^{(1)}$ K-HCO$_3$	8	4.78002E-02	9.32455E-04	6.15790E-06	1.09999E-03	-1.77636E-14	3.76E-09
$\beta^{(1)}$ Mg-HCO$_3$	8	8.04741E-01	3.20321E+06	2.99272E+04	-9.27779E+01	4.77642E-02	3.78E-08
$\beta^{(1)}$ Ca-HCO$_3$	8	3.00039E-01	2.64922E+04	1.83132E+02	-3.72588E-01	8.96910E-05	2.52E-09
$\beta^{(1)}$ Na-CO$_3$	8	1.51207E+00	4.41251E+03	4.45821E+01	-9.98912E-02	1.73372E-12	1.14E-08
$\beta^{(1)}$ K-CO$_3$	8	1.43300E+00	1.17874E-03	5.96046E-08	4.36000E-03	2.84217E-14	7.20E-09

Tab. 2 (continued)

Parameters	Ref.	c_1 [25°C]	c_2	c_3	c_4	c_5	σ
$\beta^{(2)}$ Mg-SO$_4$	3	-3.27753E+01	-9.97546E+05	-1.10152E+04	4.17733E+01	-2.72804E-02	1.26E-02
$\beta^{(2)}$ Ca-SO$_4$	1	-1.00108E+01	-1.52588E-03	-9.53674E-07	4.00431E-01	1.81899E-12	7.45E-07
C^{ϕ} Na-Cl	1	1.40794E-03	3.51122E+02	2.74184E+00	-7.33724E-03	3.31819E-06	1.36E-05
C^{ϕ} K-Cl	1	-7.87999E-04	9.12712E+01	5.86450E-01	-1.29808E-03	4.95714E-07	3.46E-11
C^{ϕ} Mg-Cl	3	6.50689E-03	4.02331E-07	1.16415E-09	-2.49949E-04	2.41831E-07	1.39E-10
C^{ϕ} Ca-Cl	1	2.34218E-03	1.95526E+03	1.65554E+01	-4.68850E-02	2.20465E-05	2.39E-04
C^{ϕ} Sr-Cl	10	-0.000314996	4.512230143	0	4.54747E-13	1.11022E-16	4.72E-11
C^{ϕ} Ba-Cl	10	-0.030466886	1140.833151	8.093666674	-0.018660049	7.03389E-06	4.78E-06
C^{ϕ} Na-SO$_4$	1	6.29815E-03	-3.89497E+02	-5.66272E+00	2.12197E-02	-1.19512E-05	2.49E-04
C^{ϕ} Mg-SO$_4$	3	2.79193E-02	1.64736E+03	1.38756E+01	-3.90051E-02	1.78328E-05	3.33E-05
C^{ϕ} Na-OH	4	3.61013E-03	-3.58353E+02	-3.42983E+00	1.04541E-02	-5.16145E-06	1.76E-04
θ Na-K	1	-3.20349E-03	1.40213E+01	-2.32831E-10	1.36424E-12	-2.22045E-16	1.57E-10
λ Na-H$_2$CO$_3$	8	8.14744E-02	1.09399E+05	1.04702E+03	-3.32657E+00	1.75320E-03	8.17E-10
λ K-H$_2$CO$_3$	8	4.49422E-02	-5.59542E+04	-5.46074E+02	1.76701E+00	-9.48700E-04	2.90E-09
λ Mg-H$_2$CO$_3$	8	1.44733E-01	3.58947E+03	1.04345E+02	-5.41843E-01	3.88120E-04	7.27E-11
λ Ca-H$_2$CO$_3$	8	1.64379E-01	2.45542E+05	2.45251E+03	-8.10156E+00	4.42472E-03	9.12E-09
λ Cl-H$_2$CO$_3$	8	2.04804E-02	-3.31596E+04	-3.15828E+02	9.96433E-01	-5.21220E-04	1.03E-09
λ SO$_4$-H$_2$CO$_3$	8	1.39973E-01	-3.39278E+04	-4.57016E+02	1.82709E+00	-1.14272E-03	7.76E-09
λ Na-SiO$_2$	9	0.092500052	-961.1272755	-7.832730047	0.020632273	-9.11756E-06	1.86E-05
λ K-SiO$_2$	9	0.032239999	749.3664763	5.637908168	-0.014390482	6.13277E-06	7.95E-06
λ Mg-SiO$_2$	9	0.292481636	-2227.025299	-19.62053046	0.057037085	-2.82131E-05	1.60E-04
λ Ca-SiO$_2$	9	0.292481636	-2227.025299	-19.62053046	0.057037085	-2.82131E-05	1.60E-04
λ SO$_4$-SiO$_2$	9	-0.139627272	-2183.086038	-18.11408045	0.051102344	-2.4427E-05	6.81E-05
λ HCO$_3$-SiO$_2$	9	0.015998085	831.1168259	7.037341963	-0.0201729	9.6735E-06	3.52E-05
ξ H-Cl-H$_2$CO$_3$	8	-4.70519E-03	1.63344E+04	1.52384E+02	-4.70474E-01	2.40526E-04	9.39E-11
ξ Na-Cl-H$_2$CO$_3$	8	-5.71530E-04	6.87903E+03	7.37451E+01	-2.58005E-01	1.47823E-04	6.20E-11
ξ K-Cl-H$_2$CO$_3$	8	-1.20697E-02	6.85326E+03	7.37998E+01	-2.57891E-01	1.47333E-04	4.04E-10
ξ Ca-Cl-H$_2$CO$_3$	8	-1.41310E-02	5.25684E+03	2.73775E+01	-1.80020E-02	-2.47349E-05	6.18E-10
ξ Mg-Cl-H$_2$CO$_3$	8	-9.84695E-03	2.77268E+04	2.53623E+02	-7.72286E-01	3.91603E-04	4.87E-10
ξ Na-SO$_4$-H$_2$CO$_3$	8	-3.74535E-02	-1.39908E+06	-1.26303E+04	3.79305E+01	-1.89473E-02	1.36E-08
ξ K-SO$_4$-H$_2$CO$_3$	8	-3.57567E-04	3.07569E+04	6.11376E+02	-2.86076E+00	1.95109E-03	4.91E-09
ξ Mg-SO$_4$-H$_2$CO$_3$	8	-4.15864E-02	1.43163E+05	1.41230E+03	-4.60833E+00	2.48921E-03	5.15E-09

Thus a data base of Pitzer interaction parameters is provided for the system Na-K-Mg-Ca-Ba-Sr-Si-H-Cl-SO$_4$-OH-(HCO$_3$-CO$_3$-CO$_2$)-H$_2$O which is valid for temperatures from 0 to 150°C. The Pitzer treatment is largely based on data of Greenberg and Moller (1989). Data for the incorporated carbonic acid system (in parentheses in the list above) are valid for temperatures from 0 to 90°C only, according to He and Morse (1993). The data base is extended with Pitzer interaction parameters for the elements Ba and Sr as well as Si published by Monnin (1999) and Azaroual et al. (1997), respectively.

Tab. 3:   Pitzer interaction parameters applied in the numerical code without temperature dependence.

Parameters	Ref.	Value	Parameters	Ref.	Value	Parameters	Ref.	Value
$\beta^{(0)}$ MgOH-Cl	7	-0.1	$\beta^{(0)}$ H-Cl	7	0.1775	$\beta^{(0)}$ Ca-SO$_4$	1	0.015
$\beta^{(0)}$ H-SO$_4$	7	0.0298	$\beta^{(0)}$ Na-HSO$_4$	7	0.0454	$\beta^{(0)}$ K-HSO$_4$	7	-0.0003
$\beta^{(0)}$ Mg-HSO$_4$	7	0.4746	$\beta^{(0)}$ Ca-HSO$_4$	7	0.2145	$\beta^{(0)}$ H-HSO$_4$	7	0.2065
$\beta^{(0)}$ K-OH	7	0.1298	$\beta^{(0)}$ Ca-OH	7	-0.1747	$\beta^{(1)}$ MgOH-Cl	7	1.658
$\beta^{(1)}$ Ca-SO$_4$	1	3.0	$\beta^{(1)}$ Na-HSO$_4$	7	0.398	$\beta^{(1)}$ K-HSO$_4$	7	0.1735
$\beta^{(1)}$ Mg-HSO$_4$	7	1.729	$\beta^{(1)}$ Ca-HSO$_4$	7	2.53	$\beta^{(1)}$ H-HSO$_4$	7	0.5556
$\beta^{(1)}$ K-OH	7	0.32	$\beta^{(1)}$ Ca-OH	7	-0.2303	$\beta^{(2)}$ Ca-OH	7	-5.72
C^{ϕ} H-Cl	7	0.0008	C^{ϕ} H-SO$_4$	7	0.0438	C^{ϕ} K-OH	7	0.0041
C^{ϕ} K-HCO$_3$	7	-0.008	C^{ϕ} Na-CO$_3$	8	0.0052	C^{ϕ} K-CO$_3$	8	0.0005
θ Mg-Na	3	0.07	θ Ca-Na	1	0.05	θ Sr-Na	10	0.051
θ H-Na	7	0.036	θ Ca-K	1	0.1156	θ H-K	7	0.005
θ Ca-Mg	7	0.007	θ H-Mg	7	0.1	θ H-Ca	7	0.092
θ SO$_4$-Cl	1	0.07	θ HSO$_4$-Cl	7	-0.006	θ OH-Cl	3	-0.05
θ HCO$_3$-Cl	7	0.03	θ CO$_3$-Cl	7	-0.02	θ OH-SO$_4$	3	-0.013
θ HCO$_3$-SO$_4$	7	0.01	θ CO$_3$-SO$_4$	7	0.02	θ CO$_3$-OH	7	0.1
θ CO$_3$-HCO$_3$	7	-0.04	λ HSO$_4$-H$_2$CO$_3$	7	-0.003	ψ Na-K-Cl	1	-0.0037
ψ Na-K-SO$_4$	1	0.0073	ψ Na-K-HCO$_3$	7	-0.003	ψ Na-K-CO$_3$	7	0.003
ψ Na-Ca-Cl	1	-0.003	ψ Na-Ca-SO$_4$	1	-0.012	ψ Na-Mg-Cl	3/7	-0.0149
ψ Na-Sr-Cl	10	-0.0021	ψ Na-Ba-Cl	10	0.0128	ψ Na-Mg-SO$_4$	7	-0.015
ψ Na-H-Cl	7	-0.004	ψ Na-H-HSO$_4$	7	-0.0129	ψ K-Ca-Cl	1	-0.0432
ψ K-Mg-Cl	3/7	-0.0264	ψ K-Mg-SO$_4$	7	-0.048	ψ K-H-Cl	7	-0.011
ψ K-H-SO$_4$	7	0.197	ψ K-H-HSO$_4$	7	-0.0265	ψ Ca-Mg-Cl	7	-0.012
ψ Ca-Mg-SO$_4$	7	0.024	ψ Ca-H-Cl	7	-0.015	ψ Mg-MgOH-Cl	7	0.028
ψ Mg-H-Cl	7	-0.011	ψ Mg-H-SO$_4$	7	-0.0178	ψ Cl-SO$_4$-Na	1	-0.0091
ψ Cl-SO$_4$-K	1	-0.0016	ψ Cl-SO$_4$-Ca	1	-0.018	ψ Cl-SO$_4$-Mg	3	-0.008
ψ Cl-HSO$_4$-Na	7	-0.006	ψ Cl-HSO$_4$-H	7	0.013	ψ Cl-OH-Na	3/7	-0.0091
ψ Cl-OH-K	7	-0.006	ψ Cl-OH-Ca	7	-0.025	ψ Cl-HCO$_3$-Na	7	-0.015
ψ Cl-HCO$_3$-Mg	7	-0.096	ψ Cl-CO$_3$-Na	7	0.0085	ψ Cl-CO$_3$-K	7	0.004
ψ SO$_4$-HSO$_4$-Na	7	-0.0094	ψ SO$_4$-HSO$_4$-K	7	-0.0677	ψ SO$_4$-HSO$_4$-Mg	7	-0.0425
ψ SO$_4$-OH-Na	3/7	-0.0126	ψ SO$_4$-OH-K	7	-0.05	ψ SO$_4$-HCO$_3$-Na	7	-0.005
ψ SO$_4$-HCO$_3$-Mg	7	-0161	ψ SO$_4$-CO$_3$-Na	7	-0.005	ψ SO$_4$-CO$_3$-K	7	-0.009
ψ OH-CO$_3$-Na	7	-0.017	ψ OH-CO$_3$-K	7	-0.01	ψ HCO$_3$-CO$_3$-Na	7	0.002
ψ HCO$_3$-CO$_3$-K	7	0.012						

Partial validation is given for the data base built this way (see below), but it has to be stated, that merging Pitzer interaction parameters from different sources can lead to inconsistent parameter sets and significant accuracy deterioration. As such the present model has to be verified by potential users calculating systems different from the ones specified below.

2.2 PERMEABILITY-POROSITY RELATIONSHIP

The relationship between permeability and porosity, implemented in the numerical code, is based on the assumption that the shape of the internal surface of rock pores follows a rule of self-similarity. Thus the theory of fractals can be applied. The fractal relationship between permeability k and porosity ϕ was expressed by Pape et al. (1999) as a general three-term power series of porosity:

$$k = A \phi + B \phi^{exp1} + C (10)^{exp2} , \qquad (4)$$

where the exponents exp1 and exp2 depend on the fractal dimension of the internal surface of the pore space. The coefficients A, B, and C need to be calibrated for each type of sedimentary basin or pore space modification, i.e. porosity change, due to chemical reactions. In average sandstone Pape *et al.* (1999) found $k = 31 \phi + 7463 \phi^2 +$ 191 $(10)^{10}$. Equation (4) reflects the fact that in different porosity regions different processes are responsible for porosity and permeability changes. In SHEMAT this is approximated by equation (5) and the possibility for the modeler to prescribe different exponents for three porosity intervals. k_0 and ϕ_0 denote the initial values containing the same information as the coefficients in equation (4):

$$k = k_0 (\phi / \phi_0)^{exp*} . \qquad (5)$$

If the simulated processes change the geometry of the pore space in the same way as diagenesis, then exp* gets the values of 1, exp1, and exp2 in the intervals $\phi < 0.01$, $0.01 < \phi < 0.1$, and $0.1 < \phi$, respectively. However in technical operations, precipitation of small crystals and also dissolution which generates rough surfaces and preferential pathways create a large fractal dimension. Then exp* in equation (5) is independent of the exponents exp1 and exp2 in equation (4) which nevertheless describes the initial state.

2.3 NUMERICAL COUPLING OF THE MODULES

In the simulation code SHEMAT the partial differential equations for flow, heat transfer and species transport are solved numerically by a finite difference approximation. A pure up-wind scheme, the Il'in-flux-blending scheme (Clauser and Kiesner, 1987), and the diffusion corrected Smolarkiewicz method (Smolarkiewicz, 1983) are implemented for solving the equations for heat and species transport.

All processes involved, i.e. flow, heat transfer, species transport, and geochemical reactions, are coupled (i) via the dependence of the material and thermodynamic properties on temperature, pressure and species concentrations, and (ii) via terms containing these prognostic variables when solving simultaneously the flow, transport, and reaction equations in sequence. The first kind of coupling accounts for the dependence of

- fluid properties (viscosity, density, heat capacity, thermal conductivity, compressibility) on temperature;
- fluid density on species concentrations;
- rock thermal conductivity on temperature;

- the chemical reactions on temperature and species concentration;

and for

- changes in porosity due to the precipitation and dissolution of minerals;
- fractal relationship between permeability and porosity changes.

The second kind of coupling accounts for the fact, that chemical reactions act as sources or sinks for the transported dissolved species. In the modeling code this is approximated at each time step by calculating subsequently the transport of dissolved species, the change in their concentrations due to mineral reactions between the reservoir rock and the fluid, the concurrent porosity changes, and the resulting changes in permeability.

3. Mineral precipitation and dissolution in the vicinity of an injection well

3.1 CONCEPTUAL MODEL

Because well injectivity is most sensitive to changes in the formation near the injection well (± 1-10 m, Boisdet et al., 1989), SHEMAT was applied for such a conceptual study. The objective of the presented study was to concentrate on mineral reaction processes and resulting changes of formation properties near the well. In this special case, fluid was studied as a function of temperature as well as of chemical reactions of the minerals anhydrite and barite. Anhydrite was identified in reservoirs of Northern Germany as the reason for a substantial reduction of permeability, and barite as a mineral with a potential risk of precipitation during the reinjection process. Further, anhydrite and barite were studied because of their different behavior, namely - retrograde and prograde solubility, respectively. The investigated zone around the injection well comprises an area of 51 m * 51 m with a layer thickness of 30 m. This is discretised into an equidistant grid with 1 m an element. For (two-dimensional) simulations a model was set up corresponding to a "typical" north German aquifer, at depth of 1500 m, with the appropriate properties (Table 4). The catchment area of the injection well is assumed to be homogeneous and isotropic. The aquifer is assumed impermeable at the top and the bottom and to have constant hydraulic head at all lateral boundaries corresponding to an aquifer depth of 1500 m. The injection well is at the center of the model area. Injection was at the rate of 50 m³/h and at a temperature of 20°C.

In the first case considered, the sandstone formation is assumed to be cemented with 0.8 weight % of anhydrite. This amount corresponds to 8.16 g anhydrite per kg of bulk rock. This is used as an initial value, and a homogeneous distribution is assumed for over the entire simulation domain. A sensitivity analysis of the permeability-porosity relationship was performed for anhydrite by simulations carried out with different exponents exp* in equation (5) reflecting the fractal dimension of the pore space structure. From petrophysical reasons when using a simple model with smooth capillaries, an exponent of 3.0 has been frequently used in basin analysis (Ungerer, 1990). However, this model only fits for the rare type of clean sandstone with smooth equal sized grains represented by the quartz cemented Fontainebleau sandstone with porosity $\phi > 8$ % (Bourbie and Zinszner, 1985). An exponent 4.85 was derived from a petrophysical data set of a sandstone cemented by anhydrite, corresponding to the initial diagenetically state. In addition, an exponent of 12 was used which was determined for dissolution and subsequent precipitation of anhydrite during a core flooding experiment (Bartels *et al.*, in prep.), corresponding to the dislocation of anhydrite during the operations of the doublet installation. In the simulations, a dispersion length of 0.5 m is used for the transport, and the injection well is assumed to be fully penetrating. The injected brine is a solution of 1.7 mol/L sodium chloride with a pH of 7.

Tab. 4: Formation properties (isotropic)

Property	Value	Dimension
Anhydrite	0.8	[Weight %]
Barite	0.004	[Weight %]
Porosity	0.15	[-]
Permeability	$0.5*10^{-12}$	[m^2]
Thermal Capacity	0.7	[J/kgK]
Therm. Conductivity	2.5	[W/mK]
Temperature	100	[°C]
Inner Surface	1.2	[m^2/g]

In the second case studied, the active mineral phase is barite instead of anhydrite and the same boundary conditions and the same brine are used as before. The sandstone is assumed to be cemented with 0.004 weight % barite. This amount corresponds to $4.2*10^{-2}$ g barite per kg of bulk rock and is used as initial value. Beside sodium and chloride, barium and sulfate were also reinjected at a concentration of 0.24 mmol/L each. This represents an amount of dissolved barite which corresponds to the thermodynamic equilibrium in a sodium chloride solution of 1.7 mol/L at 100°C, the assumed formation temperature. The solubility product of barite and the minerals investigated in the following are listed in Table 5. There is a lack of kinetic coefficients for barite in the literature. Therefore, we chose for our - temperature independent - specific reaction rate a roughly averaged value of known data for carbonates of $1*10^{-5}$ mol/m^2/s (Nagy *et al.*, 1990). This allows at least to represent the role of reaction kinetics interacting with transport in an approximately correct order of magnitude. The internal surface of the sandstone was assumed to be 1.2 m^2/g (Table 4), a value corresponding to a typical formation sandstone.

Initial simulations showed that non reactive flow reached a steady state after an injection time of 10 hours in our model. Flow velocities vary between 7.27 m/day near the well and 0.33 m/day near the model boundaries. After 4 days the front of a non-reactive tracer reaches the model boundaries. Therefore, simulation time is limited to this value.

3.2 SPECIFICATION OF THE CHEMICAL MODEL

3.2.1 Temperature dependent mineral solubilities
Due to potentially insufficient data the chemical model and its input data set were studied to accuracy and possible limits for temperature and concentration of the mineral phases anhydrite and barite in sodium chloride solutions. A verification was performed comparing model calculations with available experimental data. First the ternary sodium-potassium-chloride system was checked because most of the groundwaters and surface waters are dominated by NaCl. In Figure 1 the solubilities of halite and sylvite are shown in the ternary system Na-K-Cl. The modeling results match the laboratory data of Cornec and Krombach (1932), Blasdale (1918), Ackoumov and Wassilijew (1932), and Bergmann and Vlassov (1949).

The chemical model is used not only to calculate solubilities in the binary or ternary system that make up the parameterization data base but also to predict solubilities in more complex systems. In Figure 2 the solubility of anhydrite is shown for varying sodium chloride concentrations (system $Na-Ca-Cl-SO_4-H_2O$). The modeling results for the Pitzer formalism are comparable with laboratory data published by Blount and Dickson (1969) and Bock (1961). For mineral barite the calculated solubilities agree with laboratory data too (Figure 3). The experimental data in Figure 3 are from Templeton (1960) and Blount (1977).

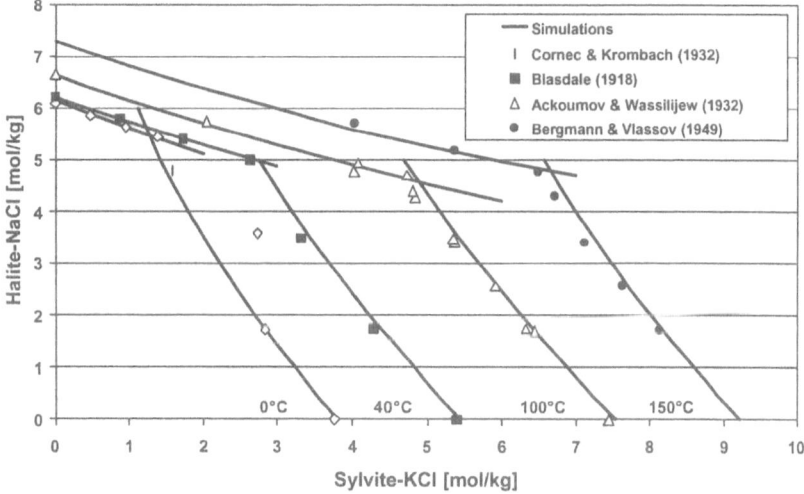

Fig. 1: Solubility of halite in potassium chloride and sylvite in sodium chloride solutions. Model calculations compared to experimental data for 0, 40, 100, and 150 °C. On the one hand the system is limited by the mineral halite (towards the top) and on the other by sylvite (towards the right side).

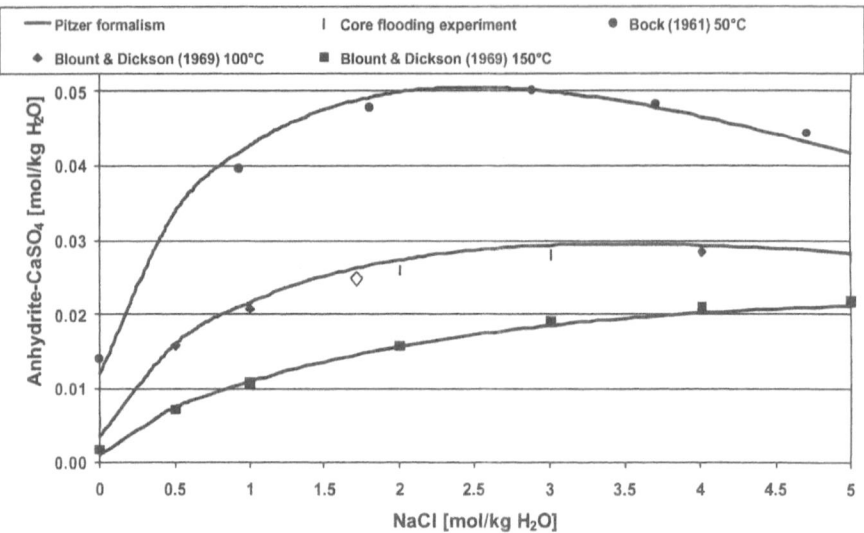

Fig. 2: Solubility of anhydrite as a function of sodium chloride concentration. The symbols
        are data of Blount and Dickson (1969) and Bock (1961) for 50, 100 and 150°C. The
        calculations with the Pitzer formalism coincide very well with the experimental
        data.

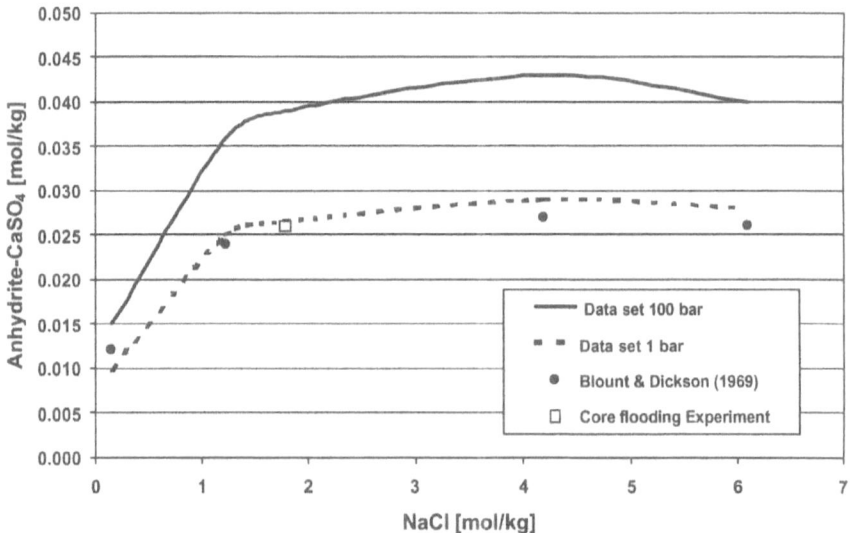

Fig. 3: Solubility of barite as a function of sodium chloride concentration. The calculated
        results are shown compared to laboratory results of Templeton (1960) and Blount
        (1977). The Pitzer formalism provides proper data.

Tab. 5: Solubility products

Mineral	Equation	log K [25°C]
Halite	$NaCl \leftrightarrow Na^+ + Cl^-$	1.582
Sylvite	$KCl \leftrightarrow K^+ + Cl^-$	0.911
Anhydrite	$CaSO_4 \leftrightarrow Ca^{2+} + SO_4^{2-}$	-4.326
Anhydrite (500 bar)	$CaSO_4 \leftrightarrow Ca^{2+} + SO_4^{2-}$	-4.079
Barite	$BaSO_4 \leftrightarrow Ba^{2+} + SO_4^{2-}$	-10.005

3.2.2 Pressure dependent mineral solubilities

Besides the influence of temperature that of pressure needs to be studied for chemical brine-rock interaction in geothermal reservoirs. For reactions between a solution and minerals significant changes of the solubility equilibria can be expected (Aggarwal *et al.*, 1990). The solubility of the minerals generally increases with pressure. Monnin and Ramboz (1996) showed, that the dependence on pressure in the range 500 - 1000 bar of a solution of anhydrite ($CaSO_4$) at 62°C is on the same order as the dependence on temperature. During heat mining in the geothermal installation the fluid undergoes a pressure change from several hundred bars within the formation to 1-10 bar at the surface. A supersaturation of mineral phases in the water can be caused by this pressure reduction. This can trigger precipitation of minerals during the operations in the surface installation. At the next step, when the water is reinjected, pressure increases to the formation pressure. Thus, the saturation indices decline, and also the potential risk of mineral precipitation.

In the literature there is a lack of pressure dependent Pitzer coefficients in general and of solubility experiments investigating the pressure range 1-250 bar in particular. Monnin (1990) quantified the variations of the activity coefficients of the aqueous solutes with respect to temperature and pressure in order to study their relevance. He tested a solution containing 3 moles of NaCl and 0.01 moles of $CaSO_4$ under a confining pressure of 300 bar and a temperature of 140°C (oil field conditions). This results show, that the variation of activity with pressure varies between 4 % and 14 %, depending on the salt. The change in the distribution of species is very small between 1 bar and 300 bar. Thus, Monnin concludes that the use of pressure dependent solubility constants is sufficient for this particular pressure range.

The solubility constants used for the numerical calculations were adapted using available anhydrite solubilities at 500 bar. Figure 4 shows the results for anhydrite solubilities as a function dependent on temperature at atmospheric pressure and at 500 bar. The Pitzer coefficients were used as above. The pressure has a great influence on the amount of dissolved anhydrite in the range 1-500 bar. In contrast the amount of anhydrite dissolved during a core flooding experiment under a confining pressure of 100 bar and a temperature of 100°C, as well as the laboratory results of Blount and Dickson (1969, 1973) coincide well with the numerical results for atmospheric pressure (see Figure 5). Due to the fact, that in a geothermal reservoir assumed to be in 1500 m depth the pressure will be about 150 bar, we decided to neglect the influence of pressure in the following studies.

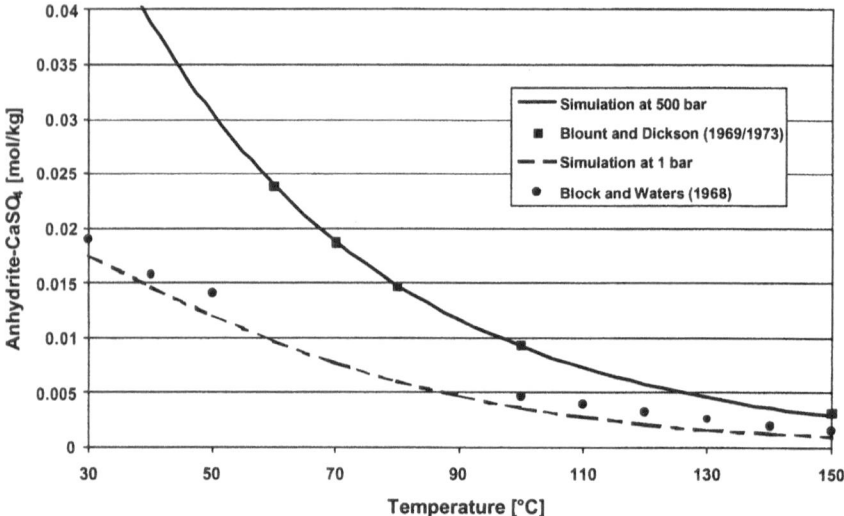

Fig. 4: Solubility of anhydrite as a function of temperature for atmospheric pressure and 500 bar. The pressure exerts a great influence on the solubility at 500 bar compared to 1 bar.

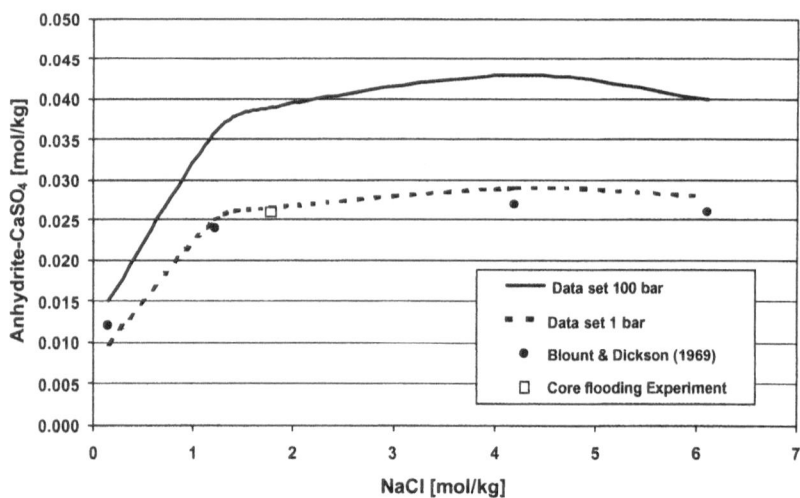

Fig. 5: Solubility of anhydrite determined with solubility constants adapted for confining pressures of 1 and 100 bar (symbols = experimental data of Blount and Dickson (1969) and a core flooding experiment, both performed at 100 bar).

3.3 RESULTS

3.3.1 Anhydrite

Figure 6 shows, that the temperature front which propagates from the injection to the production well, tails behind the tracer front. The temperature decreases from an initial value of 100°C in the formation to 20°C injection temperature over the first 2.5 days. In less than one day the tracer concentration increases from 0 to 50 mmol/L, the concentration at the injection well.

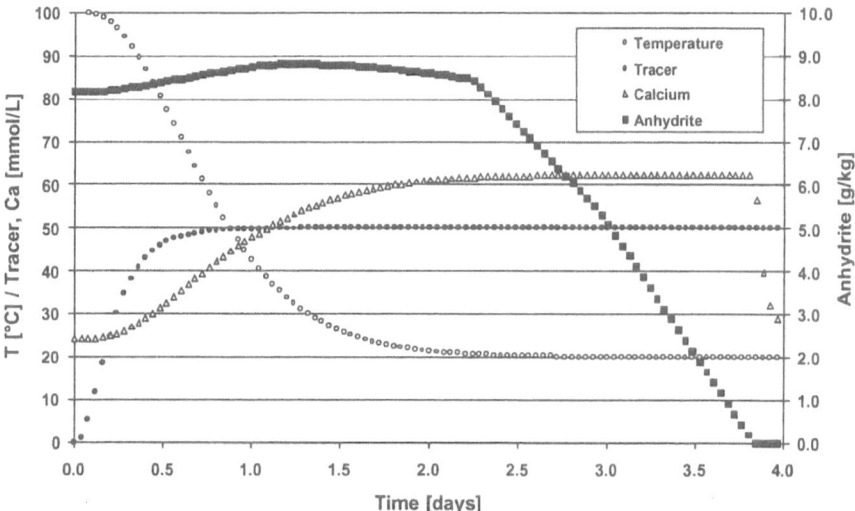

Fig. 6:   Temperature, tracer, calcium, and anhydrite concentration as a function of time at an observation point 4 m away from the injection well.

The calcium concentration of the water at an observation point 4 m away from the well increases from 25 to 63 mmol/L over the first 2.5 days. The amount of anhydrite at this point increases during this period, followed by a decrease until it reaches a value of zero after about 4 days. When this occurs, the calcium concentration drops sharply. Thus anhydrite is completely dissolved within a radius of 4 m around the well (the distance of the observation point from the well). Outside this radius anhydrite is still enriched compared to the initial value (Figure 7). Figure 8 shows the hydraulic head at the injection well as a function of time. It reflects the excess pressure necessary for reinjection. The hydraulic head rises from the initial value of 1500 m, depending on the applied fractal exponent (see equation 5), but does not reach steady state during the first 4 simulation days. Simulations were performed both for non-reactive flow and for reactive flow with exponents of 3, 4.85, and 12. The higher the fractal exponent the lower the resulting hydraulic head.

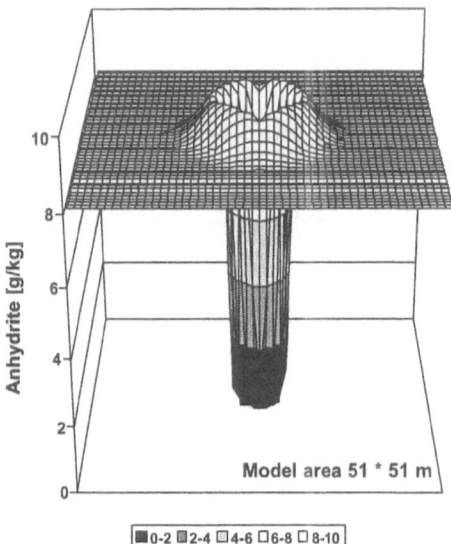

Fig. 7: Anhydrite distribution after a time period of 4 days. The amount of anhydrite is reduced to zero in a radius of about 4 m around the well. The initial value of 8.16 g/kg anhydrite is enriched in an annulus of almost 5 m width.

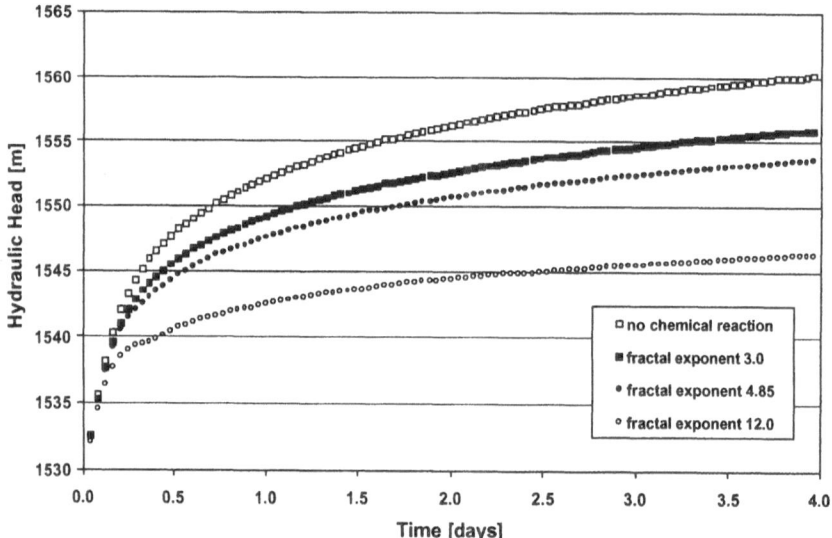

Fig. 8: Hydraulic head at the well as a function of time corresponding to reinjection pressure for non-reactive flow and for different fractal exponents in the case of reactive flow. With increasing exponent of fractal dimension the hydraulic head required to inject 50 m³/h decreases compared to non-reactive flow.

3.3.2 Barite

Figure 9 shows the concentration of barium and the change in barite concentration (actual amount of barite in the rock is normally unknown because of its small concentration) versus time for the observation points 3 m and 6 m away from the well plotted as mmol per liter pore volume.

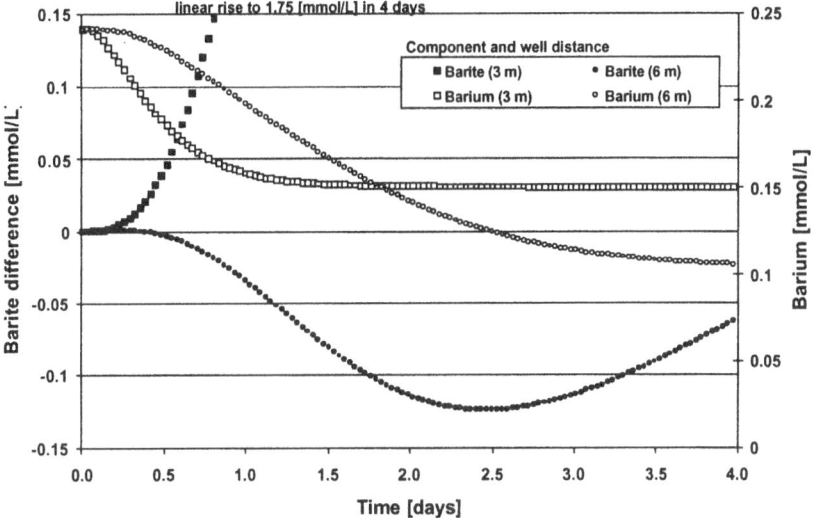

Fig. 9: Barium concentration and change of barite concentration as a function of time at observation points 3 and 6 m away from the well.

At the distance of 3 m from the injection well the amount of barite increases from the start of the injection. After less than one day the concentration of the mineral already exceeds 0.15 mmol/L and rises up to 1.75 mmol/l within 4 days. At the location the amount of barium in the solution decreases from the initial concentration of 0.24 mmol/L to a constant value of 0.15 mmol/L after about 2.5 days. At 6 m from the well the barium concentration decreases to a stationary value of 0.11 mmol/L within 4 days. At this location the amount of barite decreases by about 0.125 mmol/L during the first 2.5 days. From then on, the concentration of barite increases continuously. This leads to an enrichment of barite around the injection well and a depletion in an area directly in front of it (Figure 10).

In contrast to the anhydrite scenario the hydraulic head at the injection well does not change significantly compared to the non-reactive flow.

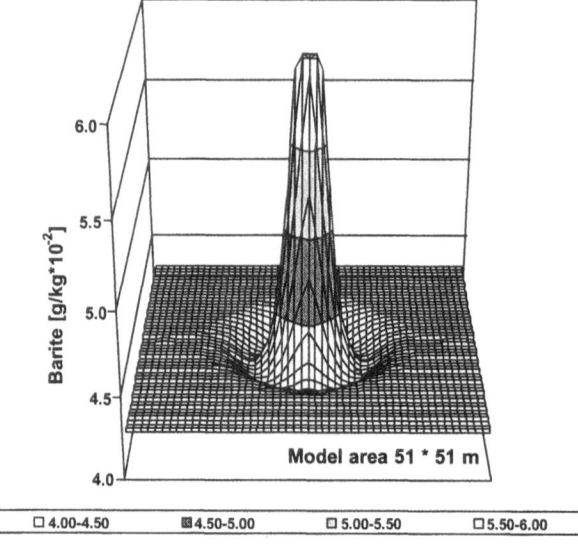

Fig. 10: Barite distribution after a time period of 4 days. The initial amount of barite of $4.2*10^{-2}$ g/kg is enriched around the well with an area of depletion in front of it.

3.4 DISCUSSION

The simulations for the non-reactive flow showed, that the propagation of the temperature front is much slower than the propagation of the tracer front (Figure 6). This is due to the thermal capacity of the rock. The excess pressure necessary for the injection strongly depends on the injection temperature. With increasing temperature the viscosity η of the water decreases, and the hydraulic conductivity k_f of the formation layer increases according to

$$k_f = (k \; \rho \; g) / \eta, \qquad\qquad (6)$$

where ρ is the density of water , and g the gravity constant. This is why the hydraulic head does not reach a steady state during the simulation period (Figure 8). The volume filled with cold water increases with time and thus the hydraulic conductivity of the entire aquifer decreases continuously.

The temperature regime during the injection process controls the chemical reactions. As anhydrite is more soluble in cold than in hot water (retrograde solubility) Figure 6 shows that the calcium concentration increases with decreasing temperature. During the first 2.5 days the equilibrium concentration increases from about 25 mmol/L at 100°C to 63 mmol/L at 20°C. After 4 days anhydrite is totally dissolved and the calcium concentration rapidly decreases.

Injecting 20°C water into an aquifer with a formation temperature of 100°C causes dissolution of anhydrite at the injection well, and transport of the dissolved elements (Ca and SO_4) through the aquifer. The species transport is faster than that of the lower temperature through the aquifer (Figure 6). Therefore water at 20°C in equilibrium with larger amounts of Ca and SO_4 eventually reaches the warm temperature front from upstream. At the higher temperature less $CaSO_4$ is soluble in the brine. Therefore, anhydrite precipitates at the warm temperature front (Figure 7). As a consequence permeability is reduced. When the cooling front propagates further and a stationary temperature of 20°C is approached the relocated anhydrite is dissolved again. That means that the region of anhydrite enrichment and permeability reduction is propagating through the aquifer together with the temperature front. As illustrated in Figure 8 the hydraulic head evolution at the injection well depends on the relationship between porosity and permeability: the larger the fractal exponent, the lower the resulting hydraulic head. As a result of the dissolution of anhydrite around the injection well, the injectivity of the layer increases compared to the non-reactive flow. In the period considered, the negative effect of the cold water on the injectivity is partially compensated by the relocation of anhydrite in the formation. In conclusion, the influence of mineral dissolution around the well dominates over the precipitation at the temperature front.

While anhydrite is an example for retrograde mineral solubility, barite reacts in a prograde way. Brines, saturated with barite at 100°C, are supersaturated for an injection temperature of 20°C. Thus, during reinjection precipitation of barite occurs in the vicinity of the borehole as shown in Figure 9. At a distance of 3 m and 6 m from the well the barium concentration reaches constant values in 1.5 days and almost 4 days, respectively, after the temperature front has passed. Both concentrations do not represent the thermodynamic equilibrium of $BaSO_4$, which is 0.095 mmol/L at 20°C. This can be explained with the kinetics of barite precipitation. At both observation points the reaction does not reach thermodynamic equilibrium.

At the observation point 6 m from the well the barite concentration decreases at first. As the injected solution has lost most of its barium sulfate content in the immediate surrounding of the well, the water is close to equilibrium at a low temperature on its path towards point 6 m. At point 6 m, the concentration of the solution at this place becomes undersaturated, due to the fact, that the transport process is faster then the spreading of the temperature front and the water eventually reaches the front and temperature increases. Therefore dissolution of barite takes place and barite depletion evolves in the area of the temperature front (Figure 10). When the cold water front has passed, barite precipitates at the observation points. At the point 3 m from the well it looks as if the precipitation of barite starts immediately with the reinjection.

Figure 10 shows, that the near surrounding of the well is subjected to barite precipitation. The radius of this region depends on the reaction rate. In our case the reaction constant k_r is chosen in an order as for calcite for which mineral data exist from literature. True barite reaction rates need to be determined in future experiments. Permeability changes induced by barite precipitation are negligible during the first 4 days of heat production because the precipitated amounts of mineral are too small.

Summary

The prediction of hydraulic and thermal regime conditions of an operated geothermal aquifer requires a numerical simulation. Reliable results can be obtained only if the fluid flow, heat transfer, species transport, and chemical reactions, are all fully coupled. The simulation code SHEMAT, presented here (Bartels *et al.*, 2000), ensures the necessary coupling via the temperature and concentration dependence of fluid and matrix properties, and above all the feedback of porosity changes on the flow field due to the precipitation and dissolution of several mineral phases.

The program includes a new chemical module which is based on the geochemical simulation code PHRQPITZ (Plummer *et al.*, 1988). The Pitzer virial coefficient approach for the correction of activity coefficients is used to calculate aqueous speciation and mineral solubilities in geothermal brines. Code and data base of PHRQPITZ were extended for the handling of temperature dependent Pitzer coefficients valid now for temperatures from 0 - 150°C in the system Na-K-Mg-Ca-Ba-Sr-Si-H-Cl-SO_4-OH-(HCO_3-CO_3-CO_2)-H_2O. Data for the incorporated carbonic acid system (set in parentheses) are valid for temperatures from 0 to 90°C, only.

The temperature of the water injected into a geothermal reservoir has a great influence on the hydraulic conductivity of the aquifer and thus on the pressure development in the well. The lower the temperature, the lower the injectivity of the formation layer. A constant pressure cannot be reached because the cold region around the wells is continuously growing.

The dissolution of anhydrite around the well induced by temperature changes increases the permeability, whereas the precipitation of anhydrite at the warm temperature front reduces it. The negative effect of the temperature on the injectivity is partially compensated by the anhydrite reactions, as the permeability increase predominates the decrease.

In contrast to anhydrite, with retrograde solubility, barite has prograde solubility. The reinjection of cold brine leads to supersaturation of the solution in this of barite. Although barite precipitates around the injection well no significant permeability change or hydraulic effect is observed during the simulated time.

Acknowledgements

The research reported in this paper was supported by the German Federal Ministry for Education, Science, Research, and Technology (BMBF) under grant 032 69 95. Special thanks are given to M. Azaroual and Ch. Monnin for the highly constructive reviews.

References

Ackoumov, E. I., and Wassilijew, B. B. (1932): *Zh. Obshch. Khim.*, **2**, 271-289.

Aggarwal, P. K., Gunter, W. D., and Kharaka Y. K. (1990): Effect of Pressure on Aqueous Equilibria. In: *ACS Symposium Series, Chemical Modeling of Aqueous Systems II* (D. C. Melchior und R. L. Bassett, eds.), **No. 416**, Chap. 7, American Chemical Society, Washington DC, 87-101.

Akaku, K. (1990): Geochemical study on mineral precipitation from geothermal waters in the Fushime field, Kyushu, Japan. *Geothermics*, **19** (5), 455-467.

Azaroual, M., Fouillac, C., and Matray, J. M. (1997): Solubility of silica polymorphs in electrolyte solutions, I. Activity coefficients of aqueous silica from 25° to 250°C, Pitzer's parameterisation. *Chemical Geology*, **140**, 155-165.

Bartels, J., Kühn, M., Pape, H., and Clauser, C. (2000): A New Aquifer Simulation Tool for Coupled Flow, Heat Transfer, Multi-Species Transport and Chemical Water-Rock Interactions. Proceedings of World Geothermal Congress 2000, Kyushu - Tohuku, Japan, May 28 - June 10, (2000) pp 3997-4002

Bartels, J., Kühn, M., Pape, H., and Clauser, C. (in prep.). Change of permeability distribution during flooding of sandstone cores: Experiment and simulation. *Geophysical Research Letters*.

Bergmann, A.G., and Vlassov N. A. (1949): *Izv. Sektora Fiz.-Khim. Analiza*, Inst. Obsch. Neorgan. Khim., Akad. Nauk SSSR, **17**, 312-337.

Blasdale, W.C. (1918): *Ind. Eng. Chem.*, **10**, 344-347.

Blount, C. W., and Dickson F. W. (1969): The solubility of anhydrite (CaSO₄) in NaCl-H₂O from 100 to 450°C and 1 to 1000 bars. *Geochimica et Cosmochimica Acta*, **33**, 227-245.

Blount, C. W., and Dickson F. W. (1973): Gypsum-Anhydrite Equilibria in Systems CaSO₄-H₂O CaCO₃-NaCl-H₂O. *American Mineralogist*, **58**, 323-331.

Blount, C.W. (1977): Barite solubilities and thermodynamic quantities up to 300°C and 1400 bars. *American Mineralogist*, **62**, 942-957.

Bock, E. (1961): On the Solubility of Anhydrous Calcium Sulphate and of Gypsum in Concentrated Solutions of Sodium Chloride at 25°C, 30°C, 40°C, and 50°C. *Canadian Journal of Chemistry*, **39**, 1746-1751.

Boisdet, A., Cautru, J. P., Czernichowski-Lauriol, I., Foucher, J. C., Fouillac, C., Honnegger, J. L., and Martin J. C. (1989): Experiments on reinjection of geothermal brines in the deep Triassic sandstones. *European Geothermal update, Proceedings of the 4th International Seminar on the results of EC Geothemal Energy*, Florence, 27-30 April 1989, 419-428.

Bourbie, T., and Zinszner, B. (1985): Hydraulic and acoustic properties as a function of porosity in Fontainebleau sandstone. *J. Geophys. Res.*, **90**, 11524-11532. Clauser, C. and Kiesner, S. (1987): A conservative, unconditionally stable, second-order three point differencing scheme for the diffusion convection equation. *Geophys. J. R. Astr. Soc.*, **91**, 557-568.

Brown, K. L., Freeston, D. H., Dimas, Z. O. & Slatter, A. (1995): Pressure drop due to silica scaling. In: *Proceedings of the 17th New Zealand Geothermal Workshop 1995* (M. P. Hochstein, ed.), Geothermal Institute, University of Auckland, 163-167.

Clauser, C. and Kiesner, S. (1987): A conservative, unconditionally stable, second-order three point differencing scheme for the diffusion convection equation. *Geophys. J.R. Astr. Soc.*, **91**, 557-568.

Clauser, C. and Villinger, H. (1990): Analysis of conductive and convective heat transfer in a sedimentary basin, demonstrated for the Rheingraben. *Geophys. J. Int.*, **100** (3), 393-414.

Cornec, E., and Krombach, H. (1932): *Compt. Rend.* **194**, 714-716.

Darwis, R.S., Tampubolon, T., Simatupang, R., and Asdassah, D. (1995): Study of water reinjection on the Kamojang Geothermal Reservoir Performance, Indonesia. In: *Proceedings of the 17th New Zealand*

Geothermal Workshop 1995 (M. P. Hochstein, ed.), Geothermal Institute, University of Auckland, 185-192.

Greenberg, J. P., and Moller, N. (1989): The prediction of mineral solubilities in natural waters: A chemical equilibrium model for the Na-K-Ca-Cl-SO4-H2O system to high concentration from 0 to 250 °C. *Geochim. Cosmochim. Acta*, **53**, 2503-2518.

Harvie, C. E., and Weare, J. H. (1980): The prediction of mineral solubilities in natural waters: The Na-K-Mg-Ca-Cl-SO4-H2O system from zero to high concentrations at 25°C. *Geochim. Cosmochim. Acta* **44**, 981-997.

Harvie, C. E., Moller, N., and Weare, J. H. (1984): The prediction of mineral solubilities in natural waters: The Na-K-Mg-Ca-H-Cl-SO4-OH-HCO3-CO3-CO2-H2O system to high ionic strengths at 25 °C. *Geochim. Cosmochim. Acta*, **48**, 723-751.

He, S. and Morse, J. W. (1993): The carbonic acid system and calcite solubility in aqueous Na-K-Ca-Mg-Cl-SO4 solutions from 0-90 °C. *Geochim. Cosmochim. Acta*, **57**, 3533-3554.

Holmes, H. F., Baes, C. F. B. Jr., and Mesmer, R. E. (1987): The enthalpy of dilution of $HCl_{(aq)}$ to 648 K and 40 MPa: Thermodynamic properties. *J. Chem. Thermodyn.*, **19**, 863-890.

Holmes, H.,F. and Mesmer, R. E. (1992): Isopiestic studies of $H_2SO_{4(aq)}$ at elevated temperatures - Thermodynamic properties. *J. Chem. thermodyn.*, **24**, 317-328.

Kühn, M., Frosch, G., Kölling, M., Kellner, T., Althaus, E., and Schulz, H. D. (1997): Experimentelle Untersuchungen zur Barytübersättigung einer Thermalsole (in German with English abstract). *Grundwasser*, 3 (2), 111-117.

Kühn, M., Vernoux, J.-F., Isenbeck-Schröter, M., Kellner, T., and Schulz, H. D. (1998): On-site experimental simulation of brine injection into a clastic reservoir as applied to geothermal exploitation in Germany. *Applied Geochemistry*, **13** (4), 477-490.

Monnin, C. (1990): The influence of pressure on the activity coefficients of the solutes and on the solubility of minerals in the system Na-Ca-Cl-SO$_4$-H$_2$O to 200°C and 1 kbar, and to high NaCl concentration. *Geochimica et Cosmochimica Acta*, **54**, 3265-3282.

Monnin, C. (1999): A thermodynamic model for the solubility of barite and celestite in electrolyte solutions and seawater to 200°C and to 1 kbar. *Chemical Geology*, **127**, 141-159.

Monnin, C., and Ramboz, C. (1996): The anhydrite saturation index of the ponded brines and sediment pore waters of the Red Sea deeps. *Chemical Geology*, **153**, 187-209.

Nagy, K. L., Steefel, C. I., Blum, A. E., and Lasaga, A. C. (1990): Dissolution and precipitation kinetics of kaolinite: Initial results at 80°C with application to porosity evolution in a sandstone. In: *Prediction of Reservoir Quality through Chemical Modeling* (I. D. Meshri, and P. J. Ortoleva,eds.), AAPG Memoir, **49**, 546-550.

Pabalan, R. T., and Pitzer, K. S. (1987a): Thermodynamics of concentrated electrolyte mixtures and the prediction of mineral solubilities to high temperatures for mixtures in the system Na-K-Mg-Cl-SO$_4$-OH-H$_2$O. *Geochim. Cosmochim. Acta*, **51**, 2429-2443.

Pabalan, R. T., and Pitzer, K. S. (1987b): Thermodynamics of $NaOH_{(aq)}$ in hydrothermal solutions. *Geochim. Cosmochim. Acta*, **51**, 829-837.

Pape, H., Clauser, C., and Iffland, J. (1999). Permeability prediction based on fractal pore-space geometry. *Geophysics*, **64** (5), 1447-1460.

Pitzer, K. S. (1973): Thermodynamics of electrolytes. 1. Theoretical basis and general equations. *J. Physical Chemistry*, **77**, 268-277.

Pitzer, K. S. (1975): Thermodynamics of electrolytes. V. Effects of higher-order electrostatic terms. *Journal of Solution Chemistry*, **4** (3), 249-265.

Pitzer, K. S. (1991): Ion interaction approach: Theory and data correlation, in: *Activity Coefficients in Electrolyte Solutions* (K. S. Pitzer, ed.), CRC Press Boca Raton, 2nd Edition, 76-153.

Pitzer, K. S., and Kim, J.J. (1974): Thermodynamics of electrolytes. IV. Activity and osmotic coefficient for mixed electrolytes. *Journal of American Chemical Society,* **96** (18), 5701-5707.

Pitzer, K. S., and Mayorga, G. (1973): Thermodynamics of electrolytes. II. Activity and osmotic coefficient for strong electrolytes with one or both ions univalnet. *Journal of Physical Chemistry,* **77** (19), 2300-2308.

Pitzer, K. S., and Mayorga, G. (1974): Thermodynamics of electrolytes. III. Activity and osmotic coefficients for 2-2 electrolytes. *Journal of Solution Chemistry,* **3** (7),539-546.

Plummer, L. N., Parkhurst, D. L., Fleming, G. W., and Dunkle, S. A. (1988): *A computer program incorporating Pitzer's equations for calculation of geochemical reactions in brines.* U.S. Geological Survey Water-Resources Investigations, Report 88-4153, 310 p.

Potter, J. M., Dibble, W. E., and Nur, A. (1981): Effects of temperature and solution composition on the permeability of St. Peters sandstone-Role of iron (III). *Journal of Petroleum Technology,* **33** (5), 905-907.

Sanyal, S. K, McNitt, J. R., Klein, C. W., and Granados, E. E. (1985): An investigation of wellbore scaling at the Miravalles Geothermal Field, Costa Rica. In: *Proceedings of the 10th Workshop Geothermal Reservoir Engineering, Jan. 22-24, 1985* (H. J. Ramey, P. Kruger ,F. G. Miller, R. N. Horne, W. E. Brigham, and J. S. Gudmundsson, eds.), 37-44.

Smolarkiewicz, P. K. (1983): A simple positive definite advection scheme with small implicit diffusion. *Monthly Weather Rev.,* **111,** 479-486.

Spencer, R. J., Moller N., and Weare, J. H. (1990): The prediction of mineral solubilities in natural waters: A chemical equilibrium model for the system Na-K-Ca-Cl-SO$_4$-H$_2$O at temperatures below 25°C. *Geochim. Cosmochim. Acta,* **54,** 575-590.

Templeton, C. C. (1960): Solubility of Barium Sulfate in Sodium Chloride Solutions from 25° to 95°C. *Journal of Chemical and Engineering Data,* **5** (4), 514-516.

Ungemach, P. (1983): Drilling, production, well completion and injection in fine grained sedimentary reservoirs with special reference to reinjection of heat depleted geothermal brines in clastic deposits. Report of an extended contractors meeting held in Brussels on 23 March 1983.

Ungerer, P., Burrus, J., Doligez, B., Chénet, P. Y., and Bessis, F. (1990): Basin evaluation by integrated two-dimensional modeling of heat transfer, fluid flow, hydrocarbon generation, and migration: *AAPG Bull.,* **74,** 309-335.

Vernoux, J.F. & Ochi, J. (1994). Aspects relative to the release and deposition of fines and their influence on the injectivity decrease of a clastic reservoir. *Geothermics 94 in Europe,* Int. Symp. held in Orléans, France, 8-9 Feb., 291-302.

Vinchon, C., Matray, J. M., and Rojas, J. (1993): Textural and mineralogical changes in argillaceous sandstone, induced by experimental fluid percolation; *Geofluids'93,* International Conference, Torquay, UK, May (1993), 233-236.

Weres, O. (1988): Environmental protection and the chemistry of geothermal fluids; *Geotherm. Sci. & Tech.,* **1** (3), 253-302.

Wolery, T. J., and Jackson, K. J. (1990): Activity coefficients in aqueous salt solutions. In: *ACS Symposium Series, Chemical Modeling of Aqueous Systems II* (D. C. Melchior und R. L. Bassett, eds.), **No. 416,** Chap. 2, American Chemical Society, Washington DC, 16-29.

Water-Rock Reactions in a Barite-Fluorite Underground Mine, Black Forest (Germany)

Ingrid Stober[1], Yinian Zhu[2] & Kurt Bucher[2]

[1]Geological Survey, Albertstr. 5, D-79104 Freiburg, Germany
(stober@lgrb.uni-freiburg.de)
[2]Institute of Mineralogy, Petrology and Geochemistry, Universität
Freiburg, Albertstr. 23b, D-79104 Freiburg, Germany
(bucher@ruf.uni-freiburg.de)

Abstract

Mineralization in the more than 700 m deep underground barite and fluorite mine Clara occurs in 3 major vein systems within the central gneiss complex of the Black Forest, SW-Germany. They are associated with quartz, carbonate, various sulfides, natural alloys and about 250 different species of secondary minerals. Water transport is strongly focused within the mineral veins owing to their high fracture and cavity-related permeability. Water from 25 different sampling points in and around the mine were collected and analyzed monthly during one year in order to elucidate the interaction of surface water with the various rock types. On the basis of the relative proportions of the major ions 3 types of water could be distinguished: $Ca-HCO_3-$, $Ca-SO_4-HCO_3-$ and $Ca-SO_4-Cl-$water. The source of the solutes is related to fluid-rock reaction, especially feldspar weathering. All waters within the mine are very close to quartz saturation. They are saturated or slightly undersaturated with respect to calcite after relatively short flow distances through the fracture pore space. All waters are undersaturated regarding dolomite or gypsum. On pertinent activity diagrams, all waters fall into the stability field of kaolinite which is an observed major alteration product in veins and rocks.

I. Stober and K. Bucher (eds.), Water-Rock Interaction, 171–187.
© 2002 *Kluwer Academic Publishers.*

1.   Geological settings

The underground mine Clara is situated in the Central Gneiss Complex of the Black Forest, SW-Germany (fig. 1). Geologically, the Black Forest area represents an erosional window through the post-Variscian cover in the Central European continental crystalline basement. The exposed basement rocks stretch in the N-S direction about 130 km and the W-E extension is about 50 km. The uplift and exhumation history of the basement is related to the formation of the Upper Rhine rift system in the Tertiary. The uplift of the Black Forest resulted in erosion that striped away the post-Variscian cover and exposed the basement of the European continental crust. A complete section of cover rocks is exposed today to E of the Black Forest window (Thury et al. 1994, Groschopf et al. 1996). Post-Variscian sediments, Buntsandstein, are still preserved in the northern and eastern parts on top of the highest hills of the Black Forest. Fig. 1 shows the general geology of the basement and its major cover units. The basement is predominantly composed of a) granitic and mafic gneisses that are locally migmatized and b) post-deformational granitic plutons of moderate size. The gneisses contain locally granite veins.

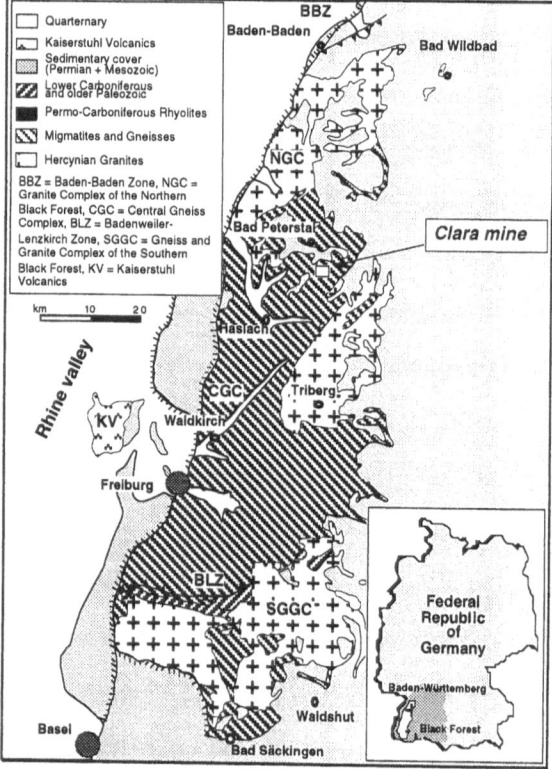

Fig. 1: Geological map of the Black Forest area (simplified after Groschopf et al. 1996) showing the location of the Clara mine

The Black Forest area has a long and famous mining tradition with small Cu-Ag-Pb-Zn veined gneisses (Bliedtner & Martin 1986). About 2000 years ago Celts and Romans were the first to establish a real mining industry in the Black Forest. During the Middle Ages considerable amounts of silver and lead were produced. Today only few mines are still in operation. One of them is the Clara Mine in the central part of the Black Forest. First reports of mining activities at the site of the later Clara Mine location are dated from 1625. The history of the Clara Mine mine started in 1850 when commercial exploitation of barite began. Since 1978 the mine also produces fluorite (Huck 1986). Fig. 2 shows a cross section through the Clara mine.

The rocks in the mine area are predominantly banded, coarse grained biotite-plagioclase-gneisses, locally with garnet. Subordinate garnet-amphibolite and garnet-rich gneiss is present as well. The gneiss consists mainly of plagioclase (An_{20}-An_{40}), K-feldspar, biotite, quartz, amphibole and accessory minerals. Small fractures are filled with calcite, gypsum, pyrite, arsenopyrite, pyrrhotine and other secondary minerals. The crystalline basement is partly covered by about 35 m thick Triassic sandstone. The barite-fluorite veins continue from the basement into the sandstone (fig. 3). Hence, the minearalization is post lower Triassic. The barite and fluorite mineralization occurs in 3 major vein systems (Huck 1986): i) Clara barite veins, ii) Stollen barite veins and iii) quartz veins. The Clara veins are the most important barite and fluorite veins of the mine. They strike approximately 150°NW-SE with a dip of 70-80°NE; they are about 650 meters long and several meters thick. The Stollen barite veins strike in a N-S direction. The third vein system consists mainly of quartz veins, but it bears barite and fluorite too. This later vein system strikes in an E-W direction (fig. 3). Today the Clara Mine has reached a depth of about 650 m and a lateral extension of about 200 x 500 m. The actual mean production rate is about 165000 t a^{-1} raw ore; the barite/fluorite ratio is about 2 : 1.

Mining in the underground produces cavities. To stabilize the overburden rocks these cavities were in former times filled with waste rock from the mining production. Later, this gneissic debris had to be stabilized by concrete. In recent years some cavities within the Clara mine were filled with fly-ash from coal firing power plants.

2.   Hydrogeological setting and water samples

In the crystalline basement natural water flows along distinct geological discontinuities. Water transport within the Clara mine is strongly focused within fracture systems and mineral veins for the simple reason of their high permeability. Water within the rock matrix is very strongly attached to mineral surfaces and consequently from a hydraulic point of view immobile (Biehler 1995, Stober 1995).

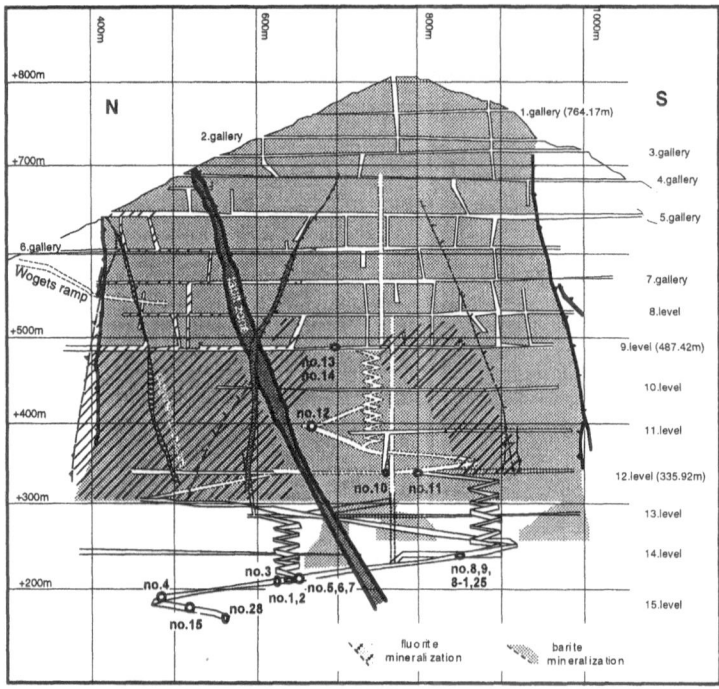

Fig. 2: Cross section through the Clara mine and location of water sample points

Water samples from more than 20 different localities in and around the barite and fluorite mine Clara were collected monthly between July 1997 and October 1998. Surface water samples were taken from streams, creeks and springs, groundwater samples were collected from exploration drillholes and open fractures, cavities or veins. We collected snow (December 1997, January 1998) and rain water (September 1998) too. The scope of the work was to elucidate the interaction of surface water with various rock types and minerals. We expected an increase of TDS (total dissolved solids) and a change of water-type with increasing depth. Therefore we took water samples at different levels of the 650 m deep mine. We collected water samples as well out of barite-veins, fluorite-veins and out of the gneisses to see if there are any significant differences caused by water circulation within different mineralogical surroundings.

Figure 2 shows the water sample points within the Clara mine. The depth of each sample point is given on the cross section. Most water samples came from drillholes (no. 1, 2, 3, 4, 5, 6, 8-1, 9, 10, 11, 14, 28) and some samples were collected from water conducting open fractures and veins (no. 8, 12, 13, 15, 26, 29). No. 7 and 25 are samples taken from fast flowing water in open trenches in the mine.

On figure 3, the sampling points outside the mine are shown. Water samples from points no. 20, 21, 22 and 23 are spring waters. Most springs occur within the gneisses; spring no. 22 originates from the Triassic sandstone (Buntsandstein).

Water sample no. 16 is taken from a small river, Rankachbach, and no. 16-1 from a nearby situated waste water pool filled with mine water. Water sample no. 24 is as well taken from a small river used as drinking water.

Water temperature, pH, electrical conductivity, Eh and O_2-concentration were measured in-situ. Major cations (Li^+, Na^+, K^+, Ca^{2+}, Mg^{2+}) and major anions (Cl^+, SO_4^{2-}, NO_3^-, F^-, Br^-) were analyzed in laboratory by IC, trace elements by Flame AAS and by hollow graphite furnace AAS. Alkalinity was determined by Gran titration. Only analyses with E.N. (electro neutrality) better than 5% are used in this study.

The Clara mine has an extensive water management system. Water needs to be pumped out from the mine continuously with an average rate of about $20 \, l \, s^{-1}$. The monthly quantity of water pumped form the mine is clearly related to the amount of precipitation. As a result of continuos pumping of water out of the mine structure, it can be expected that waste water influenced by mining activities is efficiently removed. Consequently. water collected from the deep sampling points, especially from drillholes, are assumed to represent pure natural waters.

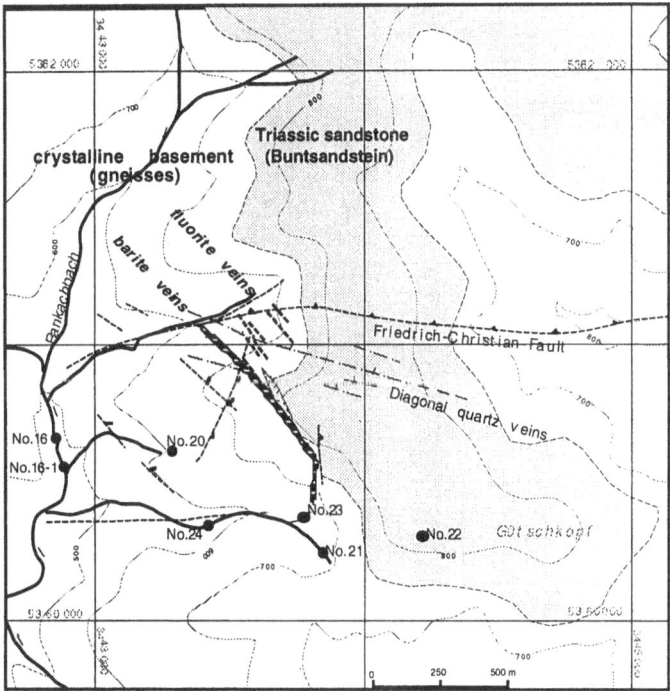

Fig. 3: Detailed geological map of the area of the Clara mine with the location of surfacewater sample points

3   Time series

The temperature within the mine increases with depth. Therefore measured water temperature from deeper locations is higher than for near surface sampling localities. In addition, groundwater temperature depends on infiltration conditions. The natural infiltration conditions within the mine are strongly disturbed by different activities within the mine itself. Relevant operations and structures include:

* continuous pumping of large amounts of water out of the mine,
* remnants of former open-pit mining on top of the present day underground mine causing increased infiltration and
* artificialy increased permeability due to generation of cavities.

Regardless, the water temperature increases with depth by about 2.1°C per 100m. Water samples with direct connections to the surface typically show a slightly higher temperature during the summertime and lower temperatures during the winter months. Some waters are characterized by a delayed response to seasonal temperature variations. For instance, the temperature is high in the fall and low in spring. Some locations within the mine are virtually independent of seasonal effects (fig. 4). The three types of temperature vs time curves show no clear depth dependence. The seasonal effect is not simply decreasing with depth. Therefore, vertical flow velocities of infiltrating meteoric water must be varying and contact times of descending meteoric water with the rock matrix differs significantly from point to point.

In situ measured pH varies between 4.75 (melted snow) and 8.30. Water samples within the mine itself vary between pH=6 and pH=8. pH of some spring waters outside the mine are sporadically slightly lower than pH=6.

Three different groups of data patterns can be observed on a diagram showing selected water parameters vs time (Fig. 4):

#1:   Data independent of seasonal effects: The total of dissolved solids (TDS) increases gradually with time.
#2:   Higher concentrations during summer, lower values during winter months.
#3:   Lower concentrations during fall, higher values during spring months

All time series of TDS and the major ion concentrations fall into these three groups. Group #1 represents water samples from boreholes in low permeability gneiss far away from barite or fluorite veins. Accordingly, water within these boreholes is stagnant. Such boreholes are typically about 100 m deep and drilled with a small diameter. Clean low TDS water of drinking water quality was used as drilling fluid. This was the "starting chemistry". The composition of the water pumped from the boreholes for our investigations evolved with time. TDS increased with time, probably as a result of direct reaction with the surrounding rock matrix and also by diffusion of solutes from nearly stagnant high-TDS water in micro-fissures of the surrounding rocks. Fig. 4 shows data from sample point no. 2. The concentration of all major ions increased during the observation period.

Nevertheless delayed seasonal temperature signals could be observed with lowest values (13.8°C) during spring and highest values (15.4°C) during fall. The concentrations of Ca and SO_4 increased faster with time than that of the other ions probably due to weathering of plagioclase or dissolutionn of calcite or gypsum from vein fillings. With about 63 meq%, Ca is the dominant cation and SO_4 the dominant anion, about 65 meq% .

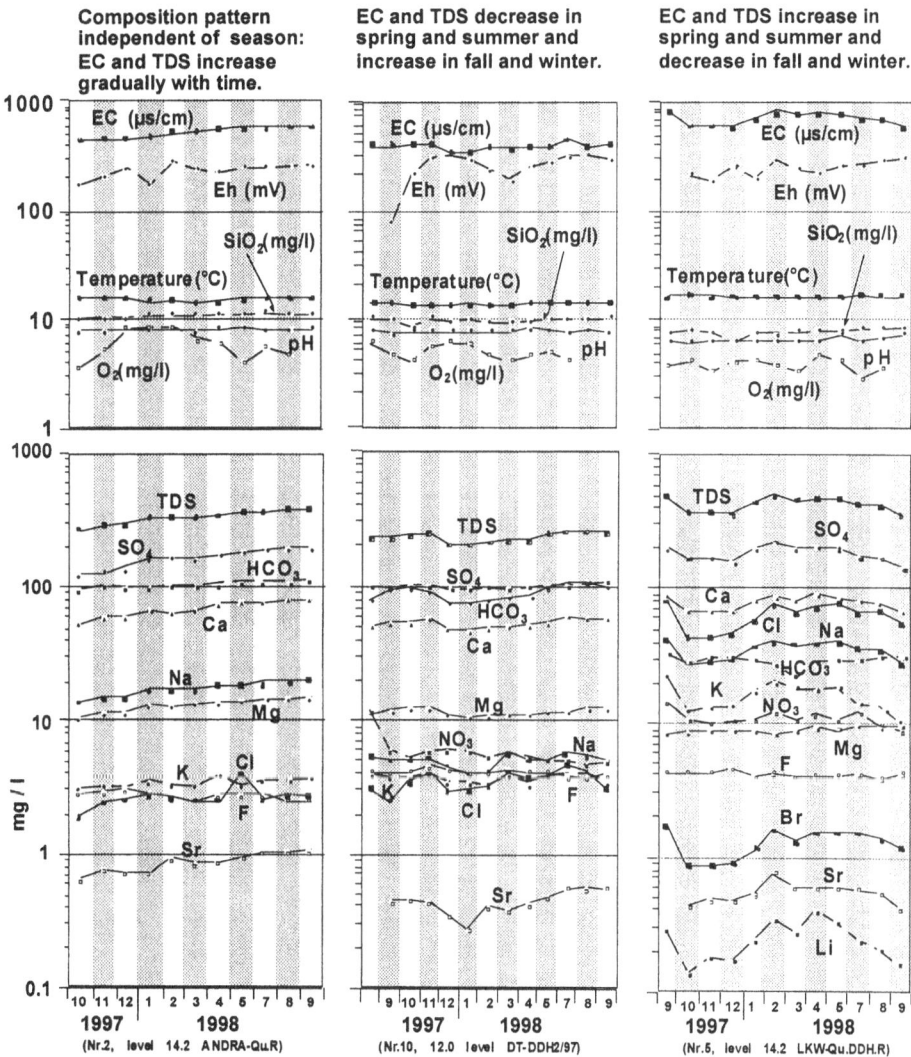

Fig. 4: Time series of water analyses data

Water samples of group #2 show typical seasonal effects. These samples were collected from higher levels of the mine. Most samples were taken from boreholes drilled into veins or fracture systems. The data of sample #10 (Fig. 4) are from water of a borehole drilled mainly in quartz veins of the third vein system. The permeability is higher and downward water transport quite fast. The observations suggest that the high precipitation, which typically begins in November, arrives at most of group #2 sampling localities with a time delay of about two month. Consequently, waters of group #2 have lower solute concentrations during the following winter months. TDS of this group is much lower than that of the other groups. Temperature fluctuations vary in the range of about 0.7°C with a maximum in late summer and a minimum in late winter months. The major cation is again Ca with about 68 meq% and the major anion is SO_4 (~48 meq%) closely followed by HCO_3 (~42 meq%).

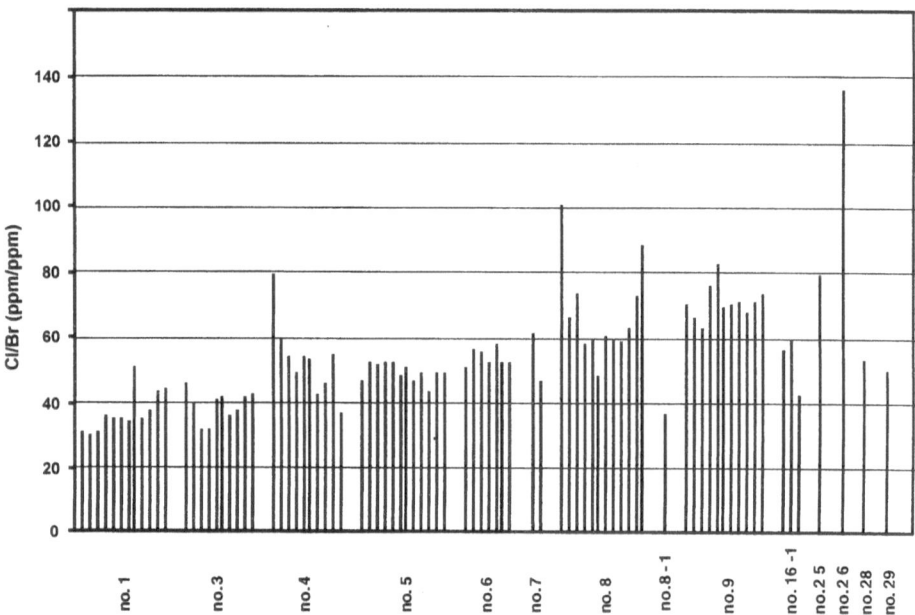

Fig. 5: Time series of Cl/Br-values of the water samples

Most water samples of group #3 were taken from the deepest parts of the mine. These waters are upwelling waters from deeper sources. The upflow is induced by the continuous pumping within the mine. Hence, TDS is much higher than in the other groups. Ca is the major cation with about 60 meq% of total cations and SO_4 is the dominant anion (~58 meq%). The value of HCO_3 (~6 meq%) is dramatically lower than in the other two groups. Chloride is, with about 26 meq%, much higher than in other sample groups. The upwelling waters bring along all typical attributes of deep water components (Stober & Bucher 1999a). Bromide could easily be detected within these samples. Nevertheless, weak seasonal effects could be observed as well. But the delay of the precipitation response is much longer, at least 10 months. The seasonal effects on the water composition are probably a consequence of a large-scale water circulation. The flow system starts with precipitation at the surface, followed by water infiltration into rock volumes outside of the actual mine area, followed by downward flow along the hydraulic gradient. The water finally enters the mine primarily as upwelling waters from below after a long circulation path within the fractured gneisses outside the mine area.

4   Origin of salinity

Chloride was the third, often the second abundant anion in all waters (fig. 6). The origin of abundant chloride is not obvious in a basement gneiss aquifer. In order to gain some insights to the source of salinity we determined chloride-bromide-ratios where possible. In low TDS waters with low chloride concentration bromide was not detectable. The Cl/Br-wt ratio of seawater is about Cl/Br = 288. In contrast, Cl/Br ratios of other possible sources of salinity show distinctly different Cl/Br ratios: Water derived from dissolved Tertiary halite deposits of the rift valley is in the order of Cl/Br = 2400 and water from dissolved Muschelkalk halite deposits has values of about Cl/Br = 9900 (Stober & Bucher 1999b). Leaching experiments on crystalline rocks on the other hand show that the average Cl/Br ratio of crystalline rocks is below Cl/Br = 100, meaning that the origin of salinity is dictated by water-rock interaction (Bucher AND Stober, this volume). Measured Cl/Br-ratios are typically in the range of 40 to 100. Two major sources of Cl are: i) weathering of biotite or amphibole with release of Cl from the crystal-lattice (Edmunds, 1996; Kullerud, 2000), ii) saline fluids or brines from cracked and leached fluid-inclusions (Böhlke & Irwin, 1992; Yardley et al., 1992; Bucher and Stober, this volume).

Figure 5 shows all measured Cl/Br-values (e.g. Br-values) of all water samples in the Clara Mine. No time dependence could be observed. The values show that in all samples the observed Cl/Br-ratio is in the range of Cl/Br = 40 to Cl/Br = 80. This strongly suggests that the salinity is not related to an ancient seawater intrusion or imported Cl from dissolution of halite deposits in the (former) sedimentary cover. The chloride has a local source and originates exclusively from water-rock interaction with the surrounding gneiss matrix.

Waters from fractured high-permeability zones and from drill holes in low-permeability gneiss matrix have major solute characteristics typical of water-rock reaction. They are controlled by incongruent dissolution of plagioclase and biotite. In addition, high-TDS waters are distinctly enriched in a Na-Cl component. The plagioclase dissolution process is briefly outlined below (see also Bucher & Stober, 2001a). The contribution of biotite alteration will be described in a companion paper. As already mentioned, the Cl/Br ratio of the water samples is low (about Cl/Br = 50. All Cl/Br-values of the water samples from the mine (fig. 7) are below the value of seawater (Cl/Br = 288). We did as well some leaching experiments with different rocks of the mine; the detailed experiments and results will be described in a forthcoming paper. The Cl/Br-values of most leaching experiments (fig. 8) are in the range of 50-70. The comparison of figure 7 with figure 8 indicates a clear internal origin of salinity in the groundwater of the mine.

Groundwater contains up to 150 mg/kg Cl of rock origin, antropogenic sources can be neglected in the chosen samples.

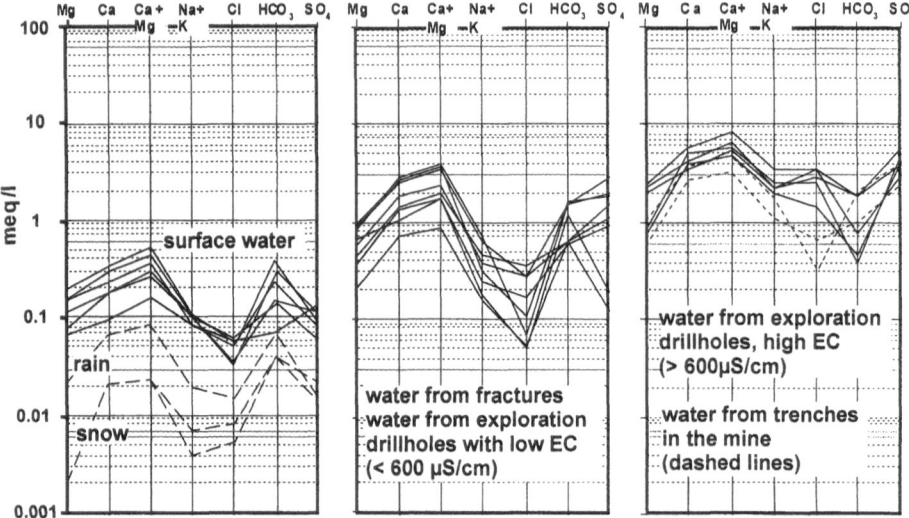

Fig. 6: Schoeller-Diagramms of water analyses: The diagram on the l.h.s. shows all surface waters (fig. 3) and the collected snow and rain waters. Waters from inside the mine collected from fractures and exploration boreholes (fig. 2) are shown on the diagram in the centre. The diagram on the r.h.s. shows water from exploration boreholes in the deepest parts within the mine. They are plotted together with water data from waste water pools in the mine.

5   Depth dependence

The concentrations of major ions of selected water analyses from all sampling localities are shown on three Schoeller-Diagrams (Fig. 6). On such diagrams, genetically related waters with different TDS are characterized by series of similar shaped parallel curves. The diagrams (Fig. 6) from left to right clearly demonstrate the general TDS increase with depth. The major components of selected waters show that each water type has its distinct pattern. The compositional variation is larger for the anions than for the cations. In all waters (Ca+Mg) is greater than (Na+K), and the concentration of Ca or Na is higher than that of Mg or K respectively. The anion concentrations also show a clear trend. In low-TDS waters HCO_3 is the dominant anion unlike to high-TDS waters which are generally rich in SO_4 and Cl. On the basis of the relative proportion of the major ions three types of waters could be distinguished:

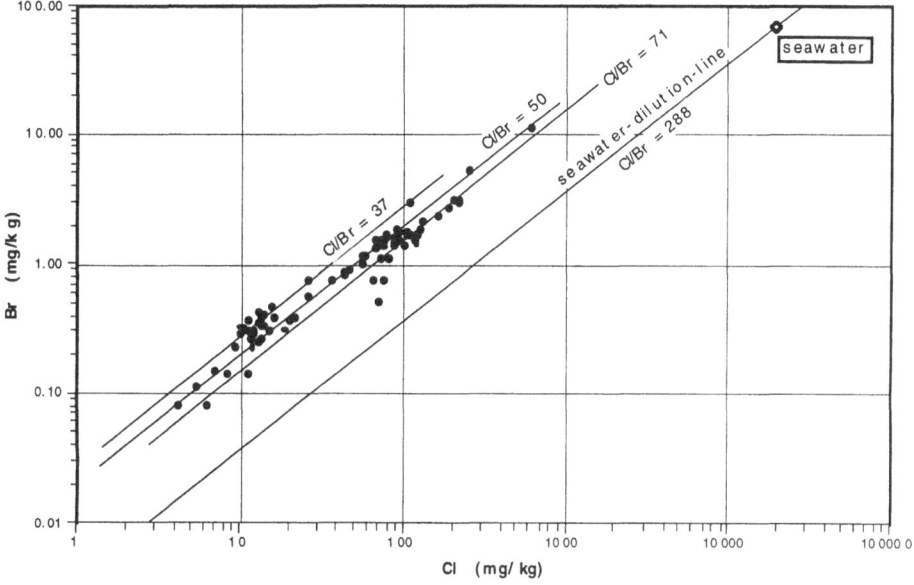

Fig. 7: Cl- versus Br-concentrations from sample points within the mine

(1) Ca-HCO₃-water

Precipitation, like rain water and snow, most surface waters and some other low TDS-waters from exploration drillholes in the mine belong to this group (Fig. 6). HCO_3 is with more than 70 meq% the dominant anion. There is not much difference between precipitation-, surface- or waters from these drillholes.

(2) Ca-SO₄-HCO₃-water

These waters show high Ca concentrations and very low Na concentrations. SO_4 is the dominant anion but HCO_3 is important too. The Cl concentrations are very low. These waters were collected out of fractures and exploration drillholes. TDS is still quite low but significantly higher than in group #1.

(3) Ca-SO₄-Cl-water

These waters were collected in the deepest parts of the mine out of exploration drillholes. TDS is significantly higher (log-scale) than in all other samples. Na concentrations are with 20-40 meq% significantly higher than in any other group before. The main anions are SO_4 and Cl with together about 60 meq%. HCO_3 remained nearly unchanged compared to group #2.

Figure 6 clearly shows that with increasing depth the concentration of all main water components is increasing and consequently TDS increases too. Relative to the other ions the most significant increase is observed for Na and Cl. Thus, the water type changes with depth. The salinity increases with depth due to interaction with the surrounding rock matrix. The complete set of water analyses data is given in Table 1.

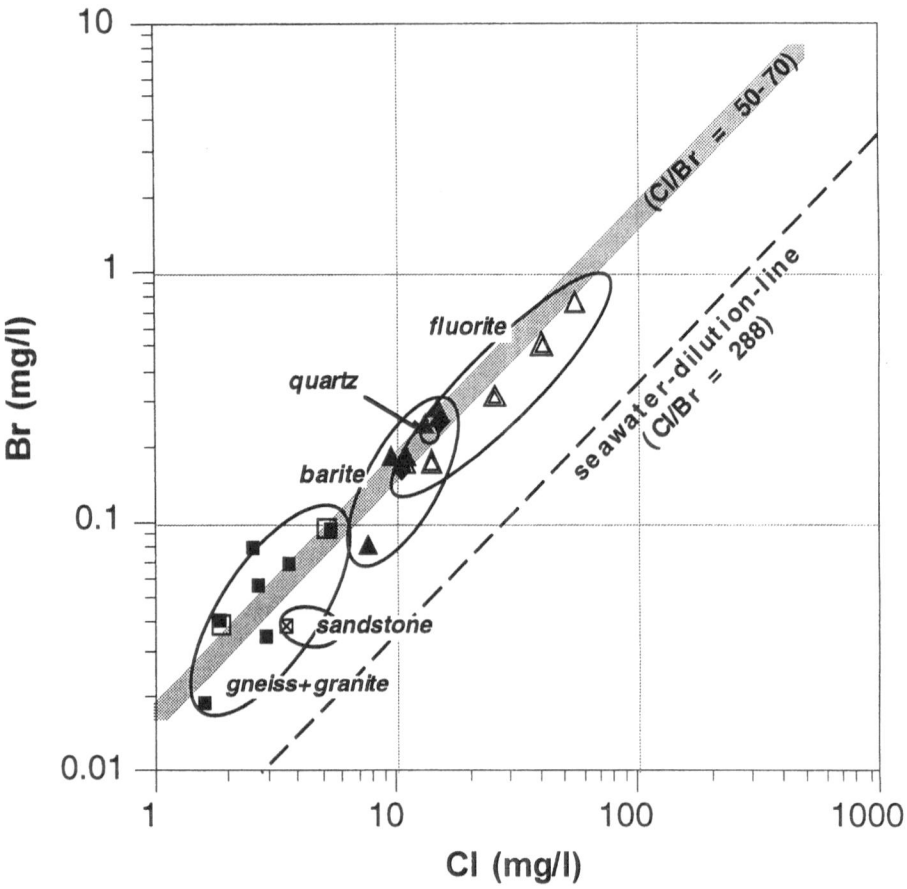

Fig. 8: Cl- versus Br-concentrations from leaching experiments

6   Thermodynamic   calculations

We calculated the activities of the components and the saturation state (SI) of the sampled groundwater using the code PHREEQE (Parkurst et al., 1990). Generally, the waters of the Clara mine are weakly oversaturated with respect to quartz ($0.1 <$ SI < 0.5). The saturation with respect to carbonates depends on TDS. Calcite saturation rapidly increases with increasing TDS and reaches equilibrium when TDS exceeds 200 mg/l. With respect to dolomite a similar correlation can be observed, however, few waters reach dolomite saturation. All waters are undersaturated with respect to gypsum. The saturation indices increase parabolicaly with increasing TDS.

Figure 9 shows selected saturation indices (SI) of water samples, mostly taken during December 1997, inside and outside the mine. Snow and rain waters are not shown because they were dramatically undersaturated with regard to all minerals. The SI of chalcedony are all slightly lower than that of quartz and are not shown on Fig. 9.

The water samples #16 and #20-24 were collected from surface waters, mainly springs and two from a small brook (#16, #24). All waters are undersaturated with respect to most minerals; they are saturated or slightly oversaturated with respect to quartz. The springs are also saturated with respect to barite. We assume that the springs arise from high-permeability veins filled with quartz and barite.

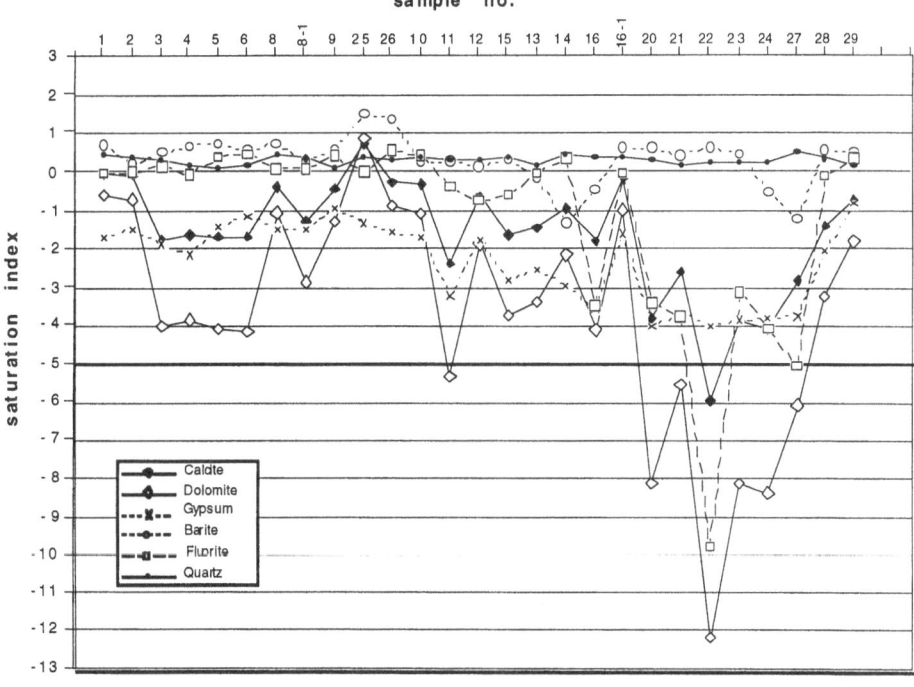

Fig. 9: Selected saturation indices (SI) of the water samples

Water #16-1 was collected from a waste water pool filled with mine water located just outside the mine. This water is saturated with respect to fluorite and calcite in addition to quartz and barite. Undersaturation in regard to dolomite and gypsum is minimal. #25 are water samples from a pool filled with mine water within the mine. Also this water represents a collection of different water sources from somewhere within the mine. The water in this pool is oversaturated with respect to calcite, dolomite, quartz, especially barite and saturated relative to fluorite and albit.

The remaining saturation indices shown on figure 9 are from water samples from boreholes and water conducting fractures. All waters are saturated or slightly oversaturated with respect to quartz, as might be expected for waters in crystalline basement rocks. The SI patterns of well waters and fracture waters show no difference (Fig. 9). Also, there is little or no depth dependence of the SI patterns. Except for #14 and #28, all other waters are close to barite saturation, perhaps not surprising in a barite mine. The saturation index of dolomite was always slightly lower than that of calcite. The SI of dolomite and calcite was lowest in waters from the major barite, fluorite and quartz veins. In contrast, waters from gneiss (also often associated with small barite and/or fluorite veins) were significantly closer to calcite and dolomite saturation.

Along the flow path of infiltrating meteoric water, water-rock interaction produces secondary minerals, predominantly kaolinite and illite, from primary minerals of the gneiss, predominantly plagioclase and biotite. The feldspar alteration can be described by the reactions:

$3 \ KAlSi_3O_8 + 2 \ CO_2 + H_2O = \ KAl_3Si_3O_{10}(OH)_2 + 6 \ SiO_2 + 2 \ K^+ + \ 2 \ HCO_3^-$
(K-feldspar)   (illite)

$4 \ NaAlSi_3O_8 + 4 \ CO_2 + 6 \ H_2O = 2 \ Al_2Si_2O_5(OH)_4 + 8 \ SiO_2 + 4 \ Na^+ + \ 4 \ HCO_3^-$
(albite)   kaolinite)

$CaAl_2Si_2O_8 + 2 \ CO_2 + 3 \ H_2O = Al_2Si_2O_5(OH)_4 + Ca^{2+} + 2HCO_3^-$
(anorthite)   kaolinite)

Figure 10 shows the stability relations among feldspars and their alteration products (Bowers et al. 1984, Berman 1988). The dashed lines represent quartz and amorphous silica saturation, respectively. All waters are weakly oversaturated with respect to quartz. Silica concetration increases with increasing TDS. All waters are consistent with kaolinite as the prime Al-rich residue of the feldspar and mica alteration. Mine waters with high pH values, however, fall near the boundary of equilibrium with K-feldspar.

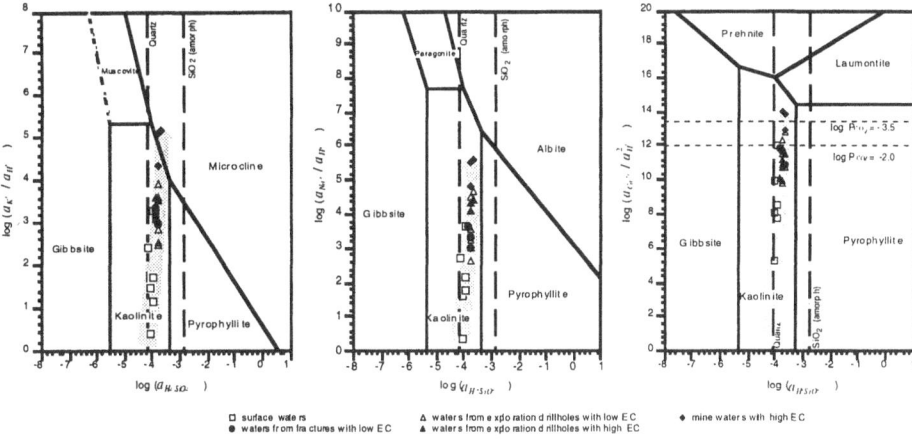

Fig.10: Stability relations among feldspars and their alteration products (Bowers et al. 1984, Berman 1988). The dashed lines represent quartz and amorphous silica saturation, respectively

7. Conclusions

Water from the Clara mine is predominantly local meteoric water. Its initial near surface chemical evolution is controlled by water-rock reactions. Feldspar (and biotite) alteration and dissolution of soluble minerals, e.g. calcite, gypsum are important. The plagioclase alteration reaction forms abundant kaolinite and creates a Ca-HCO$_3$-type water if CO$_2$ is present. This water further descends along fractures and through the highly permeable barite and fluorite veins. Ca and sulphate increases rapidly as a result of dissolution of secondary calcite and gypsum. The water changes to a Ca-SO$_4$-HCO$_3$-type. Finally in the deeper part of the mine, the water grades into a Ca-SO$_4$-Cl-type and Na increases in proportion. The systematic of the measured Cl/Br ratios of the water samples strongly suggest that chloride is mostly contributed by opened fluid inclusions in primary minerals and by alteration of Cl-bearing minerals.

Acknowledgments: We thank Dr. K.-H.Huck from the Clara mine for support in the field and Sigrid Hirth-Walther and Erika Lutz for assistance with the analytical work. Peter Möller and Margot Isenbeck-Schröter areacknowledged for their valuable comments on the manuscript.

References

Berman R. G. (1988): Internally-Consistent Thermodynamic Data for Minerals in the System: Na2O- K2O- CaO- MgO- FeO-Fe2O3- Al2O3- SiO2- TiO2- H2O- CO2. Journal of Petrology 29, 445-522.

Biehler, D. (1995): Kluftgrundwässer im kristallinen Grundgebirge des Schwarzwaldes - Ergebnisse von Untersuchungen in Stollen.- Dissertation an der Universität Tübingen, TGA, C22, 103 S., Tübingen.

Bliedtner, M. & Martin, M., (1986): Erz- und Minerallagerstätten des Mittleren Schwarzwaldes. - Geologisches Landesamt Baden-Württemberg,   782 p., Freiburg iBr.

Böhlke, J.K. & Irwin, J.J. (1992): Laser microprobe analyses of Cl, Br, I, and K in fluid inclusions: Implications for sources of salinity in some ancient hydrothermal fluids.- Geochemica et Cosmochemica Acta, v. 56, p. 203-225.

Bowers, T.S., Jackson, K.J., Helgeson, H.C. (1984): Equilibrium Activity Diagrams for Coexisting Minerals and Aqueous Solutions at Pressures and Temperature to 5kb and 600°C. Springer-Verlag Berlin Heidelberg.

Bucher, K., Stober, I. (2001a): Does plagioclase control the composition of groundwater in the crystalline basement?.- Water-Rock Interaction (WRI-10), ed. R. Cidu, p. 149-152, Balkema publishers.

Bucher, K., Stober, I. (2001b): Water-rock reaction experiments with Black Forest gneiss and granite, this volume

Edmunds, W.M. (1996): Bromine geochemistry of British groundwaters.- Mineralogical Magazine, Vol.60, pp.275-284.

Groschopf, R., Kessler, G., Leiber, J., Maus, H., Ohmert, W., Schreiner, A., Wimmenauer, W. (1996): Erläuterungen zur Geologischen Karte von Baden-Württemberg Freiburg i.Br. und Umgebung.- Geologisches Landesamt B.-W., 3. Aufl., 364 S., 27 Abb., Freiburg.

Huck, K.-H. (1986): Clara am Schwarzbruch. In: Bliedtner, M. & Martin, M.: Erz- und Minerallagerstätten des Mittleren Schwarzwaldes. - Geologisches Landesamt Baden-Württemberg, Freiburg iBr., 366-399.

Kullerud, K. (2000): Occurrence and origin of Cl-rich amphibole and biotite in the Earth's crust – implications for fluid composition and evolution.- p.205-226, In: Stober, I., K. Bucher (eds.): Hydrogeology of Crystalline Rocks, KLUWER Academic Publishers, 275 p., Dordrecht Boston London.

Parkhurst D. L., Thorstenson D. C., Plummer L. N. (1980): PHREEQE - a computer program for geochemical calculations. Water Resources Investigations, 80-96, 210 pp., U.S. Geological Survey.

Stober, I., (1995): Die Wasserführung des kristallinen Grundgebirge. Enke Verlag, Stuttgart. 191pp.

Stober, I., Bucher, K. (1999a): Deep groundwater in the crystalline basement of the Black Forest region. Appl. Geochem. 14, 237-254.

Stober, I., Bucher, K. (1999b): Origin of salinity of deep groundwater in crystalline rocks.- Terra Nova, vol. 11, no. 4, 181-185.

Thury, M., Gautschi, A., Mazurek, M., Müller, W. H., Naef, H., Pearson, F. J., Vomvoris, S., Wilson, W. (1994): Geology and Hydrogeology of the Crystalline Basement of Northern Switzerland. Synthesis of Regional Investigations 1981-1993 within the Nagra Radioactive Waste Disposal Programme. Nagra, Technical Report 93-01, 1-347.

Yardley, B.W.D., Banks, D.A., Munz , L.A. (1992): Halogen composition of fluid inclusions as tracers of crustal fluid behaviour.-Water-Rock Interaction, Kharaka & Maest (eds.), pp. 1137-1140.

Extensional Veins and Pb-Zn Mineralisation in Basement Rocks: The Role of Penetration of Formation Brines

S.A. GLEESON* & B.W.D. YARDLEY

School of Earth Sciences, University of Leeds, LS2 9JT, U.K.
*Present address: Dept. of Earth and Atmospheric Sciences, University of Alberta, Edmonton T6G 2E, Canada

Abstract
The importance of sedimentary formation waters in the genesis of sediment-hosted ore deposits has been widely recognised for many years, and similar fluids are also responsible for many, generally small, orebodies in basement rocks that have been overlain by basins in the past. It is shown that, even where ores are hosted by sediments, as in the case of the Irish Pb-Zn deposits, the circulation of basinal fluids into underlying basement rocks along fractures plays an essential part in ore genesis. Often it is at this stage that metals are leached from crystalline rocks and enter solution as chloride complexes in brines that have been generated by the evaporation of seawater at surface. The zone near the interface between basement and basin has particular ore-forming potential because it is an interface between rocks of different chemistry, mineralogy and permeability structure. These differences may lead to changes in redox and pH, and hence in sulphur speciation, as well as changes in temperature gradient due to different flow regimes.

1. Introduction

The relationship between sedimentary basins and low temperature base metal deposits has long been recognised. Many world class ore districts are hosted by sedimentary successions and some models for ore genesis involve fluid movement along porous and permeable pathways in a "closed" sedimentary basin with no flow from the base of the basin to the basement (e.g. Garven and Freeze, 1984). However, basement rocks, spatially associated with sedimentary basins, are often host to base-metal mineralised and unmineralised carbonate, quartz, barite and fluorite veins. These systems are commonly sub-economic in present-day terms but have been exploited in the past. Such occurrences are the best evidence that in certain tectonic environments, basins do not act as closed systems but leak fluid into basement rocks, and that in some cases the basement has a fundamental role to play in the formation of sediment-hosted ore deposits. This contribution highlights the presence of base-metal mineralised and unmineralised veins sourced from basinal fluids but hosted by older, basement rocks

I. Stober and K. Bucher (eds.), Water-Rock Interaction, 189–205.

and the role the basement plays in one type of world-class basin-hosted deposit. This paper is not intended to give a thorough review of these vein occurrences but seeks to point out some general features using some examples of base metal sulphide mineralised and non-mineralised veins from Ireland, Southwest England, Massif Central, Iberia and the Soultz-sous-Forêt borehole, a modern system in France (Fig. 1).

Figure 1: Sketch map of Western Europe showing some of the locations of samples mentioned in the text and summarised in Table 1. Locations are as follows: 1. Tynagh Pb-Zn mine 2. Fluorite veins in the Galway Granite 3. Fulmort Quarry 4. Lower Palaeozoic hosted veins; 5. Silvermines Pb-Zn mine 6. Porthleven, Cornwall, U.K. 7. Menheniot, Cornwall, U.K. 8. Soultz-sous-Forêt borehole, France; 9. Saint Salvy, Massif Central, France. 10. Spanish Central System, Spain.

2.   Modern day formation waters

Sedimentary formation waters can broadly be classified into three types; dilute Na-HCO_3 and Na-acetate waters; Cl-dominated, halite unsaturated brines with salinities of less than 300,000 mgl^{-1} and Cl-dominated, halite saturated brines with salinities of greater than 300,000 mgl^{-1} (Hanor, 1994). Ultimately all these aqueous fluids are surfaced derived and can be sourced from either meteoric- or sea- water, however, commonly present day formation water have mixed origins. In many sedimentary basins there is a relationship between increasing salinities and depth. In most saline formation waters, the chloride component is either derived at surface by the evaporation of seawater or Cl-bearing groundwaters or by the dissolution of halide-bearing evaporites.

Cl-dominated brines can contain significant quantities of divalent cations, particularly Ca. Analyses of brines for transition metals such as Pb and Zn are rarely reported in the literature although elements such as Fe and Mn are more commonly documented. Nevertheless, ore-forming concentrations of Pb and Zn in some saline formation waters have been reported from some regions (e.g. Carpenter et al., 1974; Kharaka et al., 1987; Land, 1995) and thus, these fluids have been suggested as the favoured solutions for the formation of low temperature base metal deposits such as the carbonate-hosted deposits of the Mississippi Valley (Sverjensky, 1986). The pH of these evolved brines is often low and chloride plays an important role in the complexation and transport of metals (Sverjensky, 1984).

3. Pb-Zn mineralising fluids

3.1 IRISH ZN-PB DEPOSITS

3.1.1. Nature of the Irish Zn-Pb deposits
The Lower Carboniferous rocks of the Irish Midlands host one of the major ore districts in Europe including the Navan, Lisheen and Galmoy deposits, which are being mined today and the Silvermines and Tynagh deposits which are now closed. These carbonate-hosted deposits lie along strong NE-SW Caledonian lineaments and appear to be spatially associated with the intersection of these structures with E-W basement structures and weaker NNE-SSW lineaments (Everett et al., 1999b). Mineralisation is found associated with normal faults and is normally situated at the point of maximum throw of the faults with minor mineralisation occurring at major fault intersections and in ramp relay zones between faults (Hitzman, 1995). Generally, in these deposits, mineralisation occurs preferentially in the stratigraphically lowest non-argillaceous carbonate unit, is broadly stratabound and displays complex sulphide textures. However, some mineralisation and alteration does extend along fractures into the basement rocks underlying the deposits (e.g. Andrew, 1986; Mallon, 1997). Mineralisation is dominated by sphalerite, galena and (variably) Fe-sulphides. Hydrothermal dolomites are the most common gangue minerals associated with the deposits although barite and siderite are also found in some deposits.

The age of these deposits is contentious with most workers accepting a Lower Carboniferous age (e.g. Ashton et al., 1992) although some consider the deposits to be younger (e.g. Hitzman, 1995; Peace and Wallace, 2000). Due to the complexity of the textures these deposits have been described as being syn-diagenetic/epigenetic (e.g. Hitzman, 1995, Anderson et al., 1998), or epigenetic with a syngenetic or exhalative component (e.g. Boyce et al., 1983; Banks, 1985; Russell et al., 1989).

Fluid inclusion studies on these deposits suggest that a moderately hot (200°C), intermediate salinity fluid (12-15wt%) metal bearing fluids mixes at the site of mineralisation with a low temperature, high salinity pore fluid (e.g. Eyre, 1998; Samson and Russell 1987; Banks and Russell 1992). Sulphur isotopes suggest that the metal bearing end-member contains a small amount of reduced sulphur whereas the dominant sulphur source in the deposits is marine sulphate reduced by the action of sulphate reducing bacteria (e.g. Caulfield et al., 1986). This sulphur reservoir is likely to be associated with the low temperature brine end member.

3.1.2.   Role of the basement in the formation of the Irish Deposits

Isotope studies carried out on the Irish deposits suggest that, in marked contrast to S, the Pb is sourced from the basement rocks that lie under the deposits ie. the basement has a fundamental role as a source for metals for the ore district (O'Keefe, 1986; LeHuray et al., 1987).

There is another equally fundamental role for the basement underneath the deposits. The high temperatures recorded by fluid inclusions (up to 220°C) in many of the deposits suggest that the mineralising fluids, trapped at shallow levels, must be sourced from depths of at least 3km (Everett et al., 1999) with many authors suggesting depths of ≈10-12 km, even with the higher heat flows expected in a rift environment (Russell, 1978; 1986). Hydrogeologic modelling suggest these high temperatures cannot be achieved wholly by topographic flow within the Upper Palaeozoic sedimentary succession, which would be analogous to the model commonly invoked for the MVT deposits (Garven et al., 1999). Some fluid flow in the Lower Palaeozoic basement must be invoked in the absence of magmatic activity (Garven et al., 1999).

A recent fluid inclusion and isotopic study has identified both the high temperature and low temperature fluid end-member seen in the deposits in veins in Lower Palaeozoic rocks that were basement to the Carboniferous sequence (Everett et al., 1999b). This, along with studies of the deposits and on the Devonian Old Red Sandstone unit underlying them (Mallon, 1987), strongly supports the hypothesis that the metal bearing fluid was heated and acquired metals at depth and flowed along fractures to shallower levels in the crust where the deposits were formed as the result of mixing with a low temperature, higher salinity brine. Some samples also show evidence of mixing with a third, low temperature, low salinity fluid as well (Everett et al., 1999b).

These mixing trends are seen elsewhere in Ireland. Palaeozoic hosted veins from Galway, Ireland show a wide range in salinities, which O'Reilly et al. (1997) interpret as representing a mixing trend between a 80-125°C, 21-28wt% NaCl equiv. fluid and a hotter more dilute fluid (Figure 2). Preliminary data from unmineralised, crosscutting quartz-calcite-fluorite veins from a Lower Carboniferous quarry in the Irish Central Midlands show a similar trend in data (Figure 2).

3.1.3.   Origin of the mineralising fluids

Fluid inclusion, isotopic and geochemical studies suggest that the intermediate salinity, high temperature ore fluid has a composition which is consistent with Carboniferous seawater that has extensively interacted and equilibrated with metasedimentary rocks (Everett et al., 1999a,b and references therein). The halogen systematics of this fluid suggest that the fluid originated as seawater and was either evaporated at surface or was dehydrated in the subsurface to generate the approx. 15wt% NaCl equiv. salinities observed in fluid inclusions (Gleeson et al., 1999). The low temperature brine present in the deposits and elsewhere is dominated by $NaCl-CaCl_2$ salts, has a similar composition to highly evolved formation waters and has been interpreted as a basinal brine (Banks and Russell, 1992).

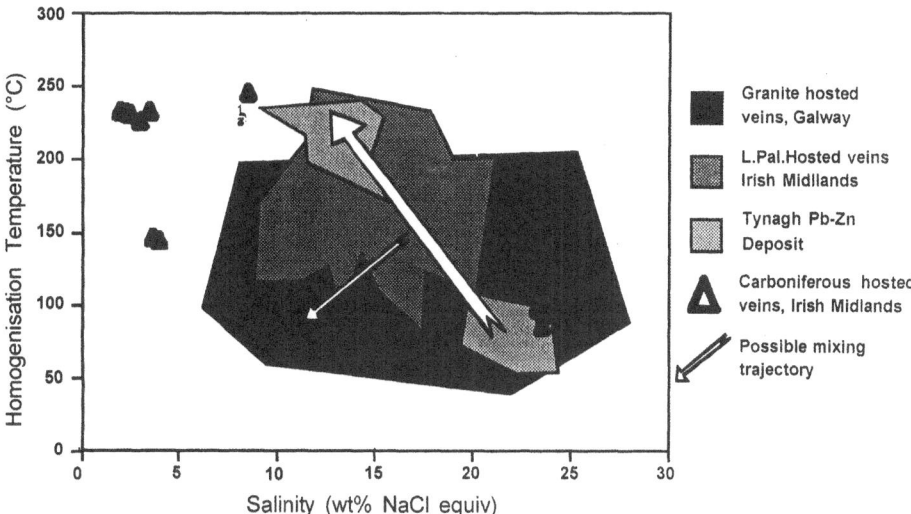

Figure 2: A summary of published microthermometric data from veins hosted by Lower Palaeozoic rocks (Everett et al., 1999b; O'Reilly et al., 1987), Tynagh, one of the Irish base-metal deposits (Banks and Russell, 1992) and some previously unpublished data from quartz-carbonate-fluorite veins hosted by Lower Carboniferous rocks in Fulmort Quarry in the Irish Midlands. All localities show evidence of mixing of a low temperature high salinity basinal brine with higher temperature, intermediate salinity fluids. Some of the Lower Palaeozoic veins also have evidence of the presence of a third, low temperature, lower salinity fluid.

Chlorine and bromine are considered by many workers to be relatively conservative elements in many crustal systems and are routinely used to distinguish between an evaporitic seawater origin and a dissolved halite source for brines. Halogen systematics have been used widely in this context to fingerprint the origins of sedimentary formation waters as well as mineralising fluids. However, if seawater is brought deep into the crust and undergoes dehydration as a result of water rock interaction at elevated temperatures and pressures (e.g. Markl et al., 1998) is it possible to tell such a subsurface process from the acquisition of salinity at the surface? If the reaction is simply one of losing OH but conserving Cl, Br and Na in the residual fluid, then molar ratio plots such as those used by Kesler et al. (1996) will not distinguish this process from seawater evaporation. The plot of absolute concentrations of Cl against Br introduced by Carpenter (1978) shows the seawater evaporation trend (SET) at surface temperatures. If the water loss is not being caused by evaporation but in fact dehydration occurring at high temperatures, the concentration point at which halite would precipitate would of course be higher (see Figure 3). In this way it should also be possible to distinguish between sub-surface dehydration reactions and surface evaporation in the generation of saline brines. In the case of the Irish orefield then this suggests that the high salinity fluid did indeed acquire its chlorinity in an evaporative environment at surface before it entered the basement rocks rather than through dehydration of the fluids at elevated temperatures. As the higher temperature ore fluid has not reached the point of halite precipitation it is not possible to distinguish between these two end-members solely on the basis of halogens. Mass balance considerations and the lack of large amounts of hydrous alteration in exposed basement rocks, however, suggest that this fluid also acquired its salinity at surface by evaporative processes.

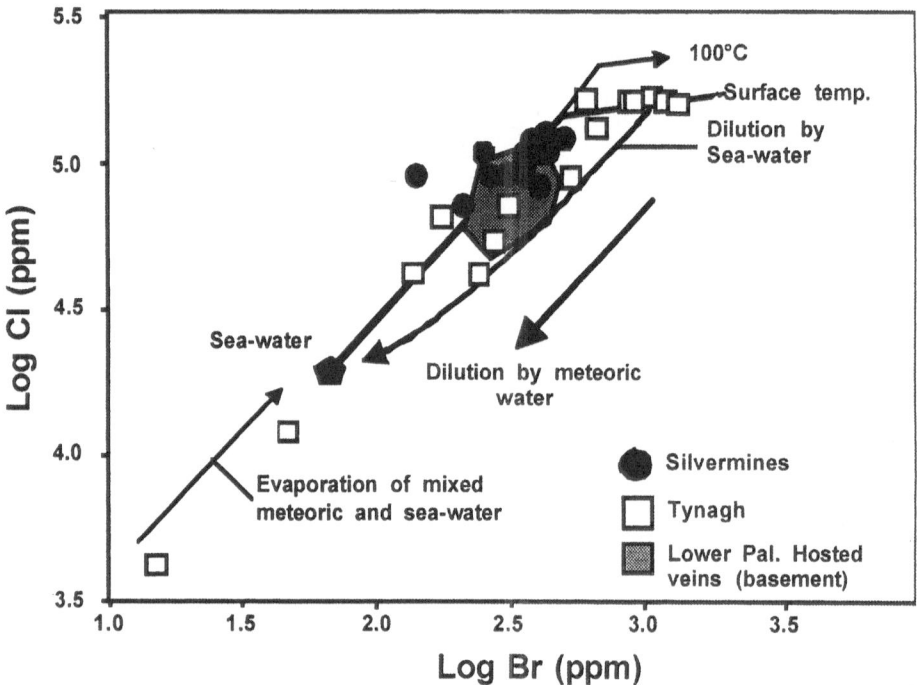

Figure 3: Plot of the absolute Cl and Br data from two Irish, Lower Carboniferous-hosted Zn-Pb deposits and data from veins in Lower Palaeozoic rocks which underlie the Carboniferous basins (after Gleeson et al., 1999). The seawater evaporation trajectory is drawn from the data compiled by Carpenter (1978) and represents the evaporation of seawater at surface temperatures. Loss of water from the fluid at higher temperatures shifts the point of halite saturation to higher chloride and bromide values.

3.2 LOW TEMPERATURE PB-ZN MINERALISATION IN WESTERN EUROPE

There are many instances of low temperature Pb-Zn (± fluorite) mineralisation found in Western Europe. Unlike the Irish deposits however, these lodes are often vein-hosted, sub-economic and are/have been spatially associated with sedimentary basins but are hosted by older rocks.

Table 1: Summary table of Pb-Zn mineralised and un-mineralised veins from Western Europe

Location	Vein-type	Minerals	Temp.[1] (°C)	Sal	Cl/Br (molar)	Age of Min.	Age host rocks	TS	Source Refs.
Porthleven, Cornwall	Cross-course	Qtz± Carb ±Sulphides	100-150	25-27	350-775	?Post L.Tri ?Early Trias.	L. Pal.	ext	Gleeson et al. 2000; Shepherd et al. 1994
Sans Salvy, France	M3	Qtz± Carb ±Sulphides	80-140	23-25	N/a	?Jurassic	L. Cam.	ext	Munoz et al., 1995
Les Malines France		Qtz± Ba±Carb ±Sulphides	<150	3-21	N/a	?Tri-Jur	Cam.	? ext	Charef & Sheppard, 1988
Spanish Central Sys.	Lw2	Qtz	90-160	25-27	703-723	?Per-Tri	Pal.	? ext	Crespo et al., 1999
Catalonian Coastal Rang.	Atrevida	Qtz F± Ba±Cart ±Sulphides	<100	21-22	N/a	L. Tri	Pal.	ext	Canals et al., 1992
Germany		Ba± F± Qtz±Carb ±Sulphides	<200	21-27	66-155	Post-Varisc.	Mostly Pal. Few PT	ext	Behr & Gerler, 1987
Central Europe		Ba± F± Qtz±Carb ±Sulphides				L. Per- L. Jur	Pal.	ext	Dill & Nielsen, 1987
Galway, Ireland	V3	Qtz± F± Ba±Cart ±Sulphides	125-205	8-28	N/a	? L.Tri	L. Pal.	ext	O'Reilly et al., 1997
Fulmort, Irel.		Qtz± F	84-232	3-26	N/a	Post Carb	Carb.	ext	New data
Midlands, Ireland		Qtz± Ba ±Carb ±Sulphides	115-226	9-18	375-825	Post- U. Pal	U. Pal.	ext	Everett et al., 1999; Gleeson et al., 1999
Soultz, France	Stage 5	Qtz	99-202	3-15	N/a	Mes	Pal.	? ext	Smith et al., 1998

[1] Fluid inclusion homgenisation temperatures. Sal = salinity in NaCl wt% equiv. PT = Permotriassic, TS = Tectonic Setting, ext = extensional, ?ext = probably extensional

Microthermometric analyses of hydrothermal veins systems can give some compositional and pressure temperature information on the nature of the mineralising fluids in these systems. Figure 4 summarises the microthermometric data from some of these areas listed in Table 1. Much of the data from the Lower-Palaeozoic hosted veins have features in common. Mineralised veins from Cornwall, Saint Salvy and the Spanish Central System are dominated by low temperature (<150°C), high salinity (>20wt% NaCl equiv.) Ca-rich fluids. These fluids are also comparable with those involved in Pb-Zn mineralisation in other provinces such as the Mississippi Valley Type deposits in the U.S.A, Canada, and Australia; and also deposits in U.K. (Pattrick and Polya, 1993), Spain (e.g. Cardallach et al., 1990; Canals et al., 1992; Velasco et al., 1994), Germany (e.g. Behr and Gerler, 1987), Central Europe (Dill & Nielsen, 1987); France (Charef & Sheppard, 1988; Bruce, 2000). All these fluids are compositionally similar to saline Cl-bearing, halite unsaturated formation waters and are considered to be sourced from local basins. Also plotted on Figure 4 are some microthermometric data from un-mineralised veins hosted by Lower Palaeozoic basement underlying the Western Approaches Basin of the English Channel. These

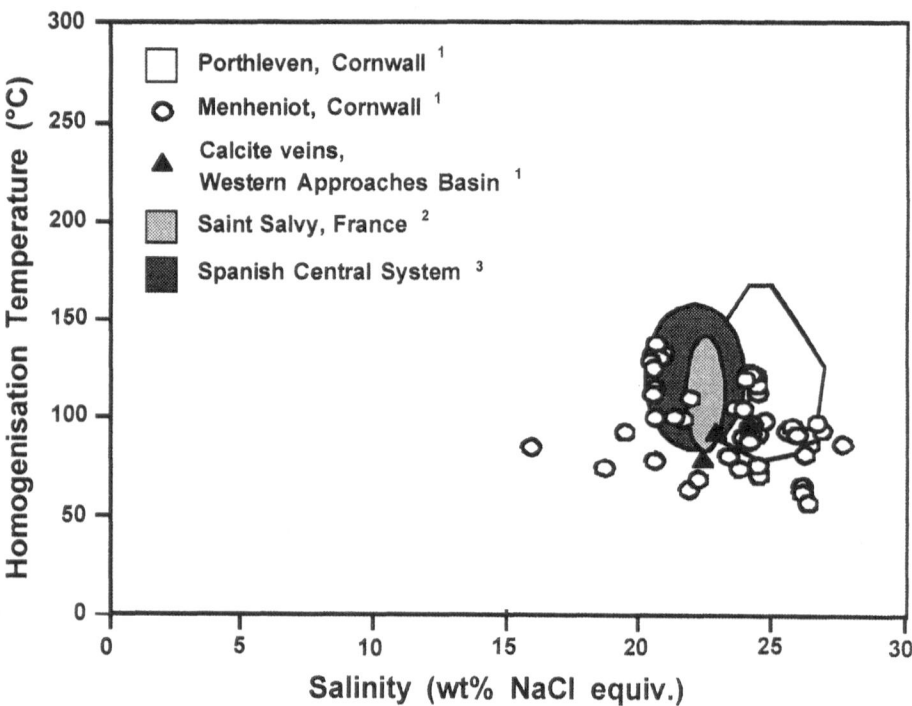

Figure 4: Microthermometric data collated from some published studies of low temperature Pb-Zn mineralised and un-mineralised veins in Western Europe. [1] Gleeson et al., (2000); [2] Munoz et al., (1994); [3] Crespo et al., (1999). This data are comparable with those obtained from many other studies on low temperature Pb-Zn deposits.

fluids clearly have similar salinities and bulk chemistries to the base metal mineralising fluid responsible for mineralisation in South Cornwall and have been interpreted as having a Permo-Triassic basinal origin for the mineralising fluids (Gleeson et al., 2000).

Table 1 summarises the results of the published halogen data for the brine end-members in several basement hosted Pb-Zn lodes. With the exception of the unmineralised veins from the Spanish Central System, the mineralised veins are dominated by fluids with Cl/Br ratios less than seawater (Cl/Br (molar) ≈ 660) ie. they are bittern brines suggesting that evaporated seawater is the source of the chlorinity in all these fluids. Like the Irish deposits there is no evidence either for a significant role for subsurface dehydration reactions or halite dissolution. These results are comparable with the studies carried out by Viets and co-workers (1996) and Kesler et al., (1996) that showed that the bulk of the Mississippi Valley Type mineralising fluids in central and western U.S. were similarly derived from seawater. These results, along with the microthermometric studies suggest that unlike modern formation waters that can have variable origins, many mineralising brines originate as evaporated seawater at surface, acquire Ca along the flow path (possibly in the basinal sequences), then migrate within the basin or into basement rocks.

4.   Mechanisms of brine migration

4.1 ROLE OF FLUID DENSITY?

In many sedimentary basins the salinity of formation waters increases with depth (Hanor, 1994). Although increased temperatures with depth reduce the density of sedimentary brines, this effect is greatly outweighed by the increase in salinity. Generally then, these saline brines found deep in basins have densities much greater than that of pure water near the surface (in the order of 1.1 g/cm^3) and are commonly at the ambient temperatures of local rocks that will be controlled by regional geothermal gradients. These dense fluids have the ability to displace other less dense fluids and will sink to the bottom of aquifers and/or into fractured basement rocks if there is sufficient permeability.

Fluids moving down into the crust are unlikely to precipitate quartz and fluorite but may precipitate carbonate and anhydrite due to the retrograde solubility of these minerals. Carbonate veins have been documented in Lower Palaeozoic rocks just underlying the basal unconformity of the Permo-Triassic Western Approaches Basins in the English Channel. Fluid inclusions in these veins suggest they form from saline brines with homogenisation temperatures of approx. 97°C. The veins may, therefore, result from formation waters leaking underneath the sedimentary basin. However, most gangue and sulphide minerals in these rocks do not have retrograde solubility and, excluding the effects of mixing and wall-rock reactions, this suggests that the fluids that formed many veins were moving down temperature and pressure and are likely to have come from deeper, hotter areas of the crust.

Density driven convective overturn is a possible mechanism for turning surface derived brines into much hotter mineralising fluids and has been suggested as the mechanism of formation of the relatively higher temperature Irish Zn-Pb deposits (e.g. Russell, 1978). One of the fundamental controls on the process is permeability.

Convective systems like those found around Mid-Ocean Ridges commonly have diffuse down-flow zones but up-flow is focussed. Strens et al. (1987) however, showed that convective cells cool relatively quickly resulting in mineral precipitation and the destruction of permeability. In order to keep a convective system operating for a significant time period it is necessary to keep propagating fractures into fresh hotter rocks. The problem therefore, is not in initiating convection in these systems but sustaining it over time. In the absence of a local magmatic heat source in the Irish Midlands in the Carboniferous, to produce the volumes of hot fluids needed at shallow levels, the fluid filled fractures would need to propagate downwards during rifting events.

4.2 ROLE OF STRUCTURE

Many genetic models for the formation of low temperature MVT-type Pb-Zn deposits invoke aquifer controlled fluid flow confined within a sedimentary basin (e.g. Garven and Freeze 1984). These models assume an essentially impermeable basement. Although the underlying basement rocks are likely to have significantly lower permeabilities and matrix flow is unlikely to occur, fracture controlled permeability may be important. This is supported by the fact that many of low tonnage Pb-Zn occurrences in Western Europe are vein-hosted deposits. Many of the examples listed in Table 1 have banded open space filling vein textures that may form as a result of episodic fluid flow along dilatant structures but equally could be the result of episodic movement on the faults promoting fluid flow. Structure also plays an important role in the localisation of the basin-hosted Irish Zn-Pb deposits. Although there are examples of formation brines forming veins in basement rocks in compressional settings (e.g. McCaig et al., 2000), and associated with local dilation in such environments, many of the vein hosted Pb-Zn occurrences in Western Europe are found in extensional regimes (Table 1).

The role of seismicity in the formation of mineral deposits has been recognised for many years. In particular, many studies on mesothermal gold deposits have given rise to models such as the fault-valve model which relates seismicity and cycling of fluid pressures from hydrostatic to lithostatic values (e.g. Sibson, 1990). However, many low temperature Pb-Zn veins form in the upper crust, at relatively shallow depths and at or close to hydrostatic pressures. Modern examples of surface derived fluids entering crystalline basement rocks at near-hydrostatic pressures are well known from continental drilling. Data from the KTB borehole suggests that in intra plate regions the brittle crust is critically stressed and pore fluid pressures are close to hydrostatic to depths of 10km (Townend and Zoback, 2000). At Soultz-sous-Forêt in the Western Rhine graben, deep drilling encountered brines in Hercynian granite basement with compositions comparable with Central Rhine Graben sedimentary formation waters (Pauwels et al., 1993). In these veins the mean homogenisation temperatures of quartz hosted fluid inclusions corresponds well to the present day fluid temperature in the borehole (Smith et al., 1998). The tightly clustered distribution of the homogenisation temperatures may be indicative of close to hydrostatically pressured systems (Yardley et al., 2000). Data from the Soultz borehole, demonstrates the pattern of fluid inclusion data that results from near-hydrostatically-pressured flow of fluid through a basement fracture system (Figure 5).

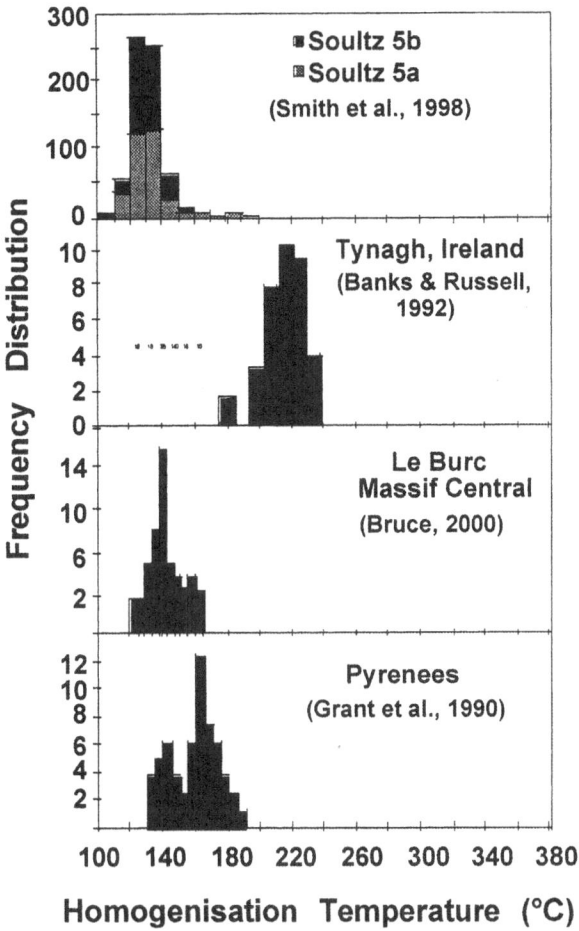

Figure 5: Histograms of homogenisation data from a modern hydrostatically pressured flow system (the geothermal system at Soultz-sous-Forêt) and examples from two mineralised palaeohydrological systems (Tynagh, Ireland and Le Burc, France). These data show a similar narrow range of homogenisation temperatures to the Soultz data. The final histogram shows data from a compressional regime. The two peaks in the histogram have been interpreted by the authors as representing trapping of a single fluid at both hydrostatic and lithostatic pressures.

One of the primary difficulties when dealing with mineralisation is the generation of high fluid fluxes in a relatively small area. If these fluid systems are at or near hydrostatic pressures then the rupturing of faults due to near lithostatic pressures (over pressures) is not likely to be widespread process. Sanderson and Zhang (1999) have shown that if a fault-fracture network changes from low differential stresses through to the critical stress state of that system then initially diffuse flow changes to highly localised flow along larger fractures and an increased hydraulic conductivity. Fluid pressures in these systems, though elevated, are sublithostatic. In a system like this stress cycling through the critical state could produce transient fluid flow which may account for incremental mineral precipitation. If the upper crust is indeed critically stressed (e.g. Townend and Zoback, 2000) then this model suggests that any tectonism could initiate large scale focussed fluid flow in major structures. Hydrogeological studies of fluid movements as the result of seismic activity suggest that normal fault earthquakes predominately expel fluids. Muir Wood (1994) suggests this occurs as dilational high angle fractures undergo partial closure during fault rupturing and as a result fluids are displaced upwards into shallower levels in the crust.

5. Discussion

Many of the low temperature Pb-Zn vein occurrences described above and listed in Table 1 are found in high angle extensional fractures formed in basement rocks, spatially associated with sedimentary basins. The age of these veins is often difficult to accurately ascertain due to the lack of minerals suitable for radiometric dating. In general, then, workers have to rely on crosscutting and structural criteria to put age constraints on the age of their formation. Although there is some variability, many of the vein-hosted Pb-Zn lodes in Western Europe are thought to have originated during Mesozoic times and associated with major phases of crustal extension; either related to Permo-Triassic basin development or to the opening of the Atlantic.

In Permo-Triassic times in Western Europe there was a major phase of crustal extension resulting in the formation of rift basins (Coward, 1995). The combination of such a tectonic setting and an arid environment produces regions where the rate of evaporation exceeds inflow of waters (Hardie, 1991), allowing the generation of evaporites. Halogen analysis of fluid inclusions suggests that many of the mineralising fluids originate as bittern brines ie. as evaporated seawater. This suggests that the brines are only generated when rifting is sufficiently well developed to allow the initial transgression of seawater into the source region. The brines are therefore, generated in an environment which will be influenced alternatively by marine and non-marine waters. The density of the brines may allow them to be displaced downwards by either meteoric waters or new influxes of seawater into the sediments or directly into fractures both in the basins and under the basins. Alternatively, the fluid flow may be confined to the sediments in the basins until renewed tectonic activity (further rifting) allows the establishment of large fractures into the basement. The downward movement of brines may be preserved as carbonate and anhydrite veins as seen in Devonian limestones under the Permo-Triassic Western Approaches Basin of the English Channel. Once in the basement rocks the brines may become heated due to the elevated geothermal gradients associated with rifting or due to perturbations of the geotherms by interaction with associated with local features such as high heat producing granites (e.g. Solomon & Henrich 1984; Gleeson et al., 2000). In many cases it is not clear to what depths the fluids have travelled but fluid temperatures suggest these low temperature vein systems are formed at maximum depths of 3-4km below the surface. If the critical stress levels of the upper crust mean that potentially focussed permeable pathways for fluid flow will be generated during any tectonic activity, then it is likely that the formation of extensional sedimentary basins may always result in structurally controlled leakage of basinal brines into basement rocks and vein formation. The paragenesis of the resulting veins may be controlled by the chemistry of the source basin and the flow-path, but also by more localised controls such as fluid mixing or wall-rock interactions at the site of mineralisation.

Such processes do not produce large ore deposits however. The Irish Zn-Pb deposits also form from Cl-rich bittern brines generated at surface and these fluids do indeed migrate into basement rocks. These deposits are significantly different from the examples mentioned above in that they have large tonnages and require much larger fluid volumes, the deposits are hosted by the overlying basins and the metal bearing fluid is much hotter. In these deposits the basement has an active role to play in the supply of metals for the mineral deposits but even more significantly, the basement must be providing sufficient permeability to allow large volumes of fluid to be heated. The nature of this permeability structure, the heat source and the architecture of the fluid flow regime are still not understood. We do not find such large deposits associated with every rifting event through geological time and the formation of these deposits is likely to involve a complex series of processes involving basement (source of metals and heat) and basinal processes (source of sulphur and site of mixing).

6. Conclusions

Hydrothermal ore bodies resulting from the passage of saline formation waters are developed both in sedimentary rocks near the bottom of basins, and in the underlying basement to the basement. Most commonly such ore bodies are vein-hosted, formed at relatively low temperatures (<150°C) and shallow depths and are sub-economic in nature. Halogen analyses suggest these fluids obtain their salinity at surface by the evaporation of seawater and they are introduced from sedimentary basins into older basement rocks during periods of extension and basin development. In some cases, modern drilling has encountered such formation waters flowing through crystalline basement rocks, as in the Rhine graben. The fluids which formed the economically significant Irish Zn-Pb carbonate-hosted deposits also originated in this manner but these surface derived fluids circulated in the basement to the deposits where they were heated and acquired base metals. These ore deposits are formed when the hot metal bearing fluids from depth interact with saline formation waters in the host-rocks. Therefore in the formation of these deposits both basement and basin play a vital role.

7. Acknowledgements

This work is funded by NERC grant GR3/11087. SG would like to acknowledge many discussions on these topics with D.A. Banks, A.J. Boyce, C.E. Everett, G.R.T. Jenkin, G.D. Sevastopulo and J.J. Wilkinson. In particular we would like to acknowledge reviews of this paper by J.R. Cann and B. Jamtveit.

References

Anderson, I.K., Ashton, J.H., Boyce, A.J., Fallick, A.E. & Russell, M.J. 1998. Ore depositional processes in the Navan Zn + Pb deposit, Ireland. Econ. Geol., 93, 535-561.

Andrew, C.J. 1986. The tecton-stratigraphic controls to mineralization in the Silvermines area, Co. Tipperary, Ireland. In: Andrew, C.J., Crowe, R.W.A., Finlay, S., Pennell, W.M. & Pyne, J. (eds.) Geology and genesis of mineral deposits in Ireland. Irish Association for Economic Geology, 377-417.

Ashton, J.H., Black, A., Geraghty, J., Holdstock, M. & Hyland, E. 1992. The geological setting and metal distribution patterns of Zn-Pb-Fe mineralisation in the Navan Boulder Conglomerate. In: Bowden, A.A., Earls, G., O'Connor, P.G. & Pyne J.F. (eds.) The Irish Minerals Industry 1980-1990. Irish Association for Economic Geology, 243-280.

Banks, D.A. 1985. A fossil hydrothermal worm assemblage from the Tynagh Pb-Zn deposit in Ireland. Nature, 313, 128-131.

Banks, D.A. & Russell, M.J. 1992. Fluid mixing during ore deposition at the Tynagh base metal deposit, Ireland. Eur. Jour. Min. 4: 921-931.

Behr, H.J. & Gerler, J. 1987. Inclusions of sedimentary brines in post-Variscan mineralizations in the Federal Republic of Germany- a study by neutron activation analysis. Chem. Geol., 61, 65-77.

Boyce, A.J., Coleman, M.L. & Russell, M.J. 1983. Formation of fossil hydrothermal chimneys and mounds from Silvermines, Ireland. Nature, 306, 545-550.

Boyce, A.J., Fallick, A.E., Fletcher, T.J., Russell, M.J. & Ashton, J.H. 1994. Detailed sulphur isotope studies of Lower Palaeozoic hosted pyrite below the giant Navan Pb+Zn mine, Ireland: evidence of mass transport of crustal S to a sediment-hosted deposit. Min. Mag., 58A, 109-110.

Bruce, S. 2000. The genesis of mineralising brines in the South West Massif Central, France. Unpublished Ph.D. thesis. University of Leeds, U.K.

Canals, A., Cardellach, E. Rye, D.M. & Ayora, C. 1992. Origin of the Atrevida Vein (Catalonian Coastal Ranges, Spain): Mineralogic, fluid inclusion and stable isotope study. Econ. Geol., 87, 142-153.

Cardellach, E., Canals, A. & Tritlla, J. 1990. Late and post-Hercynian low temperature veins in the Catalonian Coastal Ranges. Acta Geol. Hisp., 25, 75-81.

Carpenter A.B. 1978. Origin and chemical evolution of brines in sedimentary basins. Okla. Geol. Surv. Circ. 79, 60-77.

Carpenter A.B., Trout, M.L. and Picket, E.E. 1974. Preliminary report on the origin and chemical evolution of lead- and zinc-rich oilfield brines in Central Mississippi. Econ. Geol., 69, 1191-1206.

Caulfield, J.B.D., LeHuray, A.P. and Rye, D.M. 1986. A review of lead and sulphur isotope investigations of Irish sediment-hosted base metal deposits with new data from Keel, Ballinalack, Moyvoughly and Tatestown deposits. In: Andrew, C.J., Crowe, R.W.A., Finlay, S., Pennell, W.M. & Pyne, J. (eds.) Geology and genesis of mineral deposits in Ireland. Irish Association for Economic Geology, 591-615.

Charef, A. & Sheppard, S.M.F. 1988. The Malines Cambrian carbonate-shale-hosted Pb-Zn deposit; France: Thermometric and isotopic (O,H) evidence for pulsating hydrothermal mineralization. Min. Depos., 23, 86-95.

Coward, M.P. 1995. Structural and tectonic setting of the Permo-Triassic basins of northwest Europe. In: S.A.R. Boldy (ed.) Permian and Triassic rifting in Northwest Europe. Geological Society, London, Special Publications 91, 7-39.

Crespo, T.M., Lopez, J.A., Banks, D., Vindel, E., and Garcia, E. 1999. Hydrothermal fluid in barren quartz veins (Spanish Central System). A comparison with W-(Sn) and F-(Ba) veins. Bol. Soc. Esp. Min., 22, 83-94.

Dill, H. and Nielsen, H. 1987. Geochemical and geological constraints on the formation of unconfomity-related vein baryte deposits of central Europe. Jour. Geol. Soc. Lond., 144, 97-105.

Everett, C.E., Wilkinson, J.J., Boyce, A.J., Ellam R., Gleeson, S.A., Rye, D.M. and Fallick, A.E. 1999a. The genesis of Irish-type base metal deposits: Characteristics and origins of the principal ore fluid. In: Stanley, C. J. (ed.) Mineral Deposits; Processes to Processing. A.A. Balkema, Rotterdam. p. 845-848.

Everett, C.E., Wilkinson, J.J. and Rye, D.M. 1999b. Fracture-controlled fluid flow in the Lower Palaeozoic basement rocks of Ireland: Implications for the genesis of Irish-type Zn-Pb deposits. In: McCaffrey, K.J.W., Lonergan, L. & Wilkinson, J.J. (eds.) Fractures, Fluid Flow and Mineralization. Geol. Soc. Spec. Pub., 155, 247-276.

Eyre, S.L. 1998. Geochemistry of dolomitization and zinc-lead mineralization in the Rathdowney trend, Ireland. Unpublished Ph.D. thesis, University of London, U.K.

Garven, G. and Freeze, R.A. 1984. Theoretical analysis of the role of groundwater flow in the genesis of stratabound ore deposits. 1, Mathematical and numerical model. Am. Jour. Sci., 284, 1085-1124

Garven, G., Appold, M.S., Toptygina, V.I., Hazlett, T.J. 1999. Hydrogeologic modeling of the genesis of carbonate-hosted Pb-Zn ores. Hydro. Journal, 7, 108-126.

Gleeson, S.A., Wilkinson, J.J., Stuart, F.M. and Banks, D.A. (in press). The origin and chemical evolution of base metal mineralising formation waters and hydrothermal fluids, South Cornwall. Geochim. Cosmochim. Acta.

Gleeson S.A., Wilkinson J.J., Shaw H.F., and Herrington R.J. 2000. Post-magmatic hydrothermal circulation and the origin of base metal mineralisation, Cornwall, U.K. Jour. Geol. Soc. Lon., 157, 580-600.

Gleeson, S.A., Banks, D.A., Everett, C.E., Wilkinson, J.J., Samson, I.M. and Boyce, A.J. 1999. Origin of mineralising fluids in Irish-type deposits: constraints from halogen analyses. In: Stanley, C. J. (ed.) Mineral Deposits; Processes to Processing. A.A. Balkema, Rotterdam. p. 857-860.

Hanor J.S. 1994 Origin of saline fluids in sedimentary basins. In Geofluids: Origin, migration and evolution of fluids in sedimentary basins (ed. J. Parnell) Spec. Pub. Geol. Soc. Lon., 78, 151-174.

Hardie L.A. 1991. On the significance of evaporites. Ann. Rev. Earth Planet. Sci., 19, 131-168.

Hitzman, M.W. 1999. Extensional faults that localize Irish syndiagenetic Zn-Pb deposits and their reactivation during Variscan compression. In: McCaffrey, K.J.W., Lonergan, L. & Wilkinson, J.J. (eds.) Fractures, Fluid Flow and Mineralization. Geol. Soc. Spec. Pub., 155, 233-247.

Hitzman, M.W. 1995. Mineralization in the Irish Zn-Pb-(Ba-Ag) orefield. In: Anderson I.K., Ashton, J.H., Earls, G., Hitzman, M.W. & Tear, S. (eds.) Irish Carbonate-hosted Zn-Pb Deposits. Soc. Econ. Geol. Guide. Series, 21, 25-61.

Kesler S.E., Martini A.M., Appold M.S., Walter L.M., Huston T.J., and Furman F.C. 1996. Na-Cl-Br systematics of fluid inclusions from Mississippi Valley-type deposits, Appalachian Basin: Constraints on solute origin and migration paths. Geochim. Cosmochim. Acta 60, 225-233.

Kharaka, Y.K., Maest, A.S., Carothers, W.W., Law, L.M. Lamothe, P.J. and Fries, T.L. 1987. Geochemsitry of metal-rich brines from central Mississippi Salt Dome basin, U.S.A. Appl. Geochem., 2, 543-561.

Land, L.S. 1995. Na-Ca-Cl saline formation waters, Frio Formation (Oligocene), south Texas, USA: Products of diagenesis. Geochim. Cosmochim. Acta, 59, 2163-2174.

LeHuray, A.P., Caulfield, J.B.D., Rye, D.M. & Dixon, P.R. 1987. Basement controls on sediment-hosted Pb-Zn deposits: a lead isotope study of Carboniferous mineralisation in Central Ireland. Econ. Geol., 1695-1709.

Mallon, A.J. 1997. Petrological and mineralogical characteristics of the Old Red Sandstone facies rocks beneath base metal deposits as a guide to the setting of mineralisation in the Irish midlands. Ph.D. thesis. University College Cork, Ireland.

Markl, G. Ferry, J. and Bucher, K. 1998. Formation of saline brines and salt in the lower crust by hydration reactions in partially retrogressed granulites from the Lofoten Islands, Norway. Amer. Jour. Sci., 298, 705-757.

McCaig, AM, Tritlla, J., and Banks, D.A. 2000. Fluid mixing and recycling during Pyrenean thrusting: Evidence from fluid inclusion halogen ratios. Geochim. Cosmochim. Acta, ??, 3395-3412.

Muir-Wood, R. 1994. Earthquakes, strain-cycling and the mobilization of fluids. In: Parnell, J. (ed.) Geofluids: Origin, Migration and Evolution of Fluids in Sedimentary Basins. Geol. Soc. Lond. Spec. Pub., 78. 85-98

Munoz, M., Boyce, A.J., Courjault-Rade, P., Fallick, A.E. and Tollon, F. 1994. Multi-stage fluid incursion in the Palaeozoic basment-hosted Saint-Salvy ore deposit (NW Montagne Noir, southern France. Appl. Geochem., 9. 609-626.

O'Keefe, W.G. 1986. Age and postulated source rocks for mineralisation in Central Ireland, as indicated by lead isotopes. In: Andrew, C.J., Crowe, R.W.A., Finlay, S., Pennell, W.M. & Pyne, J. (eds.) Geology and genesis of mineral deposits in Ireland. Irish Association for Economic Geology, 617-624.

O' Reilly, C., Jenkin, G.R.T., Feely, M., Alderton, D.A. and Fallick, A.E. 1997. A fluid inclusion and stable isotope study of 200 Ma of fluid evolution in the Galway Granite, Connemara, Ireland. Contrib. Mineral. Petrol., 129, 120-142.

Pattrick, R.A.D. and Polya, D.A. 1993. Mineralization in the British Isles, Chapman & Hall, London

Pauwels, H., Fouillac, C., and Fouillac, A.-M. 1993. Chemistry and isotopes of deep geothermal saline fluids in the Upper Rhine Graben: Origin of compounds and water-rock interactions. Geohim. Cosmochim. Acta., 57, 2737-2749.

Peace W.M. & Wallace, M.W. 2000. Timing of mineralization at the Navan Zn-Pb deposit: A post-Arundian age for Irish mineralization. Geology, 28, 711-714.

Russell, M.J. 1986. Extension and convection: a genetic model for the Irish Carboniferous base metal and barite deposits. In: Andrew, C.J., Crowe, R.W.A., Finlay, S., Pennell, W.M. & Pyne, J. (eds.) Geology and genesis of mineral deposits in Ireland. Irish Association for Economic Geology, 545-554.

Russell, M.J. 1978. Downward excavating hydrothermal cells and Irish type ore deposits: importance of an underlying thick Caledonian prism. Trans. Inst. Mining Metall., 87, B168-171.

Russell, M.J., Hall, A.J. & Turner, D. 1989. In vitro growth of iron sulphide chimneys: possible culture chambers for origin of life experiments. Terra Nova, 1, 238-241.

Samson, I.M. and Russell, M.J. 1987. Genesis of the Silvermines lead-zinc-barite deposit Ireland: Fluid inclusion and stable isotope evidence. Econ. Geol. 82, 371-394.

Sanderson, D.J. and Zhang, X. 1999. Critical stress localization of flow associated with deformation of well-fractured rock masses, with implications for mineral deposits. In: McCaffrey, K.J.W., Lonergan, L. & Wilkinson, J.J. (eds.) Fractures, Fluid Flow and Mineralisation. Geol. Soc. Spec. Pub., 155, 69-81.

Scrivener, R.C., Darbyshire, D.P.F. & Shepherd, T.J. 1994. Timing and significance of crosscourse mineralisation in SW England. Journal of the Geological Society of London, 150, 587-590.

Sibson, R.H. 1990. Faulting and fluid flow. Min. Soc. Can. Short. Course, 18, 93-132.

Smith, M.P., Savary, V., Yardley, B.W.D., Valley, J.W., Royer, J.J. & Dubois, M. 1998. The evolution of the deep flow regime at Soultz-sous- Forêts, Rhine Graben, Eastern France: Evidence from a composite quartz vein. Jour. Geophys. Res., 103, 27223-27237.

Solomon M. and Heinrich C.A. 1992 Are high-heat producing granites essential to the origin of giant lead-zinc deposits at Mount Isa and McArthur River, Australia? Explor. Mining Geol., 1, 85-91.

Strens, M.R., Cann, D.L., and Cann, J.R. 1987. A thermal balance model of the formation of sedimentary exhalative lead-zinc deposits. Econ. Geol., 82, 1192-1203.

Sverjensky, D.A. 1986. Genesis of Mississippi Valley Type ore deposits. Ann. Rev. Earth. Planet. Sci., 14, 177-199.

Sverjensky D.A. 1984. Oil field brines as ore-forming solutions. Econ. Geol. 79, 23-37.

Townend, J. and Zoback, M.D. 2000. How faulting keeps the crust strong. Geology, 28, 399-402.

Velasco, F., Herrero, J.M., Gil, P.P., Alvarez, L. & Yusta, I. 1994. Mississippi Valley-Type, Sedex and iron deposits in Lower Cretaceous Rocks of the Basque-Cantabrian Basin, Northern Spain. Fontbote & Boni (eds.). Sediment hosted Zn-Pb ores. Spec. Pub. No. 10 S.G.A., Springer-Verlag, Berlin.

Viets J.G., Hofstra A.H., and Emsbo P. (1996). Solute compositions of fluid inclusions in sphalerite from North American and European Mississippi Valley-Type ore deposits: ore fluids derived from evaporated seawater. Soc. Econ. Geol. Spec. Pub. 4. pp. 465-482.

Yardley, B.W.D., Gleeson, S.A., Bruce, S. & Banks, D.A. 2000. Origin of retrograde fluids in metamorphic rocks. Jour. Geochem. Explor., 69-70, 281-287.

Interaction of polysilicic and monosilicic acid with mineral surfaces

MARTIN DIETZEL

Geochemisches Institut der Universität Göttingen, Goldschmidtstraße 1, D-37077 Göttingen, Germany. email: mdietze@gwdg.de

Abstract

Interaction of polysilicic and monosilicic acid was studied via adsorption experiments with lepidocrocite, hematite, feroxyhyte, goethite, akaganeite, magnetite, ferrihydrite, and gibbsite. The kinetics of monosilicic acid adsorption follows a first order reaction. At equilibrium monosilicic acid adsorption may be described by surface complexation with an adsorption maximum at pH 9.8. If polysilicic acid is adsorbed to the surface, one part is bound to the surface within a relatively short time. The other part decomposes to monomer in the solution. The polymeric silica at the surface is stabilised at pH < 6. Thus the present results show that polymerization of silica at the mineral surface has to be considered only in acidic solutions.

The adsorption experiments with monosilicic acid onto iron hydroxides result in a molar ratio of Si/Fe = 0.21 at the mineral surface. Considering natural systems it may be concluded that the silica content of recent sedimentary iron oxides (Si/Fe = 0.19) is directly related to the adsorption of silicic acid onto the primary precipitates via the formation of surface complexes. The adsorption of monosilicic acid may represent the initial step for the formation of silicates. The experimental results show that this is favoured in slightly alkaline solutions. In contrast to monomeric silica disordered linked silica polymers are expected to inhibit the crystallisation of silicates.

1. Introduction

In natural environments the mobilisation and transport of silica is closely related to the interaction between silicic acid and mineral surfaces. Silicic acid is generated by the dissolution of rocks. The composition and concentration depend on the geochemical conditions. Dissolved silica of most natural waters consists of monomeric species (Hem, 1970). The concentration of silicic acid in natural solutions near the earth surface is usually in the range form 0.083 to 0.67 _ 10^{-3} mol SiO_2 L^{-1} (5 to 40 mg SiO_2 L^{-1}) with an average value of 0.25 _ 10^{-3} mol SiO_2 L^{-1} (Davies, 1964; Siever, 1972). However, additional quantities of polysilicic acids may occur e.g. in alkaline solutions and during the primary step of silicate dissolution (Iler, 1979; Dietzel, 2000). In alkaline and hydrothermal solutions the silica content may increase up to several mmol SiO_2 L^{-1} (several hundreds of mg SiO_2 L^{-1}; White et al., 1956; Jones et al., 1967; Hem, 1970;

I. Stober and K. Bucher (eds.), Water-Rock Interaction, 207–235.

Rowe et al., 1973). In sedimentary surroundings silicic acid may be withdrawn from the solution by crystallisation of silica containing solids, and by adsorption at the surface of solids, especially oxides and hydroxides (Williams and Crerar, 1985; Stumm, 1992; Cornell and Schwertmann, 1996). Oxides and hydroxides of aluminium and iron are world-wide spreaded solids in natural systems and are major constituents of many rocks, of soils,

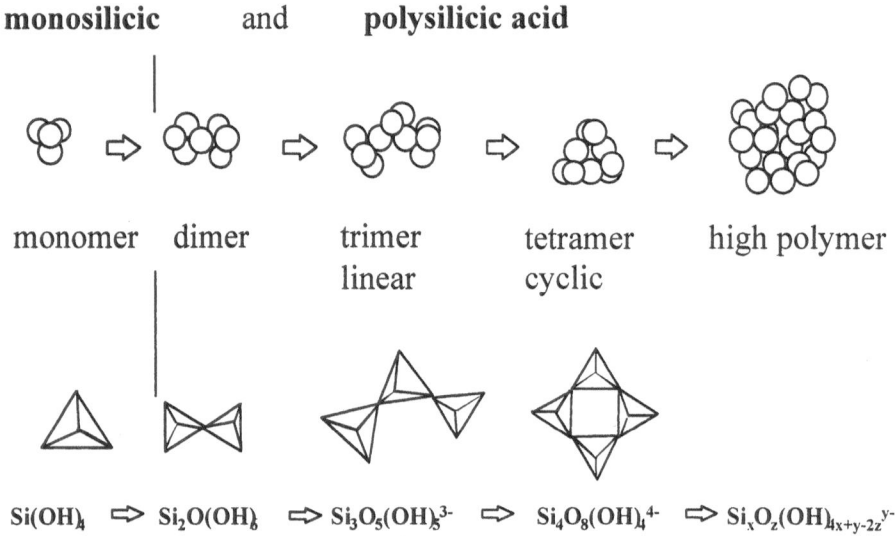

Fig. 1. Silicic acids in aqueous solution

and of sediments in rivers, streams, lakes, and the sea. They occur mostly as fine particles with high specific surface areas which favours the interaction between dissolved components and the solid surface. In many natural environments adsorption of dissolved silica limits the availability of dissolved silica, e.g. in stream sediments (Liss and Spencer, 1970), in soils (Parfitt, 1978), and in precipitation environments of iron oxides and hydroxides (Cornell and Schwertmann, 1996). In recent precipitates of iron hydroxides silica is fixed up to several weight % (e.g. Carlson and Schwertmann, 1981). Glasauer (1995) postulated that the silica is disorderly distributed within the iron hydroxide structure. General mechanism is the adsorption of dissolved silica onto the primary precipitates (e.g. Sigg and Stumm, 1981; Stumm, 1992; Hansen et al., 1994). Moreover, Schwertmann and Thalman (1976) and Cornell and Giovanoli (1987) documented that the interaction of dissolved silica with hydroxide surfaces have a significant effect on specific crystallisation rates, on morphology, and on the modification of Fe-O-OH solids. The SiO_2 fixed in the primary hydroxides and oxides represents an important source of silica for the neoformation of iron and aluminium silicates via alteration and diagenetic processes of the sediment (e.g. Füchtbauer, 1988).

The adsorption of silica onto oxide and hydroxide surfaces depends on several factors but mostly on specific surface area, surface charge, and functional groups of the solids as well as on the pH and the chemical composition of the solution (e.g. Hingston et al., 1972). With respect to monosilicic acid the adsorption may be sufficiently described by surface complexation (Sigg and Stumm, 1981; Hansen et al., 1994). In contrast to this the behaviour of polysilicic acid at the mineral surface is more complex and was insufficiently investigated in previous studies.

The adsorption behaviour of polysilicic acid may be used to study the stability of linked silica molecules at the surface of solids. Following this, it is still speculative at which boundary conditions polymerization of silica at the mineral surface has to be considered for the interaction of silica between the solid and the solution e.g. in soils. Moreover, it is claimed by Iler (1979) that neoformation of silicates requires isolated silica molecules and that disorderly linked silica molecules inhibit the arrangement of silica molecules in crystalline structures. Following this it appears that more knowledge about the stability of polymeric silica in solution and at mineral surfaces is required. The aim of this study is to decipher the boundary conditions where polymerization of silica at the mineral surface may occur.

2. System SiO_2-H_2O

In Figure 1 the species of silicic acid in aqueous solution are shown. Silicic acid may exist as isolated molecules, so called monosilicic acid, and as linked molecules, referred to as polysilicic acid. Polysilicic acid consists of silica-tetrahedrons, which are linked via silicon-oxygen-silicon bonds. Polymerization of silica may build up dimeric, trimeric, tetrameric silica, and others up to high polymeric silica.

Polysilicic acid exists especially at high pH and high concentration of dissolved silica (e.g. Eickenberg, 1990; Grenthe et al., 1992). The distribution of silica species at equilibrium with respect to amorphous silica in 0.1n Na(Cl) solution is shown in Figure 2. Monosilicic acid dissociates above a pH $\approx$ 8, and polymeric silica exists above a pH $\approx$ 9.5. Monosilicic acid concentration is related to dissociation and complexation with sodium ions. The total concentration of silica, which is the sum of dissolved components containing silica, increases in alkaline solutions. At pH = 7 and 10.5 the total concentration at equilibrium with respect to amorphous silica is about 2.0 and 47.3 $\cdot 10^{-3}$ mol SiO_2 L^{-1}, respectively.

NMR, Raman, and diffusion rate studies of dilute silica solutions by Cary et al. (1982), Alvarez and Sparks (1985), and Applin (1987), respectively, indicate the existence of disilicic acid also in neutral and acid pH regions. Considering their results Dove and Rimstidt (1994) postulated that up to 40 mol% of silica may consist of dimeric silica at equilibrium with amorphous silica. However, polysilicic acids do not only exist in alkaline solutions. They exist also in acid solutions as metastable components. Significant amounts of polysilicic acid were obtained at the primary step of silicate dissolution at pH = 3 (Dietzel, 2000).

siloxan bonds. This implies evidence about the conversion mechanism of monomer to polymer and visa versa. Whereas protonation and deprotonation of silicic acids is a fast process, the polymerization and depolymerization is slower and may be represented with $\equiv$Si-OH $+$ HO-Si$\equiv$ $=$ $\equiv$Si-O-Si$\equiv$ $+$ H_2O.

The rate of construction and destruction of polymeric silica depends mostly on pH (Baumann, 1959; Iler, 1979; Dietzel and Usdowski, 1995). The higher the pH, the faster the rate of polymerization and depolymerization. Dietzel and Usdowski (1995) showed that the depolymerization of silica may be described by a second order reaction Thus,

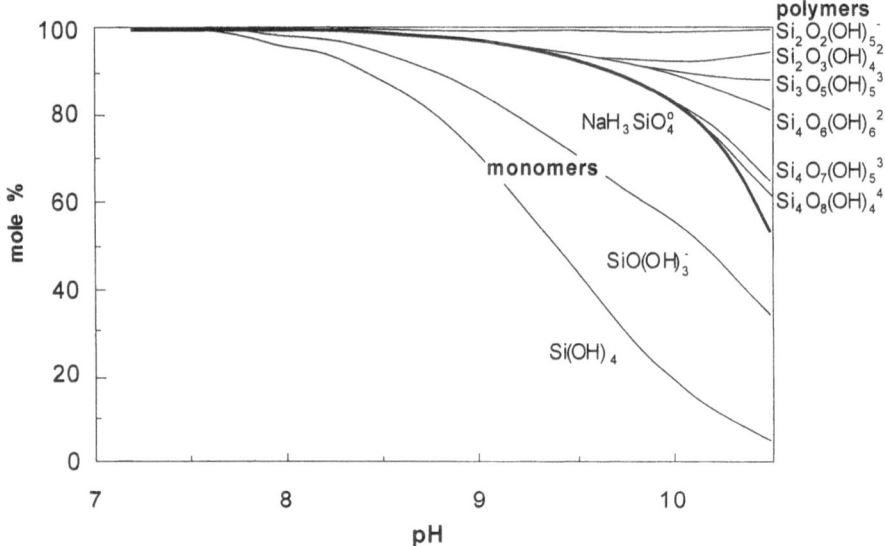

Fig. 2. Cumulative distribution of silicic acids at equilibrium with amorphous silica in 0.1 mol Na(Cl) L^{-1}. Calculation was done via the computer program MINEQL+ 4.06 (Schecher and McAvoy, 1998) at a constant ionic strength of I = 0.2. The following equations are used:

reaction			pK(25°)
$Si(OH)_4 = SiO(OH)_3^-$	$+ H^+$		9.81 ± 0.02^1
$Si(OH)_4 = SiO_2(OH)_2^{2-}$	$+ 2H^+$		23.14 ± 0.09^1
$2Si(OH)_4 = Si_2O_2(OH)_5^-$	$+ H^+$	$+ H_2O$	8.1 ± 0.3^2
$2Si(OH)_4 = Si_2O_3(OH)_4^{2-}$	$+ 2H^+$	$+ H_2O$	19.0 ± 0.3^2
$3Si(OH)_4 = Si_3O_5(OH)_5^{3-}$	$+ 3H^+$	$+ 2H_2O$	27.5 ± 0.3^2
$4Si(OH)_4 = Si_4O_6(OH)_6^{2-}$	$+ 2H^+$	$+ 4H_2O$	13.44 ± 0.2^3
$4Si(OH)_4 = Si_4O_7(OH)_5^{3-}$	$+ 3H^+$	$+ 4H_2O$	25.5 ± 0.3^2
$4Si(OH)_4 = Si_4O_8(OH)_4^{4-}$	$+ 4H^+$	$+ 4H_2O$	36.3 ± 0.5^2
$Na^+ + Si(OH)_4 = NaSiO(OH)_3^\circ$	$+ H^+$		8.66 ± 0.05^4
$Si(OH)_4 = SiO_{2(am.SiO2)} + 2H_2O$			2.72 ± 0.02^5

[1]Grenthe et al. (1992), extrapolation to I = 0 with values from Baes and Mesmer (1976), Busey and Mesmer (1977), and Sjöberg et al. (1981; 1983); [2]Grenthe et al. (1992), correction to I = 0 with data published by Sjöberg et al. (1985) via ion interaction coefficients for phosphate species of corresponding charge. [3]Baes and Mesmer (1976) for I = 0; [4]Arnorsson et al. (1982), extrapolated to 25°C with data from Seward (1974); [5]Rimstidt and Barnes (1980)

the decrease of the concentration of polysilicic acid as a function of reaction time, t, is obtained according to the equation

$$\frac{1}{[(SiO_2)_n]} = k_D \cdot t + \frac{1}{[(SiO_2)_n]_i} \tag{1}$$

Table 1. The specific surface area A (m^2 g^{-1}; BET with nitrogen gas) of the solids used in the experiments. k (10^{-3} h^{-1}) represents the reaction rate constant for the adsorption of monosilicic acid and $[Si(OH)_4]_{eq}$ (10^{-3} mol SiO_2 L^{-1}) denotes the final concentration of monosilicic acid at adsorption equilibrium after 20 days (see equation 11; at constant 34 m^2 surface area per 1 L^{-1} solution; pH = 6.2). r^2 denotes the correlation coefficient (see Figure 5). The initial concentration is $[Si(OH)_4]_i \approx 0.333 \_ 10^{-3}$ mol SiO_2 L^{-1}.

Solid	Formula	A	k	r^2	$[Si(OH)_4]_{eq}$
gibbsite[1]	γ-Al(OH)$_3$	1.8	9.17	0.978	0.187
amorphous[2]	Fe(OH)$_3$	101	10.8	0.968	0.207
ferrihydrite[2]	5Fe$_2$O$_3$·9H$_2$O	120	6.25	0.989	0.216
feroxyhyte[2]	δ-FeOOH	161	7.46	0.988	0.240
akaganeite[2]	β-FeOOH	18	7.00	0.974	0.246
lepidocrocite[2]	γ-FeOOH	153	6.38	0.952	0.258
magnetite[4]	Fe$_3$O$_4$	3.0	6.25	0.931	0.265
goethite[2]	α-FeOOH	68	10.2	0.971	0.270
hematite[3]	α-Fe$_2$O$_3$	17.7	5.54	0.985	0.271

[1]Merck (1093); [2]prepared solids (see text) [3]Riedel de Haen (1378); [4]Johnson Matthey (Alpha 976191)

In this expression $[(SiO_2)_n]_i$ and $[(SiO_2)_n]$ represent the initial and present molar concentration of silica contained in polysilicic acid, respectively, and k_D (L mol^{-1} h^{-1}) is the rate constant of the depolymerization reaction of polysilicic acid to monosilicic acid. The value of k_D is obtained from the program DEPO* which is applicable to a great variety of chemical composition of solutions. Main input parameters are temperature, pH, initial concentrations of polysilicic and monosilicic acid, and concentrations of the dissolved components. The program is based on extensive experimental studies upon the stability of polysilicic acid in aqueous solutions. It is described in detail by Dietzel (2000).

---

* The program DEPO may be obtained upon request from the author

3. Adsorption experiments

The interaction of silicic acid and solid surfaces was studied via adsorption experiments. Two types of experiments were carried out. One with initial monosilicic acid the other with initial polysilicic acid. To compare the adsorption of monosilicic and polysilicic acid both experimental sets were performed at identical boundary conditions.

The experiments were carried out in batch reactors of 1 litre polyethylene vessels. One litre of the solution contains 13 to 54 m^2 surface area of the solid under consideration.

Table 2. Concentration of monosilicic acid at adsorption equilibrium after 20 days as a function of pH (0.1n NaCl). $[Si(OH)_4]_i$ and $[Si(OH)_4]_{eq}$ denote the initial monosilicic acid concentration and the concentration at adsorption equilibrium. The concentrations are given in 10^{-3} mol SiO_2 L^{-1}.

Goethite (34 m^2 L^{-1})			Gibbsite (54 m^2 L^{-1})		
pH	$[Si(OH)_4]_i$	$[Si(OH)_4]_{eq}$	pH	$[Si(OH)_4]_i$	$[Si(OH)_4]_{eq}$
4.00	0.353	0.300	3.00	0.311	0.290
5.41	0.337	0.287	3.52	0.303	0.244
6.20	0.351	0.283	4.20	0.300	0.201
7.02	0.332	0.259	4.75	0.302	0.171
8.20	0.333	0.253	5.00	0.302	0.156
9.29	0.335	0.253	5.50	0.304	0.144
9.68	0.331	0.256	6.08	0.303	0.130
10.21	0.334	0.259	6.58	0.305	0.122
			6.98	0.298	0.105
			7.46	0.303	0.100
			7.80	0.301	0.0912
			8.50	0.301	0.0867
			9.00	0.303	0.0779
			9.70	0.301	0.0642
			10.15	0.304	0.0736
			10.50	0.320	0.0895
			10.87	0.320	0.0982
			11.40	0.324	0.132
			11.75	0.306	0.148

The solids used and their related specific surface areas are summarised in Table 1. The continuously stirred suspension was kept at a constant temperature of 25±1°C and fixed pH (±0.05). The pH was adjusted either by mixing diluted hydrochloric acid with sodium hydroxide solution at 0.1 M NaCl (3 < pH < 11.5), by acetate buffering (0.01 M NaAc; 4.5 < pH < 6.2), or buffering by carbonate (0.04 M NaHCO$_3$; pH $\approx$ 8). The solutions were prepared with double distilled water and analysis grade reagents. At the beginning of each experiment monosilicic or polysilicic acid was added to the suspension via the addition of various amounts of the stock solutions. The concentrations of silicic acid were kept below 0.5 · 10^{-3} mol SiO$_2$ L^{-1} in each experiment to avoid surpassing amorphous silica solubility ($\approx$ 2 · 10^{-3} mol SiO$_2$ L^{-1}; see Iler, 1979). Monosilicic acid was obtained from a 1.7 · 10^{-3} mol SiO$_2$ L^{-1} stock solution prepared by dissolving solid Na$_2$SiO$_3$·9H$_2$O (Baker 0310) in double distilled water. Polysilicic acid was gained from a polymer stock solution of 25.2 wt% SiO$_2$ and 8.0 wt% Na$_2$O with a polymerization degree of about 50 ((SiO$_{1.6}$(OH)$_{0.8}$)$_{50}$) containing 10 mol% monosilicic acid (Dietzel, 1993).

The content and concentration of the dissolved silica were analysed after various time periods of the experiment via the ß-silico-molybdato-method. This reaction rate method implies that the evolution of the ß-silicomolybdic acid from molybdic acid and both, polymeric and monomeric silica, occur according to an irreversible pseudo first order reaction. Thus the reaction of a mixture of polymeric and monomeric silica may be expressed by two pseudo first order parallel reactions which yield a common product. The method is described in detail by Iler (1979) and was refined by Dietzel and Usdowski (1995). The analytical error is 2% and 5% for monosilicic and polysilicic acid, respectively.

Considering 54 m^2 gibbsite and 34 m^2 goethite surface area L^{-1} (Table 2), the values of the specific surface areas summarised in Table 1, and the molecular weight of Al(OH)$_3$ and FeOOH total molar concentrations of 385 · 10^{-3} mol Al L^{-1} and 6 · 10^{-3} mol Fe L^{-1} are calculated for the present experiments with gibbsite and goethite, respectively. Dissolution experiments with gibbsite and goethite used in the present study show that at critical acid condition (pH $\approx$ 3.0) aluminium and iron concentrations are about 7 · 10^{-5} mol Al L^{-1} and 5 · 10^{-6} mol Fe L^{-1} after a reaction time of about 20 days (Böhme et al., 1999; Knollmann, 2001). Whereas, for the experiments at pH 6.2 the solubility with respect to gibbsite and goethite is only about 10^{-8} mol Al L^{-1} and 10^{-10} mol Fe L^{-1}, respectively (e.g. Stumm and Morgan, 1996; Sigg and Stumm, 1994). Following this the amount of dissolved aluminium and iron from gibbsite and goethite in the adsorption experiments is < 0.02 and < 0.08 Mol%, respectively, and may be neglected for further calculations.

4. Preparation of solids

Gibbsite, hematite, and magnetite are commercial products (see Table 1). The other solids were synthesised via crystallisation in aqueous solutions. The preparation procedures were based on experiments of Bernal et al. (1959), Brauer (1982), and Schwertmann and Cornell (1991). Goethite was obtained by titration of 5 L solution containing 0.05 M Fe^{II}(NO$_3$)$_2$·9H$_2$O with 0.1 M KOH up to pH = 13.5 and a reaction time of 30 hours at 50°C. Akaganeite was prepared in 8 days in chloridic 0.1 M Fe^{II} solution at 60°C. Synthesis of feroxyhte was done under N$_2$- atmosphere in chloridic

0.01 M Fe^{II} solution at pH = 10. Oxidation was induced by the addition of H_2O_2. After a reaction time of 30 minutes the pH was kept at 8. N_2-atmosphere was also necessary to produce lepidocrocite in chloridic 0.2 M Fe^{II} solution at 60°C. The rate of oxidation was relatively slow using $NaNO_2$. Lepidocrocite was obtained after one hour mixing with acidic hexamethylentetramine solution (2 M). The reaction of 25 g $Fe^{II}(NO_3)_2 \cdot 9H_2O$ with 2 L solution at 75°C yielded 2 L-ferrihydrite after 10 minutes. Amorphous $Fe(OH)_3$ was produced by the addition of 0.1 M $Fe^{II}Cl_2 \cdot 4H_2O$ in alkaline solution and oxidation of precipitated Fe^{II}-hydroxide under atmospheric O_2.

The solids were washed via centrifugation until the remaining solution was free of chloride or nitrate, checked by ion chromatography (Knauer, Hamilton PRP X100). Moreover, the suspended solids were purged using dialyse tubes (Nadir 38) and double distilled water. The solids were identified by X-ray diffraction and infrared spectroscopy (Perkin Elmer FTIR 1600). These analytic procedures gave the same results with respect to the identified solids before and after the experiments.

5. Distribution of surface hydroxyl groups

Hydrous oxide surfaces contain surface hydroxyl groups. In the case of gibbsite and goethite the density of surface hydroxyl groups, <≡TOTAlOH> and <≡TOTFeOH> in mol m^{-2}, was measured via potentiometric titrations. An excess of sodium hydroxide was added to a suspension of S = 35 and 68 m^2 solid surface (per 0.3 L solution) for gibbsite and goethite, respectively. After equilibration the solid was separated, and the remaining base in the supernatant solution was titrated back with 0.1 M HNO_3. The whole procedure was undertaken under nitrogen atmosphere. Considering the difference between the addition of base and the consumption of acid, ΔV in ml, the density of surface hydroxyl groups was calculated according to the expression

$$<TOT \equiv FeOH> = \frac{\Delta V \cdot m}{1000 \cdot S} \qquad (2)$$

where m is the molarity of the NaOH and HNO_3 titration solutions (0.1 M) and S is the surface area of the solid in m^2. The obtained densities of surface hydroxyl groups are <≡TOTAlOH> = $1.1 \cdot 10^{-5}$ mol m^{-2} and <TOT≡FeOH> = $9.3 \cdot 10^{-6}$ mol m^{-2}. This results in 6.9±0.6 (number of independent experiments n = 6) and 7.2±0.3 (n = 3) per nm^2 solid surface for gibbsite and goethite, respectively. These values lay within the typical range from 2 to 12 hydroxyl groups per nm^2 for metal hydroxides (e.g. Stumm, 1992).

The surface hydroxyl groups exhibit amphoteric behaviour which can be expressed by

$$\equiv FeOH_2^+ \quad = \quad \equiv FeOH^\circ + H^+ \qquad ; K_1^{app} \qquad (3)$$

$$\equiv FeOH^\circ \quad = \quad \equiv FeO^- + H^+ \qquad ; K_2^{app} \qquad (4)$$

where $\equiv FeOH_2^+$, $\equiv FeOH°$, and $\equiv FeO^-$ represent positively charged, neutral, and negatively charged surface hydroxyl groups, e.g. for goethite. K_1^{app} and K_2^{app} are the apparent surface acidity constants which include surface charge effects and hence depend on the surface ionisation. The corresponding mass law equations are

$$K_1^{app} = \frac{[\equiv FeOH°] \cdot (H^+)}{[\equiv FeOH_2^+]} \tag{5}$$

and

$$K_2^{app} = \frac{[\equiv FeO^-] \cdot (H^+)}{[\equiv FeOH°]} \tag{6}$$

where ( ) represent activities of the dissolved species and [ ] indicate concentrations of surface species in mol L^{-1}. In this equation the activity coefficients for the surface species are assumed to be equal (for detailed discussion see Dzombak and Morel, 1990; Stumm, 1992). The total concentrations of the surface species $[TOT\equiv FeOH] = [\equiv FeO^-]$ + $[\equiv FeOH°]$ + $[\equiv FeOH_2^+]$ and $[TOT\equiv AlOH]= [\equiv AlO^-]$ + $[\equiv AlOH°]$ + $[\equiv AlOH_2^+]$ in mol L^{-1} are obtained from the densities of the above surface hydroxyl groups $<TOT\equiv FeOH>$ and $<TOT\equiv AlOH>$ in mol m^{-2}, from the specific surface areas of goethite (A = 68 m^2 g^{-1}; see Table 1) and gibbsite (A = 1.8 m^2 g^{-1}), and from the concentrations of solid in the suspension in g L^{-1} for each experiment.

The values of the apparent constants K_1^{app} and K_2^{app} were obtained from potentiometric titrations with 0.1 M HNO_3 and KOH of suspensions in the presence of an inert electrolyte (0.1 to 5 M KNO_3). Titrations were carried out with a computer controlled dosage system (Methrom 702 Titrino) measuring pH via the electrodes Schott B2220 and N1180. The titrations were carried out in 0.5 L vessel with 0.3 L solution under nitrogen atmosphere at 25±0.1°C using a temperature controlled water bath (Julabo HCF18). The suspensions contain 136 m^2 solid surface per litre of solution.

In Figure 3A the titration curves for goethite are shown as a typical case. At pH $\approx$ 8.2 the effect of concentration of inert electrolyte on pH is negligible, representing the point of zero charge of the surface (pH_{PZC}). Following the procedure described in detail by Stumm (1992) the concentrations of the added acid (C_S) and base (C_B, Figure 3A) may be converted to surface charge Q in mol g^{-1} according to the expression

$$Q = ([\equiv FeOH_2^+] + [\equiv FeO^-]) \cdot \alpha^{-1} = (C_S - C_B + [OH^-] - [H^+]) \cdot \alpha^{-1} \tag{7}$$

where the factor α denotes the amount of 0.6 g goethite per 0.3 L solution. Assuming that the surface charge Q $\approx$ $[FeOH^+] \cdot \alpha^{-1}$ for pH < pH_{PZC} and Q $\approx$ $[FeO^-] \cdot \alpha^{-1}$ for pH > pH_{PZC} the values of the apparent constants are obtained according to

$$K_1^{app} = ([\equiv TOTFeOH] - Q \cdot \alpha) \cdot Q^{-1} \cdot \alpha^{-1} \cdot (H^+) \qquad \text{for pH} < pH_{PZC} \tag{8}$$

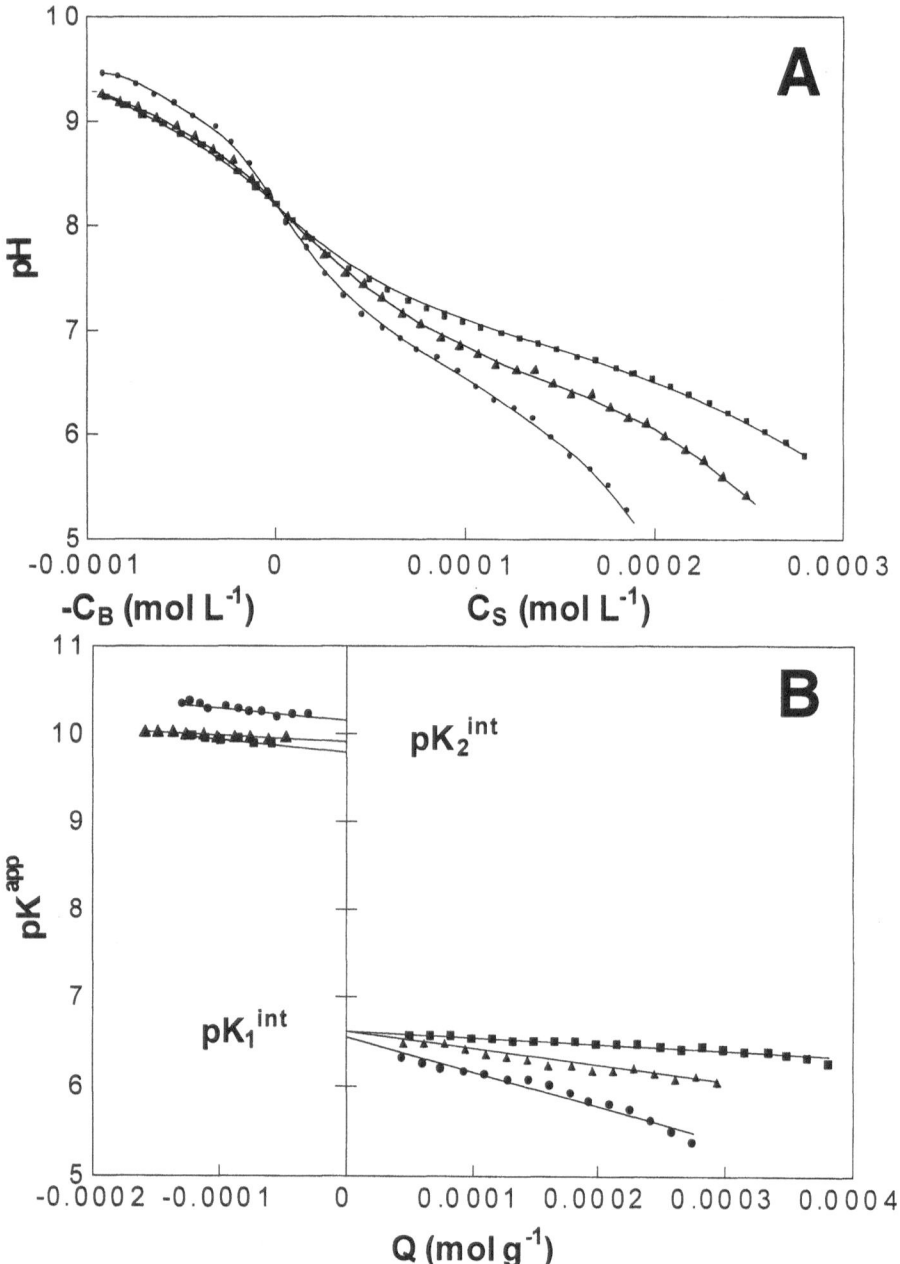

Fig. 3. A: Titration of a suspension of goethite in the absence of specific adsorbable ions and 136 m^2 surface area of goethite per 1 litre solution. C_B and C_S denote the concentrations of added base (KOH) and acid (HNO$_3$), respectively.
B: Evaluation of values of surface acidity constants according to equations 8 and 9.
■: 5 M KNO$_3$; ▲: 1 M KNO$_3$; •: 0.1 M KNO$_3$

$$K_2^{app} = ([\equiv TOTFeOH] - Q \cdot \alpha)^{-1} \cdot Q \cdot \alpha \cdot (H^+) \text{ for } pH > pH_{PZC} \qquad (9)$$

(see equation 5 and 6; Stumm, 1992). As shown in Figure 3B the intrinsic values for the surface acidity constants, which are independent of electrical surface charge, are calculated by a linear extrapolation of pK^{app} versus Q to zero charge condition ($K^{int} = K^{app}$ for Q = 0). For goethite values of pK_1^{int} = 6.60, 6.57, and 6.52 and of pK_2^{int} = 9.79, 9.89, and 10.15 are obtained which are related to 5, 1, and 0.1 M KNO_3, respectively. This results in mean values of pK_1^{int} = 6.56±0.04 and pK_2^{int} = 9.94±0.19 (pH_{PZC} = 0.5 (pK_1^{int} + pK_2^{int}) = 8.25). These values are within the range of pK_1^{int} values with 4.2, 6.0, 7.5, 6.7, and 7.1 and within the range of pK_2^{int} values with 10.5, 7.3, 9.5, 9.0, and 11.2 obtained for goethite by Yates (1975), Sigg and Stumm (1981), Lövgren et al. (1990), Müller and Sigg (1992), and Lumbsdom and Evans (1994), respectively. In analogy to the above procedure the values pK_1^{int} = 7.08±0.10 and pK_2^{int} = 10.00±0.14 are measured for gibbsite (pH_{PZC} = 8.54), which are similar to the values for Al_2O_3 (pK_1^{int} = 7.4 and pK_2^{int} = 10.0; Stumm, 1992) and for amorphous $Al(OH)_3$ (pK_1^{int} = 6.8 and pK_2^{int} = 8.7; Morel and Hering, 1993).

6. Adsorption of monosilicic acid

6.1 KINETICS

In the following the results for gibbsite are presented as a typical case. Six experiments were carried out at various concentrations of monosilicic acid in the range from 0.07 to $1.4 \cdot 10^{-3}$ mol SiO_2 L^{-1}. Figure 4 shows the evolution of silica concentration as a function of reaction time if monomer is adsorbed onto gibbsite surface. The concentration decreases continuously and the adsorption kinetics may be described according to a pseudo first order reaction, which is similar to the silica adsorption kinetics onto hematite (Dietzel and Böhme, 1997). Thus the decrease of monosilicic acid may be described by the equation

$$-\frac{d[Si(OH)_4]}{dt} = k \cdot ([Si(OH)_4] - [Si(OH)_4]_{eq}) \qquad (10)$$

In this equation $[Si(OH)_4]$ and $[Si(OH)_4]_{eq}$ represent the concentration of monosilicic acid at any time and at adsorption equilibrium, respectively. t denotes the reaction time and k is the rate constant. The integration of equation 10 yields

$$\ln\left(\frac{[Si(OH)_4]_i - [Si(OH)_4]_{eq}}{[Si(OH)_4] - [Si(OH)_4]_{eq}}\right) = k \cdot t \qquad (11)$$

where $[Si(OH)_4]_i$ is the initial concentration of monosilicic acid. In Figure 5 the results for the experiments shown in Figure 4 are presented according to equation 11. The mean value of the rate constant is $k = 9.2 \cdot 10^{-3} \pm 0.2 \cdot 10^{-3}$ h^{-1}. Similar results are obtained for the other solids of Table 1 if monosilicic acid is adsorbed (Table 2). After 20 days the concentration of monomer remains constant representing adsorption equilibrium ($\approx$ 500 h in Figure 4). As it is shown in Table 1 the adsorption of monosilicic acid increases in the order of hematite, goethite, magnetite, lepidocrocite, akaganeite, feroxyhyte, ferrihydrite, amorphous iron hydroxide, and gibbsite at identical surface areas of the solids of 34 m^2 L^{-1} and pH = 6.2. This is documented by the decrease of the

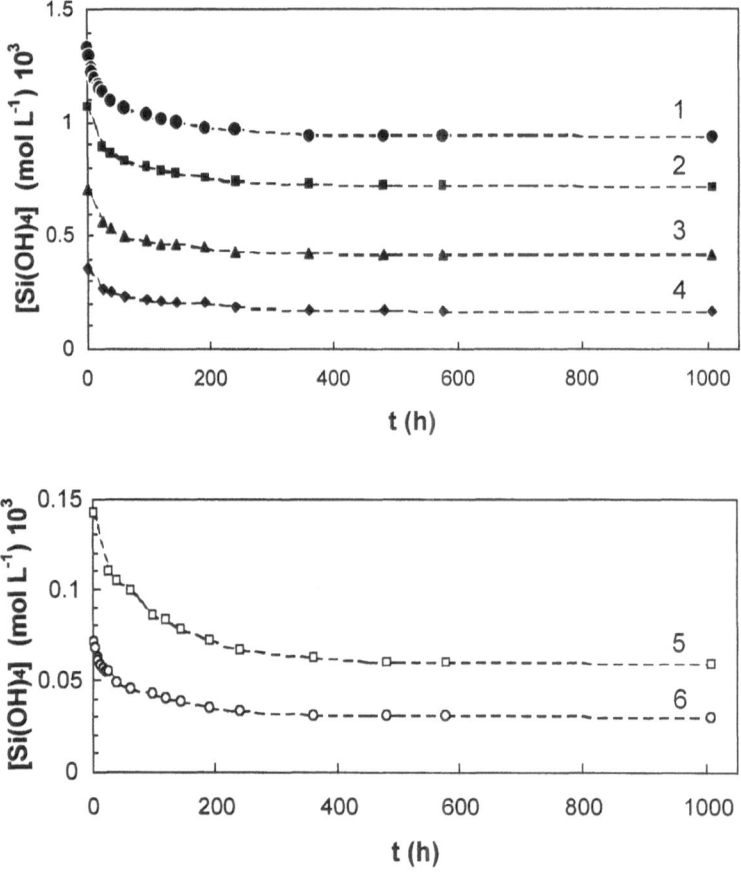

Fig. 4. The decrease of monosilicic acid as a function of reaction time as the adsorption onto the surface of gibbsite proceeds. Experiments 1 to 6 are carried out at constant pH = 6.2 (acetate buffer) and at 54 m^2 surface area L^{-1}.

concentration of monosilicic acid at adsorption equilibrium, $[Si(OH)_4]_{eq}$, in the above order.

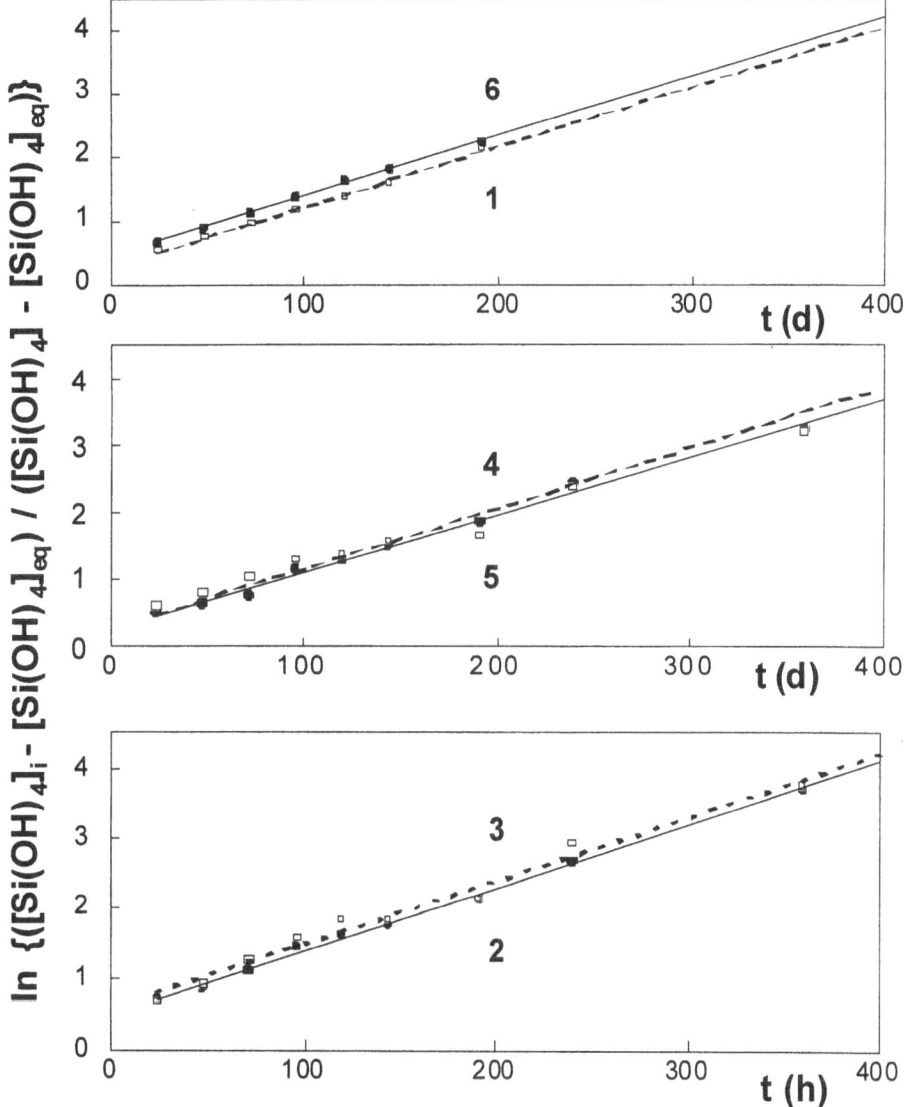

Fig. 5. Adsorption kinetics of monosilicic acid. The slopes of the regression lines represent the values of the reaction rate constant k according to equation 11. k = 9.46, 9.13, 9.29, 8.96, 8.46, and 9.38 (10^{-3} h^{-1}) for experiments 1 to 6 (see Figure 4). The related correlation coefficients are r^2 = 0.995, 0.981, 0.970, 0.977, 0.9923, and 0.998.

6.2 ADSORPTION EQUILIBRIUM AS A FUNCTION OF pH

The adsorption equilibrium with respect to monosilicic acid was investigated as a function of pH for goethite and gibbsite. Figure 6 and 7 show that at equilibrium the adsorption increases up to pH $\approx$ 9.8 (data in Table 2). This adsorption behaviour may be described by surface complex formation (see Sigg and Stumm, 1981; Davies and Kent, 1990; Dzombak and Morel, 1990; Parks, 1990; Stumm, 1992; Morel and Hering, 1993; Hansen et al., 1994). Following this concept the adsorption of silica is due to the reaction of monosilicic acid with surface hydroxyl groups and the formation of silica containing surface complexes according to the equations

$$Si(OH)_4 + \equiv FeOH^\circ \quad = \quad \equiv FeOSi(OH)_3^\circ + H_2O \qquad ;K_{S1}^{app} \qquad (12)$$

$$Si(OH)_4 + \equiv FeOH^\circ \quad = \quad \equiv FeOSiO(OH)_2^- + H_2O + H^+ \quad ;K_{S2}^{app} \qquad (13)$$

The related mass law equations are

$$K_{S1}^{app} = \frac{[\equiv FeOSi(OH)_3^\circ]}{\gamma(SiOH)_4 \cdot [Si(OH)_4] \cdot [\equiv FeOH^\circ]} \qquad (14)$$

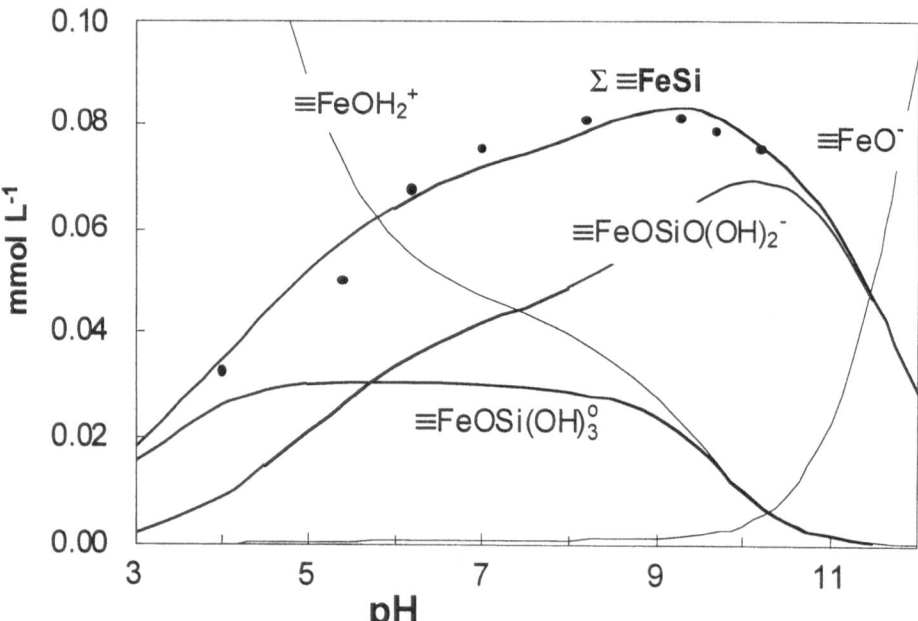

Fig. 6. Adsorption of monosilicic acid onto goethite ($\bullet$: experimental data, Tab.2; 0.5g goethite L^{-1}, representing 34 m^2 surface area L^{-1}). The solid lines denote the concentration of species at the surface of goethite. $\Sigma \equiv FeSi = [\equiv FeOSiO(OH)_2^-] + \equiv FeOSi(OH)_3^\circ]$.

$$K_{S2}^{app} = \frac{[\equiv FeOSiO(OH)_2^-] \cdot (H^+)}{\gamma(SiOH)_4 \cdot [Si(OH)_4] \cdot [\equiv FeOH^\circ]} \qquad (15)$$

where $\gamma Si(OH)_4$ represents the activity coefficient with respect to monosilicic acid which is assumed to be 1 for an uncharged species at a relatively low ionic strength. Thus total concentration of surface groups may be obtained by the concentrations of surface hydroxyl groups and those of the silica containing surface complexes according to the equation

$$[TOT\equiv FeOH] = [\equiv FeOH_2^+]+[\equiv FeOH^\circ]+[\equiv FeO^-]+[\equiv FeOSi(OH)_3^\circ]+[\equiv FeOSiO(OH)_2^-]$$

$$(16)$$

Considering the equations 5, 6, 14, and 15 expression 16 may be transformed to

$$[TOT\equiv FeOH] = [\equiv FeOH^\circ] \cdot \{(H^+) \cdot P \cdot (K_1^{int})^{-1} + 1 + K_2^{int} \cdot (P \cdot (H^+))^{-1} + K_{S1}^{int} [Si(OH)_4] + K_{S2}^{int} \cdot [Si(OH)_4] \cdot (P \cdot (H^+))^{-1}\} \qquad (17)$$

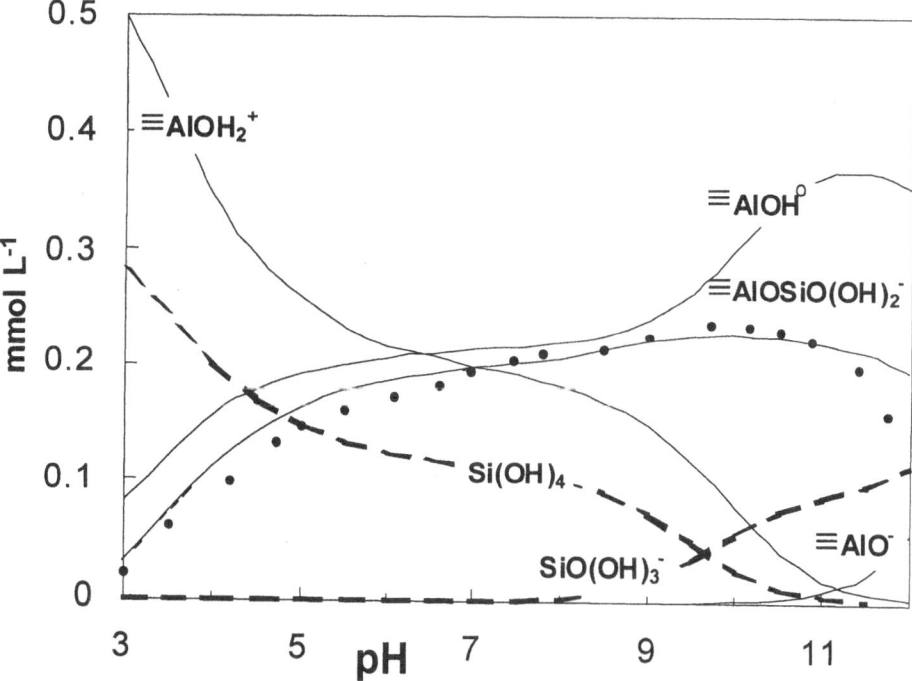

Fig. 7. Adsorption of monosilicic acid onto gibbsite (•: experimental data in Table 2; 54 m^2 surface L^{-1}). The dashed lines denote the concentrations of species in the solution and the solid line those at the surface of gibbsite.

In this equation the electrostatic correction is considered via

$$K^{app} \quad = K^{int} \cdot P^{\Delta Z} \qquad = K^{int} \cdot e^{-F\Psi/(RT)\Delta Z} \tag{18}$$

where K^{int} denotes the intrinsic constant which is independent of electrical surface charge and ΔZ is the change of charge at the surface ($\Delta Z = Z(product) - Z(educt)$). Following this, the value of ΔZ is -1, -1, 0, and -1 for the reaction 5, 6, 14, and 15, respectively. The coefficient for electrostatic correction, P, may be derived from the Faraday constant ($F = 96490$ C mol^{-1}), the gas constant ($R = 8.314$ J mol^{-1} K^{-1}), the absolute temperature (T), and the present surface potential (Ψ in mV). The value of the surface potential may be obtained from the equation

$$\Psi \quad = \sinh^{-1}(\sigma \cdot 8.518 \cdot C^{-0.5}) \cdot 0.05139 \tag{19}$$

for a symmetrical electrolyte with the molar concentration C at standard conditions (detailed derivation in Stumm (1992) and Morel and Hering (1993)). The surface charge σ (C m^{-2}) is a function of the concentrations of the charged surface groups according to the equation

$$\sigma \quad = F \cdot ([\equiv FeOH_2^+] - [\equiv FeO^-] - [\equiv FeOSiO(OH)_2^-]) \cdot \alpha^{-1} \cdot A^{-1} \tag{20}$$

where A denotes the specific surface area of the solid (Table 1) and the factor α is the amount of added grams of solid per litre solution.

Following the idea of surface complexation the mass balance of silica is expressed by

$$[TOTSi] = [Si(OH)_4] + [SiO(OH)_3^-] + [\equiv FeOSi(OH)_3^\circ] + [\equiv FeOSiO(OH)_2^-] \tag{21}$$

Considering the above equations expression 21 may be transformed into

$$[Si(OH)_4] = [TOTSi] \cdot \{1 + K_2 \cdot (H^+)^{-1} + K_{S1}^{int} [\equiv FeOH^\circ] + K_{S2}^{int} [\equiv FeOH^\circ] (P \cdot (H^+))^{-1}\}^{-1} \tag{22}$$

where K_2 denotes the first dissociation constant of silicic acid (see Figure 2; $pK_2 = 9.81$). The combination of equation 17 and 22 yields an expression as a function of $[\equiv FeOH^\circ]$ at constant pH. Assuming at first $P = 1$ and taking the individual values of K_{S1}^{int} and K_{S2}^{int} an approximated value of $[\equiv FeOH^\circ]$ is obtained. This permits to calculate the concentrations of the other components via the equations 5, 6, 14, and 15. In the next step a new value of P is obtained according to equations 18 to 20 which results in a new distribution of components for the system. The above procedure is continued until P is constant within an error of 3%.

To apply this model to the experimental results, the values of the intrinsic constants K_{S1}^{int} and K_{S2}^{int} are varied until the concentration of the surface groups fits well to the experimental data. In Figures 6 and 7 the distribution of species versus the experimental data are shown for $pK_{S1}^{int}(\equiv FeOSi(OH)_3^\circ) = -2.6$ and $pK_{S2}^{int}(\equiv FeOSiO(OH)_2^-) = 4.6$, and $pK_{S2}^{int}(\equiv AlOSiO(OH)_2^-) = 3.2$ with respect to goethite and gibbsite, respectively. It may be seen that the calculation of complexation fits the experimental data very well up to a pH of about 11. For the adsorption of monosilicic acid onto gibbsite the surface complex $\equiv AlOSi(OH)_3{}^\circ$ may be neglected because $pK_{S1}{}^{int}(\equiv AlOSi(OH)_3{}^\circ) > 1$.

7. Mechanisms concerning polysilicic acid sorption

7.1 EVOLUTION OF SILICA CONCENTRATIONS

The sorption processes are more complex if polysilicic acid is adsorbed. In contrast to monosilicic acid the interaction between dissolved silica on the form of polymers and the solid surface can not be described by a single kinetic equation or by surface complexation. This is due to the complex structures of the polymeric molecules and the variable number of silicon atoms per molecule (see Figure 1 and 2). Furthermore polysilicic acids are stable in alkaline solutions. In neutral and acid solutions polysilicic acid decomposes to monosilicic acid according to equation 1.

In Figure 8 adsorption experiments with initial polysilicic acid are compared to experiments with initial monosilicic acid for gibbsite and goethite. With monosilicic acid the concentration of total dissolved silica decreases continuously following the first order reaction 10. However in the experiments with initial polymeric silica the concentration passes through a minimum value, which corresponds to a maximum for silica adsorption. Moreover, the silica concentrations at adsorption equilibrium for initial polymeric silica are slightly higher than those for initial monosilicic acid. Figure 9 shows that similar results are obtained for the other solids used in the present study. The maximum adsorption is observed within several minutes up to one hour. The above described evolution of total dissolved silica in the experiments with initial polysilicic acid may be deciphered by analysis of the specific concentrations of monosilicic and polysilicic acid. In Figure 10 the individual concentrations of monomer and polymer are presented for gibbsite as a typical case (see Figure 8A; experimental data in Table 3). At a primary stage the concentration of polymer decreases by adsorption at the surface of the gibbsite within a relatively short time from 0.33 to $0.05 \cdot 10^{-3}$ mol SiO_2 L^{-1}. This is followed by the destruction of polysilicic acid to monosilicic acid in solution which is shown by a slow decrease of polymer concentration in the solution ($[(SiO_2)_n]$ measured; Figure 10). Thus monosilicic concentration, $[Si(OH)_4]$ measured, increases as depolymerization proceeds.

The depolymerization rate of silica in solution may be described by the kinetics in homogeneous solution, without a solid. In Figure 11 polymer concentrations are shown according to equation 1. The slope of the regression line represents the value of the rate constant $k_D = 62.7$ L mol^{-1} h^{-1} and the intersection with respect to the ordinate exhibits the theoretical value for the initial polymer concentration for the depolymerization in the solution $[(SiO_2)_n]_i = 0.0608 \cdot 10^{-3}$ mol SiO_2 L^{-1} (see equation 1). Considering the individual value of $[(SiO_2)_n]_i$ and the chemical composition of the solution a rate constant $k_D = 71.3$ L mol^{-1} h^{-1} is calculated via the program DEPO which is in close agreement to the value of 62.7 L mol^{-1} h^{-1} within the error of calculation of about 20% (Dietzel and Usdowski, 1995).

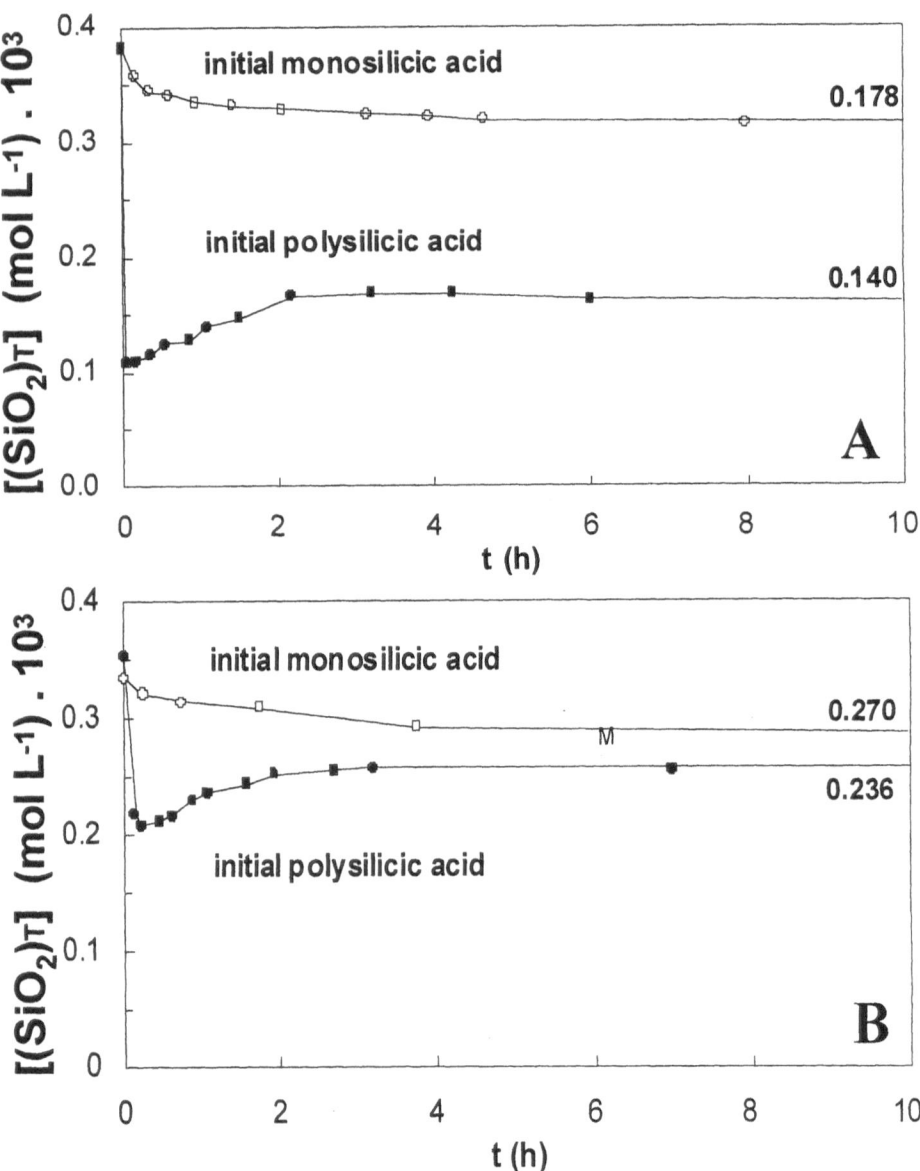

Fig. 8. Concentration of total dissolved silica (monosilicic and polysilicic acid), $[(SiO_2)_T] = [Si(OH)_4] + [(SiO_2)_n]$, as a function of reaction time (pH = 6.2). The values at the right hand side denote the concentrations of total dissolved silica in 10^{-3} mol SiO_2 L^{-1} at adsorption equilibrium. A: 44 m^2 gibbsite L^{-1}
(data in Table 3); B: 34 m^2 goethite L^{-1}

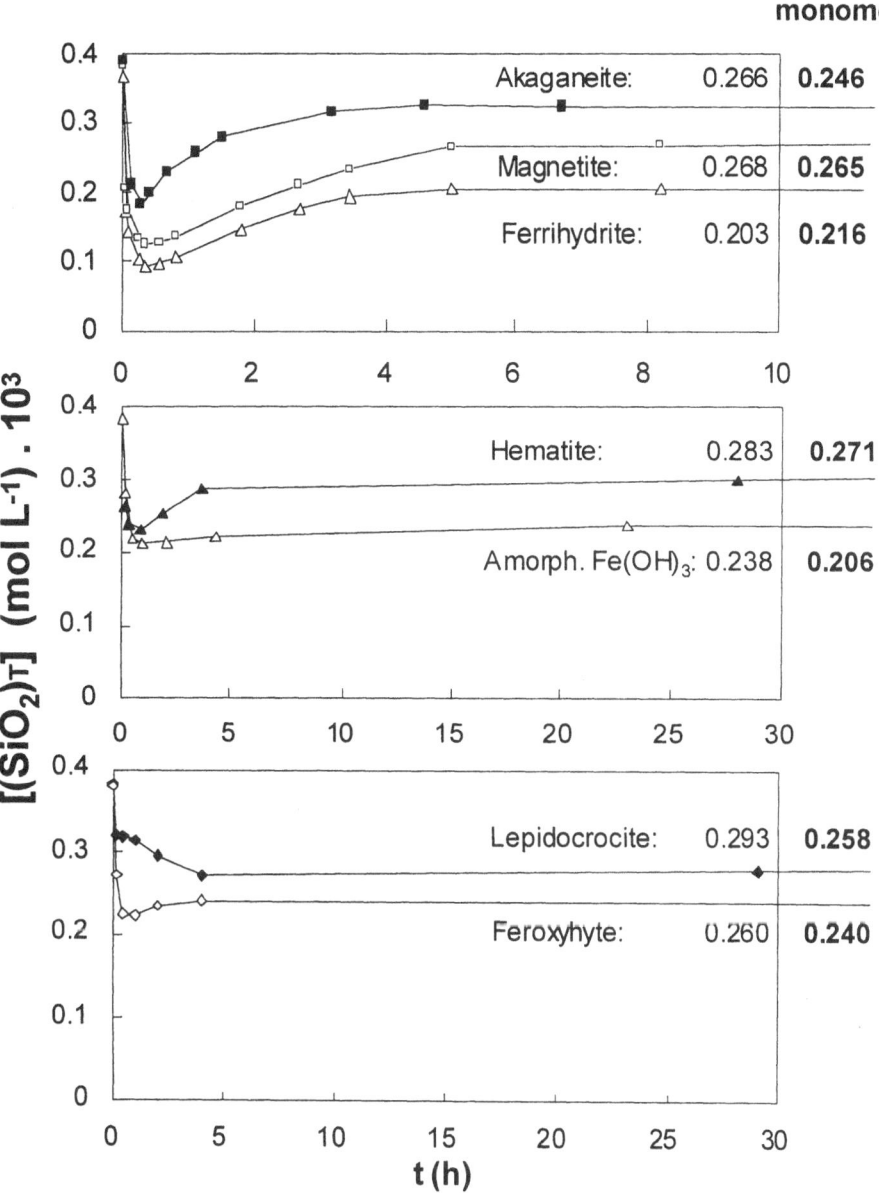

Fig. 9. Evolution of total dissolved silica (monosilicic and polysilicic acid), $[(SiO_2)_T] = [Si(OH)_4] + [(SiO_2)_n]$, for the adsorption experiments with initial polysilicic acid (pH = 6.2; 34 m^2 surface L^{-1}). Concentrations at adsorption equilibrium with respect to experiments with initial polysilicic acid are shown inside the diagrams. The bold values on the right hand side of the diagram denote 10^{-3} mol SiO_2 L^{-1} at equilibrium with respect to experiments with initial monomer (see Table 1).

Table 3. Experiment with gibbsite and initial polysilicic acid ($44 \ m^2 \ L^{-1}$; pH = 6.2). The initial concentrations of monomer and polymer are about 0.033 and 0.33 _ 10^{-3} mol $SiO_2 \ L^{-1}$, respectively (pH = 6.2; acetate buffer). $[Si(OH)_4]$, $[(SiO_2)_n]$, and $[(SiO_2)_T]$ are the present monomeric, polymeric, and total silica concentrations in 10^3 mol $SiO_2 \ L^{-1}$ at reaction time t in hours.

t	$[Si(OH)_4]$	$[(SiO_2)_n]$	$[(SiO_2)_T]$
0.05	0.0599	0.0489	0.109
0.17	0.0699	0.0393	0.109
0.35	0.0885	0.0258	0.114
0.53	0.104	0.0208	0.125
0.85	0.114	0.0146	0.129
1.08	0.127	0.0112	0.138
1.50	0.138	0.0092	0.147
2.18	0.159	0.0067	0.166
3.20	0.166	0.0004	0.166
4.23	0.168	-	0.168
12.0	0.163	-	0.163
27.2	0.154	-	0.154
65.0	0.153	-	0.153
264	0.150	-	0.150
744	0.141	-	0.141
3360	0.140	-	0.140

Moreover, the increase of monosilicic acid (see Figure 10; $[Si(OH)_4]$ measured) can not solely be described by destruction of dissolved polysilicic acid ($[Si(OH)_4]$ depolymerized). Additional amounts of monosilicic acid are gained by a release of primary adsorbed silica, which is observed by the adsorption maximum in Figure 8A. Thus the polysilicic acid decomposes at the mineral surface, liberating monosilicic acid into the solution.

7.2 ADSORPTION EQUILIBRIUM AS A FUNCTION OF PH

In Figure 12 the typical evolution for three polysilicic acid adsorption experiments with gibbsite in the pH range from 4.6 to 8.0 are presented. The figure shows that the principal behaviour during the adsorption process depends on pH. It may be seen from Table 3 for a typical adsorption experiment with polysilicic acid that at adsorption equilibrium dissolved silica consists only of monosilicic acid for both experiments with initial monosilicic and with initial polysilicic acid. Nevertheless, the quantities of adsorbed silica at various pH values are different for initial polysilicic and monosilicic acid.

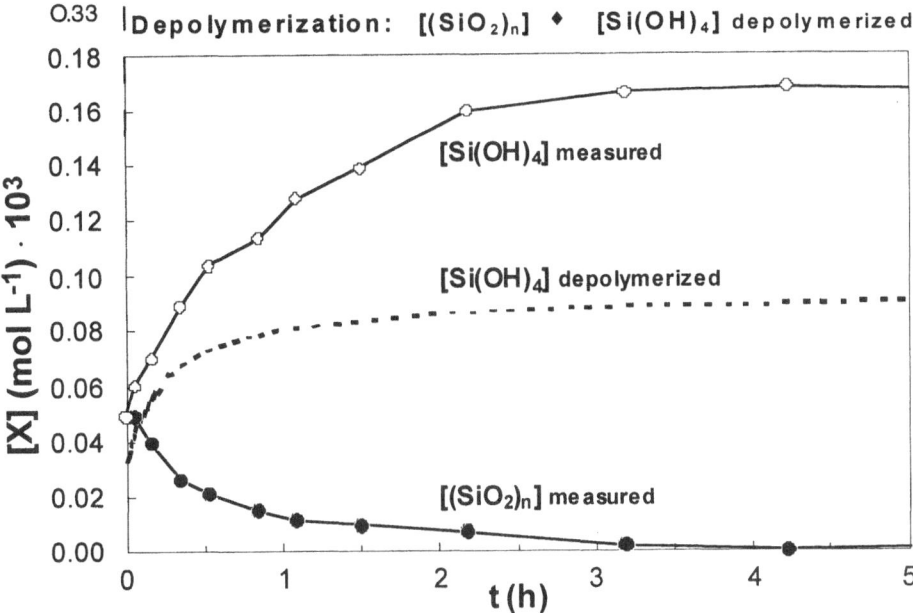

Fig. 10. Evolution of polysilicic and monosilicic acid for an experiment with initial polysilicic acid (see experiment with gibbsite in Figure 8A for total dissolved silica). Measured concentration of polysilicic acid, $[(SiO_2)_n]$, and monosilicic acid, $[Si(OH_4)]$ in Table 3. Dashed line: calculated monomeric silica concentration according to the depolymerization in homogeneous solution (eq 1 with $k_D = 62.7$ L mol^{-1} h^{-1}; Fig. 11).

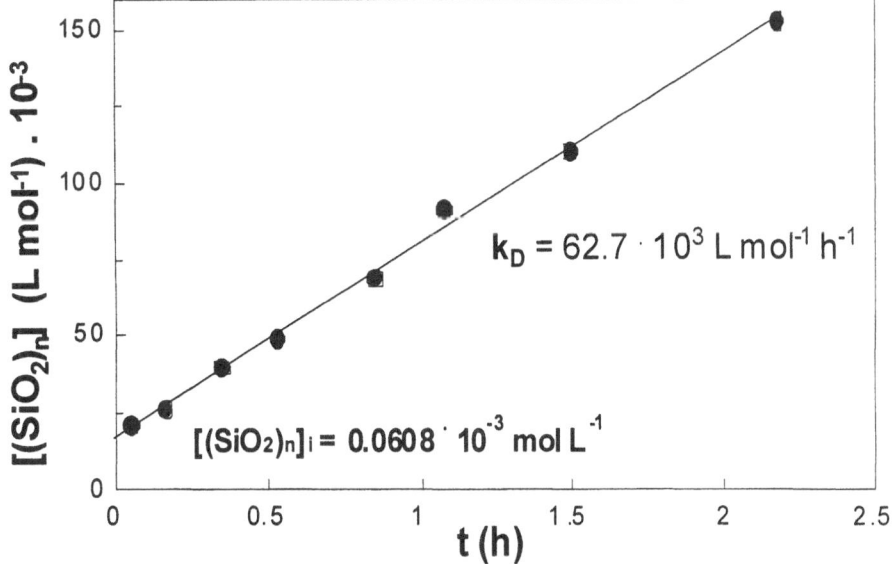

Fig. 11. Destruction of polysilicic acid to monosilicic acid in the solution (experiment of Figure 10). Regression line: $[(SiO_2)_n] = 62.7 \cdot t + 16.44$ (see equation 1).

Table 4. oncentration of total dissolved silica, $[(SiO_2)_T] = [Si(OH)_4] + [(SiO_2)_n]$, with initial polysilicic acid at adsorption equilibrium after 30 days (0.1n NaCl). The concenrations are given in 10^{-3} mol SiO_2 L^{-1}. Initial concentration is $0.383 \_ 10^{-3}$ mol SiO_2 L^{-1}.

Gibbsite (13 m² L⁻¹)		Goethite (34 m² L⁻¹)	
pH	$[(SiO_2)_T]$	pH	$[(SiO_2)_T]$
4.60	0.154	4.50	0.220
5.30	0.174	5.01	0.245
6.20	0.241	5.50	0.251
8.00	0.273	6.60	0.275
		7.40	0.275

In Table 4 the experimental results for initial polysilicic acid with respect to gibbsite and goethite as a function of pH are shown. The adsorption of silica increases as pH decreases, especially for gibbsite, whereas adsorption with initial monosilicic acid shows reversed behaviour (Figure 6 and 7). Thus polysilicic acid is stabilised at the mineral surface as pH decreases. At pH < 6 the depolymerization of primary adsorbed polymeric silica at the mineral surface is incomplete. Thus minor amounts of monosilicic acid are liberated into the solution.

8. Summary and conclusions

In Figure 13 the reactions at the mineral surface of hydroxides and in the solutions are shown. On the one side monosilicic acid is adsorbed at the mineral surface of hydroxides within two weeks (1). At equilibrium adsorption of silica may be described by surface complexation. The adsorption kinetics is described by the first order reaction 10. The obtained values for the complexation constants are $pK_{S1}^{int}(\equiv FeOSi(OH)_3^\circ) = -2.6$ and $pK_{S2}^{int}(\equiv FeOSiO(OH)_2^-) = 4.6$, and $pK_{S2}^{int}(\equiv AlOSiO(OH)_2^-) = 3.2$ with respect to goethite and gibbsite, respectively (see equations 14 and 15). It is shown in Figure 6 and 7 that maximum adsorption of monosilicic acid is observed at pH ≈ 9.8 as documented by high concentrations of $[\equiv FeOSi(OH)_3^\circ] + [\equiv FeOSiO(OH)_2^-]$ and $[\equiv AlOSiO(OH)_2^-]$ at the mineral surface of goethite and gibbsite, respectively.

The adsorption of monosilicic acid increases in the order of hematite, goethite, magnetite, lepidocrocite, akaganeite, feroxyhyte, ferrihydrite, amorphous iron hydroxide, and gibbsite. This may be seen from Table 1 by the decrease of monosilicic acid concentration of the solution at adsorption equilibrium in the above order of solids. It is important to note that the adsorption of monosilicic acid at the surface of the solid is a full reversible process. This is confirmed by desorption experiments with monosilicic acid and gibbsite (Dietzel, 1998).

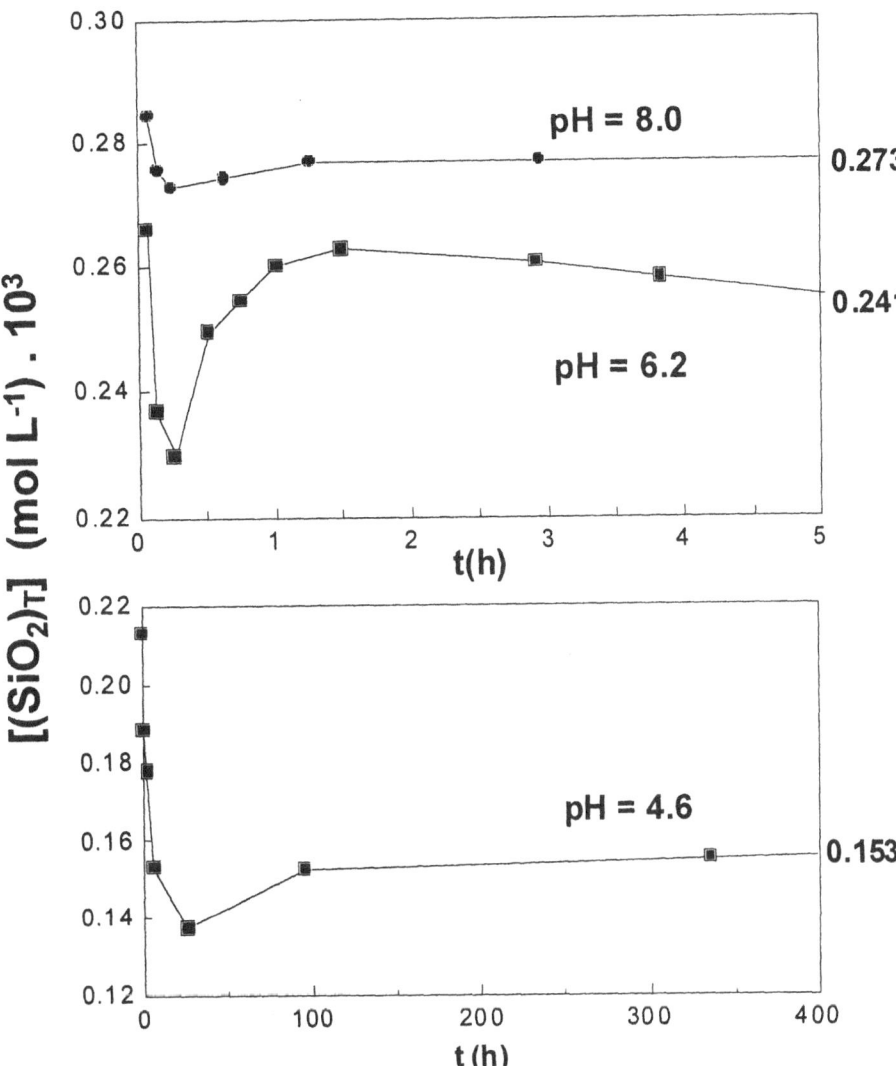

Fig. 12. Evolution of total dissolved silica (monosilicic and polysilicic acid), $[(SiO_2)_T] =$ $[Si(OH)_4] + [(SiO_2)_n]$, for adsorption experiments with polysilicic acid at various pH (13 m^2 gibbsite L^{-1}). The values on the right hand side of the diagram denote total dissolved silica concentrations in 10^{-3} mol SiO_2 L^{-1} at adsorption equilibrium (Table 4). The initial silica concentration is $0.383 \_ 10^{-3}$ mol SiO_2 L^{-1}.

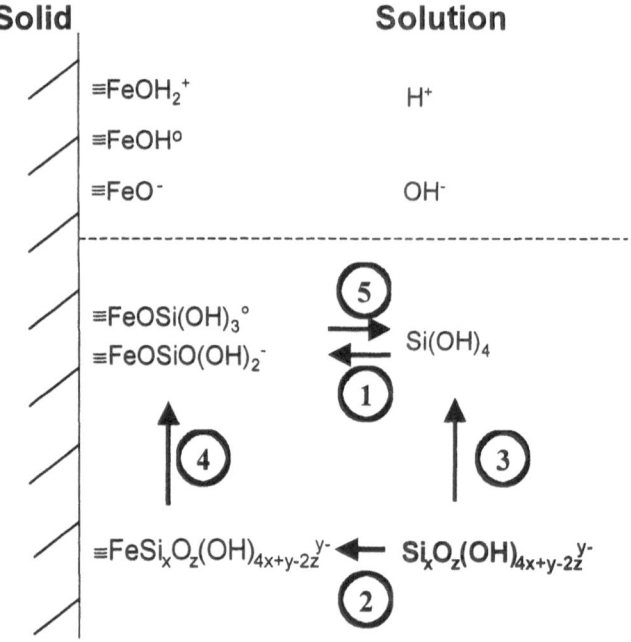

Solid   **Solution**

Fig. 13. Interaction of polysilicic and monosilicic acid with the surface of hydroxides (shown for goethite). Adsorption of polysilicic and monosilicic acid ($\leftarrow$), destruction of polysilicic acid in the solution and at the mineral surface ($\uparrow$), and desorption of monomeric silica ($\rightarrow$).

On the other side polysilicic acid is adsorbed at the surface within a short time (2). The reaction time is some minutes (see Figure 10). Simultaneously polysilicic acid decomposes slowly to monosilicic acid in the solution (3). The depolymerization rate follows equation 1 and the value of the rate constant k_D may be calculated by the program DEPO for various compositions of a solution. The decrease of polysilicic acid via destruction to monosilicic acid according to equation 1 is presented in Figure 11. Moreover, it is shown in the present study that polysilicic acid decomposes not only in the solution but also at the surface of hydroxides (4). The depolymerization of polymeric silica at the mineral surface liberates monosilicic acid into the solution within several days (5). The amount of released monosilicic acid decreases as pH decreases (see Figure 12). Thus polysilicic acid is adsorbed as a metastable component and is stabilised at the mineral surface, especially in acid solutions (reverse behaviour with respect to monomer). At pH > 6 adsorption equilibrium with respect to initial monosilicic acid is obtained which is documented e.g. by the silica adsorption equilibrium with respect to monosilicic and polysilicic acid for goethite in Table 2 and 4, respectively.

If these results are applied to natural systems silicic acids, monosilicic acid and polysilicic acid, are adsorbed at the mineral surface of iron oxides and gibbsite. Monosilicic acid is the most abundant species in natural solutions and its adsorption is lower in comparison to polysilicic acid. Nevertheless, adsorption of monosilicic acid may result in a significant decrease of dissolved silica concentration in natural environments e.g. in soils (McKeague and Cline, 1963; Mott, 1969; Obihara and Russell, 1972; Parfitt, 1978) and in sediments (Liss and Spencer, 1970; Carlson and Schwertmann, 1981; Cornell and Schwertmann, 1996). In solutions containing polysilicic acid the adsorption magnitude is higher and the adsorption mechanisms are more complex.

It is shown here that monosilicic acid forms specific surface complexes within a short time compared to time scales of natural processes. The molar ratio of total concentrations of silica containing surface complexes versus that of hydroxyl groups for goethite is about 0.21 at a pH between 5 and 9 (see Figure 6). This ratio corresponds well with the molar ratio Si/Fe = 0.19 of recent iron oxides shown in Figure 14 representing silica concentrations of the solid up to several weight %. In the presented iron oxide precipitations of Figure 14 no evidence for the existence of iron silicate minerals were obtained from the cited authors by X-ray diffraction or infrared studies. This result is conform with adsorption theory and corresponds to various studies dealing with the fixation of dissolved compounds and especially silicic acid onto iron oxides surfaces (Dzombak and Morel, 1980; Sigg and Stumm, 1981; Stumm, 1992; Glasauer, 1995; Cornell and Schwertmann, 1996). Thus it may be concluded that the silica content of sedimentary iron oxides is directly related to the adsorption of silicic acid onto the primary precipitates via the formation of the above described surface complexes. Nevertheless, the silica amount of iron oxides may be an important source for the formation of iron silicates via post sedimentary processes (e.g. Iler, 1979).

Parfitt et al. (1992) postulated by infrared studies on natural ferrihydrites that both, Fe-O-Si and linked Si-O-Si, exist at the mineral surface of iron oxide. The existence of Fe-O-Si bonds are described by surface complexation reactions according to equations 12 and 13, whereas the boundary conditions for polymerization of silica, Si-O-Si, at the mineral surface were speculative in former studies. The experiments in the present study show that polymerization (Si-O-Si bonds) at the mineral surface of hydroxides is favoured in acid solutions. Above a pH of about 6 polymeric silica bonded at the surface is not stable and the structure of the disordered linked silica molecules at the mineral surface decompose liberating monomer into the solution. Thus, linked $Si(OH)_4$ molecules onto hydroxide and oxide surfaces may only occur in acid environments with respect to soil and interstitial solutions.

Further, the observed polymerization behaviour at the surface of minerals may be seen in relation to the results obtained from silicate dissolution experiments by Casey et al. (1993), who identified polymeric silica at the surfaces of the minerals in acid solutions. In analogy, the results of the present study show that polymeric silica is stabilised at the surface of solids especially in more acid solutions. On another point of view, it is shown in the present study that adsorption of monosilicic is favoured in slightly alkaline solutions and that at this pH polymerization of silica at the surface of solids is neglected, which may inhibit crystallisation of silicates (Iler, 1979).

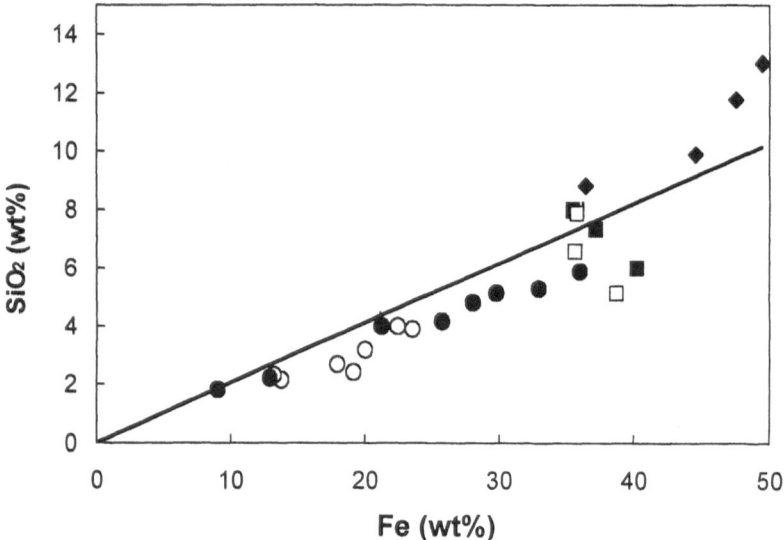

Fig. 14. Silica versus iron content of recent iron oxides precipitated from natural waters. Samples were prepared via acid oxalate extraction method (see Schwertmann, 1964). The slope of the regression line is 0.205 corresponding to a molar ratio Si/Fe = 0.19.
●: precipitation from spring water (Harz Mountain, Northern Germany);
○: precipitation from ground water (Northern Hessia, Germany);
■: corrosion products at iron tubes in water supply systems (Göttingen, Northern Germany);
□: precipitation from spring water in Finland (Carlson and Schwertmann, 1981);
♦: streambeds in New Zealand and paddy race in Japan (Parfitt et al., 1992).

Acknowledgements

The author wishes to thank E. Usdowski and H. Schwecke at the Geochemisches Institut der Universität Göttingen for a preview of the manuscript. The constructive comments by V. Ragnarsdottir and one anonymous reviewer are greatly appreciated.

References

Alvarez R. and Sparks D. L. (1985) Polymerization of silicate anion in solutions at low concentrations. *Nature* **318**, 649-651.

Applin K. R. (1987) The diffusion of dissolved silica in dilute aqueous solution. *Geochim. Cosmochim. Acta* **51**(9), 2147-2151.

Arnorsson S., Sigurdsson S., and Svavarsson H. (1982) The chemistry of geothermal waters in Iceland. I. Calculation of aqueous speciation from 0° to 370°C. *Geochim. Cosmochim. Acta* **46**, 1513-1532.

Baes C. F. and Mesmer R. E. (1976) The hydrolysis of cations. *Wiley-Interscience, New York*, 489p.

Baumann H. (1959) Polymerisation und Depolymerisation der Kieselsäure unter verschiedenen Bedingungen. *Kolloid Zeitschrift* **162**(1), 28-35.

Bernal J. D., Dasgupta D. R., and Mackay A. L. (1959) The oxides and hydroxides of iron and their structural interrelationship. *Clay Min. Bull.* **4**, 15-30.

Böhme G., Dietzel M., Heinrichs H., Heydemann A., Schlabach S., and Usdowski E. (1999) Wechselwirkungen zwischen Lösungen und Festkörpern in offenen Systemen und strömenden Medien. *In: Wechselwirkungen an geologischen Grenzflächen - SFB 468 Arbeits- und Ergebnisbericht, Universität Göttingen.* A4-1 - A4-45.

Brady P. V. and House W. A. (1996) Surface controlled dissolution and growth of minerals. *In: Physics and Chemistry of Mineral Surfaces (eds.: Brady, P.V.) Chapter 4. CRC Press, Boca Raton New York London Tokyo*, 225-305.

Brauer G. (1982) Handbuch der präparativen anorganischen Chemie. *Band 3. Enke, Stuttgart.*

Busey R. H. and Mesmer R. E. (1977) Ionization equilibria of silicic acid and polysilicate formation in aqueous sodium chloride solutions to 300°C. *Inorg. Chem.* **16**(10), 2444-2450.

Carlson L. and Schwertmann U. (1981) Natural ferrihydrites in surface deposits from Finland and their association with silica. *Geochim. Cosmochim. Acta* **45**(3), 421-429.

Cary L. W., De Jong B. H. W. S., and Dibble W. E. (1982) A ^{29}Si NMR study of silica species in dilute aqueous solution. *Geochim. Cosmochim. Acta* **46**(7), 1317-1320.

Casey W. H., Westrich H. R., Banfield J. F., Ferruzzi G., and Arnold G. W. (1993) Leaching and reconstruction at the surfaces of dissolving chain-silicate minerals. *Nature* **366**, 253-256.

Cornell R. M. and Giovanoli R. (1987) The influence of silicate species on the morphology of goethite (a-FeOOH) grown from ferrihydrite (5Fe$_2$O$_3$ 9H$_2$O). *J. Chem. Soc., Chem. Commun.*, 413-414.

Cornell R. M. and Schwertmann U. (1996) The iron oxides. Structure, properties, reactions, occurrence and uses. *VHC Verlagsgesellschaft mbH. Weinheim*, 573p.

Davies S. (1964) Silica in streams and ground water. *Am. J. Sci.* **262**, 870-876.

Davis J. A. and Kent D. B. (1990) Surface complexation modeling in aqueous geochemistry. *In: Mineral-water interface geochemistry (eds.: Hochella, M.F., White, A.F.) Chapter 5. Rev. Min.* **23**, 177-260.

Dietzel M. (1993) Depolymerisation von hochpolymerer Kieselsäure in wäßriger Lösung. *Dissertation, Universität Göttingen, Germany*, 93p.

Dietzel M. (1998) Gelöste polymere und monomere Kieselsäuren und die Wechselwirkung mit Gibbsit und Fe-O-OH-Festphasen. *Habilitations-Schrift, Universität Göttingen*, 205p.

Dietzel M. (2000) Dissolution of silicates and the stability of polysilicic acid. *Geochim. Cosmochim. Acta* **64**(19), 3275-3281.

Dietzel M. and Usdowski E. (1995) Depolymerization of soluble silicate in dilute aqueous solutions. *Colloid Polym. Sci.* **273**, 590-597.

Dietzel M. and Böhme G. (1997) Adsorption und Stabilität von polymerer Kieselsäure. *Chem. Erde* **57**, 189-203.

Dove P. M. and Rimstidt J. D. (1994) Silica-Water Interactions. *In: Silica (eds.: Heaney P.J., Prewitt C.T., Gibbs G.V.). Chapter 8. Rev. Min.* **29**, 259-308.

Dzombak D. A. and Morel F. M. M. (1990) Surface complexation modeling. Hydrous ferric oxide. *Wiley-Interscience, New York*, 393p.

Eikenberg J. (1990) On the problem of silica solubility at high pH. Nationale Genossenschaft für die Lagerung radioaktiver Abfälle, Baden (Switzland). *Technical Report* **90-36**, 54p.

Füchtbauer H. (1988) Sediment-Petrologie Teil II. Sedimente und Sedimentgesteine. *Schweizerbart'sche Verlagsbuchhandlung, Stuttgart*, 1141p.

Glasauer S. M. (1995) Silicate associated with Fe(hydr)oxides. *Dissertation, Technische Universität München-Weihenstephan*, 133p.

Grenthe I., Fuger J., Konings R. J. M., Lemire R. J., Muller A. B., Nguyen-Trung Cregu C., and Wanner H. (1992) Chemical Thermodynamics of Uranium. *Chemical Thermodynamics 1. Nuclear Energy Agency. North-Holland Elsevier*, 750pp.

Hansen H. C. B., Wechte T. P., Raulund-Rasmussen K., and Borggaard O. K. (1994) Stability constants for silicate adsorbed to ferrihydrite. *Clay Min.* **29**, 341-350.

Hem J. D. (1970) Study and interpretation of the chemical characteristics of natural waters. *Geol. Sur. Water-Supply Paper* **1473**, 363p.

Hingston F. J., Posner A. M., and Quirk J. P. (1972) Anion adsorption by goethite and gibbsite. 1. The role of the proton in determining adsorption envelopes. *J. Soil Sci.* **23**(2), 177-192.

Iler R. K. (1979) The chemistry of silica - Solubility, Polymerization, Colloid and Surface Properties, and Biochemistry. *Wiley- Interscience, New York*, 866p.

Jones B. F., Retting S. F., and Eugster H. P. (1967) Silica in alkaline brines. *Science* **158**, 1310-1314.

Knollmann S. (2001) Korrosionsprodukte in Trinkwasserrohren: Gelöste Kieselsäure und die Stabilität von Goethit. *Bachelorarbeit, Universität Göttingen*, 44p.

Liss P. S. and Spencer C. P. (1970) Abiological processes in the removal of silicate from sea water. *Geochim. Cosmochim. Acta* **34**(10), 1073-1088.

Lövgren L., Sjöberg S., and Schindler P. W. (1990) Acid/base reactions and Al(III) complexation at the surface of goethite. *Geochim. Cosmochim. Acta* **54**(5), 1301-1306.

Lumbsdom D. G. and Evans L. G. (1994) Surface complexation model parameters for goethite (a-FeOOH). *J. Colloid Interface Sci.* **164**, 119-125.

McKeague J. A. and Cline M. G. (1963) Silica in soil solutions. II. The adsorption of monosilicic acid by soil and by other substances. *Canadian Journal of Soil Science* **43**, 83-96.

Morel F. M. M. and Hering J. G. (1993) Principles and applications of aquatic chemistry. *Wiley-Interscience. New York Toronto*, 588p.

Mott C. J. B. (1969) Sorption of Anions by Soils. *S. C. I. Monograph* **37**, 40-53.

Müller G. and Sigg L. (1992) Adsorption of lead(II) on the goethite surface; voltametric evaluation of surface complexating parameters. *J. Colloid Interface Sci.* **148**, 517-532.

Obihara C. H. and Russell E. W. (1972) Specific adsorption of silicate and phosphate by soils. *J. Soil Sci.* **23**, 103-117.

Parfitt R. L. (1978) Anion adsorption by soils and soil materials. *Dep. Argon. Ser. Pap.* **1225**.

Parfitt R. L., Van der Gaast S. J., and Childs C. W. (1992) A structural model for natural siliceous ferrihydrite. *Clays and Clay Minerals* **40**(6), 675-681.

Parks G. A. (1990) Surface energy and adsorption at mineral interface: An introduction. *Mineral-water interface geochemistry (eds.: Hochella, M.F., White, A.F.) Chapter 4. Rev. Min.* **23**, 133-176.

Rimstidt J. D. and Barnes H. L. (1980) The kinetics of silica-water reactions. *Geochim. Cosmochim. Acta* **44**(11), 1683-1699.

Rowe J. J., Fournier R. O., and Morey G. W. (1973) Chemical analysis of thermal waters in Yellowstone National Park, Wyoming, 1960-1965. *Geol. Surv. Bull.* **1303**, 31p.

Schecher W. D. and McAvoy D. C. (1998) MINEQL+: A chemical equilibrium modeling system, Version 4.0 for Windows, User`s manual. *Environmental Research Software, Hallowell, Maine*, 318p.

Schwertmann U. (1964) Differenzierung der Eisenoxide des Bodens durch Extraktion mit Ammoniumoxalat-Lösung. *Z. Planzenernaehr. Dueng. Bodenkd.* **108**, 37-45.

Schwertmann U. and Cornell R. M. (1991) Iron oxides in the laboratory. *VCH Verlagsgesellschaft Weinheim*, 32p.

Schwertmann U. and Thalmann H. (1976) The influence of Fe(II), Si and pH on the formation of lepidocrocite and ferrihydrite during oxidation of aqueous $FeCl_2$ solutions. *Clay Min.* **11**, 189-200.

Seward T. M. (1974) Determination of the first ionization constant of silicic acid from quartz solubility in borate buffer solutions to 350°C. *Geochim. Cosmochim. Acta* **38**(11), 1651-1664.

Siever R. (1972) Silicon-abundance in natural waters, 14-I. *In: Handbook of Geochemistry II-2 (ed.: Wedepohl, K.H.) Springer, Berlin Heidelberg New York.*

Sigg L. and Stumm W. (1981) The interaction of anions and weak acids with the hydrous goethite (a-FeOOH) surface. *Colloids and Surfaces* **2**, 101-117.

Sigg L. and Stumm W. (1994) Aquatische Chemie: eine Einführung in die Chemie wässriger Lösung und natürlicher Gewässer. 3. *Aufl. Zürich, Verl. der Fachvereine, Stuttgart, Teubner.*

Sjöberg S., Nordin A., and Ingri N. (1981) Equilibrium and structural studies of silicon(IV) and aluminium(III) in aqueous solution. II. Formation constants for the monosilicate ions $SiO(OH)_3^-$ and $SiO_2(OH)_2^{2-}$. *Mar. Chem.* **10**, 521-532.

Sjöberg S., Hägglund Y., Nordin A., and Ingri N. (1983) Equilibrium and structural studies of silicon(IV) and aluminium(III) in aqueous solutions: V. Acidity constants of silicic acid and the ionic product of water in the medium range 0.05-2.0m Na(Cl) at 25°C. *Mar. Chem.* **13**, 35-44.

Sjöberg S., Öhman L. O., and Ingri N. (1985) Equilibrium and structural studies of silicon(IV) and aluminium(III) in aqueous solution.11. Polysilicate formation in alkaline aqueous solution. A combined potentiometric and ^{29}Si NMR study. *Acta Chem. Scand.* **A 39**, 93-107.

Stumm W. (1992) Chemistry of the solid-water interface. *Wiley-Interscience, New York*, 428p.

Stumm W. and Morgan J. J. (1996) Aquatic chemistry - Chemical equilibria and rates in natural waters. *3nd ed. Wiley-Intersience New York*, 1022p.

White D. E., Brannock W. W., and Murata K. J. (1956) Silica in hot-spring waters. *Geochim. Cosmochim. Acta* **10**(1), 27-59.

Williams L. A. and Crerar D. A. (1985) Silica diagenesis, II. general mechanisms. *J. Sed. Pet.* **55**(3), 312-321.

Yates D. E. (1975) The structure of the oxide/aqueous electrolyte interface. *Ph. D. Thesis Univ. Melbourne. Australia.*

Water Science and Technology Library

Water Science and Technology Library

Water Science and Technology Library

39. I.V. Nagy, K. Asante-Duah and I. Zsuffa: *Hydrological Dimensioning and Operation of Reservoirs*. Practical Design Concepts and Principles. 2002 ISBN 1-4020-0438-9
40. I. Stober and K. Bucher (eds.): *Water-Rock Interaction*. 2002 ISBN 1-4020-0497-4

Kluwer Academic Publishers – Dordrecht / Boston / London